STUDENT'S SOLUTIONS MANUAL

TAMSEN HERRICK

Butte College

A SURVEY OF MATHEMATICS WITH APPLICATIONS
TENTH EDITION

Allen R. Angel
Monroe Community College

Christine D. Abbott
Monroe Community College

Dennis C. Runde
State College of Florida

PEARSON

Boston   Columbus   Indianapolis   New York   San Francisco

Amsterdam   Cape Town   Dubai   London   Madrid   Milan   Munich   Paris   Montreal   Toronto

Delhi   Mexico City   Sao Paulo   Sydney   Hong Kong   Seoul   Singapore   Taipei   Tokyo

Copyright © 2017, 2013, 2009 Pearson Education, Inc.
Publishing as Pearson, 501 Boylston Street, Boston, MA 02116.

ISBN-13: 978-0-13-411220-6
ISBN-10: 0-13-411220-2

1 2 3 4 5 6   20 19 18 17 16

www.pearsonhighered.com

Table of Contents

CHAPTER ONE

CRITICAL THINKING SKILLS

Exercise Set 1.1

1. Natural

3. Counterexample

5. Inductive

7. Deductive

9. $5 \times 3 = 15$

11. 1 $5(=1+4)$ $10(=4+6)$ $10(=6+4)$
    $5(=4+1)$ 1

13.

15.

17. 9, 11, 13 (Add 2 to previous number.)

19. 5, -5, 5 (Alternate 5 and -5.)

21. $\dfrac{1}{5}, \dfrac{1}{6}, \dfrac{1}{7}$ (Increase the denominator value by 1.)

23. 36, 49, 64 (The numbers in the sequence are the squares of the counting numbers.)

25. 34, 55, 89 (Each number in the sequence is the sum of the previous two numbers.)

27. There are three letters in the pattern.
    $39 \times 3 = 117$, so the 117th entry is the second R in the pattern. Therefore, the 118th entry is Y.

29. a) 36, 49, 64

    b) Square the numbers 6, 7, 8, 9 and 10

    c) $8 \times 8 = 64$ $9 \times 9 = 81$

    72 is not a square number since it falls

    between the two square numbers 64 and 81.

31. Blue: 1, 5, 7, 10, 12   Purple: 2, 4, 6, 9, 11

    Yellow: 3, 8

1

33. a) ≈ $2050 million   b) We are using observation of specific cases to make a prediction.

35.

37. a) You should obtain the original number.
    b) You should obtain the original number.
    c) Conjecture: The result is always the original number.

    d) $n, 3n, 3n+6, \dfrac{3n+6}{3} = \dfrac{3n}{3} + \dfrac{6}{3}$

    $= n+2, n+2-2 = n$

39. a) You should obtain the number 5.
    b) You should obtain the number 5.
    c) Conjecture: The result is always the number 5.

    d) $n, n+1, n+(n+1)$

    $= 2n+1, 2n+1+9$

    $= 2n+10, \dfrac{2n+10}{2} = \dfrac{2n}{2} + \dfrac{10}{2}$

    $= n+5, n+5-n = 5$

41. $3 \times 5 = 15$ is one counterexample.

43. Two is a counting number. The sum of 2 and 3 is 5. Five divided by two is $\dfrac{5}{2}$, which is not an even number.

45. One and two are counting numbers. The difference of 1 and 2 is $1-2 = -1$, which is not a counting number.

47. a) The sum of the measures of the interior angles should be $180°$.

    b) Yes, the sum of the measures of the interior angles should be $180°$.

    c) Conjecture: The sum of the measures of the interior angles of a triangle is $180°$.

49. Inductive reasoning: a general conclusion is obtained from observation of specific cases.

51. 129, the numbers in positions are found as follows: $\begin{matrix} a & b \\ c & a+b+c \end{matrix}$

53. c

Exercise Set 1.2

(Note: Answers in this section will vary depending on how you round your numbers. The answers may differ from the answers in the back of the textbook. However, your answers should be something near the answers given. All answers are approximate.)

1. Estimation

3. $26.7 + 67.6 + 219 + 143.3$
   $\approx 30 + 67 + 220 + 143 = 460$

5. $197,500 \div 4.063 \approx 200,000 \div 4.000 = 50,000$

7. $\dfrac{405}{0.049} \approx \dfrac{400}{0.05} = 8000$

9. $51,608 \times 6981 \approx 50,000 \times 7000 = 350,000,000$

11. $22\% \times 9116 \approx 20\% \times 9000 = 0.20 \times 9000 = 1800$

13. $\dfrac{\$410}{4} \approx \dfrac{\$400}{4} = \$100$

15. $12 \, \text{months} \times \$47 \approx 12 \times \$50 = \$600$

17. $\$7.99 + \$4.23 + \$16.82 + \$3.51 + \$20.12$
       $\approx \$8 + \$4 + \$17 + \$4 + \$20 = \$53$

19. $95 \, \text{lb} + 127 \, \text{lb} + 210 \, \text{lb} \approx 100 + 100 + 200 = 400 \, \text{lb}$

21. 15% of $\$26.32 \approx 15\%$ of $\$26$
       $= 0.15 \times \$26 = \$3.90$

23. $\$595 + \$289 + \$120 + \$110 + 230$
       $\approx \$600 + \$300 + \$100 + \$100 + \$200 = \$1300$

25. $11 \times 8 \times \$1.50 \approx 10 \times 8 \times \$1.50$
       $= 10 \times \$12 = \$120$

27. 100 Mexican pesos $= 100 \times 0.068$ U.S. dollars
       $\approx 100 \times 0.07$ U.S. dollars $= 7$ U.S. dollars
       $\$50 - \$7 = \$43$

29. ≈ 90 miles

31.   a) 23% of $700 \approx 25\%$ of $700 = 0.25 \times 700 = 175$

       b) 12% of $700 \approx 10\%$ of $700 = 0.10 \times 700 = 70$

       c) 21% of $700 \approx 20\%$ of $700 = 0.20 \times 700 = 140$

33.   a) 5 million
       b) 98 million
       c) $98 \, \text{million} - 33 \, \text{million} = 65 \, \text{million}$
       d) $19 \, \text{million} + 79 \, \text{million} + 84 \, \text{million}$
          $+ 65 \, \text{million} + 33 \, \text{million} = 280 \, \text{million}$

35.   a) 85%
       b) $68\% - 53\% = 15\%$
       c) 85% of 70 million acres $= 59,500,000$ acres
       d) No, since we are not given the area of each
          state.

37.   20

39. ≈ 120 bananas

41. $150°$

43. 10%

45.   9 square units

47.   160 feet

49.-57. Answers will vary.

59.   a) Answers will vary.
       b) 11.6 days. There are
          $24 \cdot 60 \cdot 60 = 86400$ seconds in a day,
          $1,000,000 / 86400 = 11.574074.$

Exercise Set 1.3

1. $\dfrac{1 \, \text{in.}}{12 \, \text{mi}} = \dfrac{4.25 \, \text{in.}}{x \, \text{mi}}$
       $1x = 12(4.25)$
       $x = 51 \, \text{mi}$

3. $\dfrac{3 \, \text{ft}}{1.2 \, \text{ft}} = \dfrac{x \, \text{ft}}{15.36 \, \text{ft}}$
       $3(15.36) = 1.2x$
       $\dfrac{46.03}{1.2} = \dfrac{1.2x}{1.2}$
       $x = \dfrac{46.03}{1.2} = 38.4 \, \text{ft}$

5.   2.9% of $8642 $= 0.029 \times \$8642 = \$250.62$

      $\$8642 + \$250.62 = \$8892.62 \approx \$8893$

7.   a)   Ent./Misc.: 19.1% of $1950

               $= 0.191 \times \$1950 = \$372.45$

           Food: 12.7% of $1950

               $= 0.127 \times \$1950 = \$247.65$

         $\$372.45 - \$247.65 = \$124.80$

     b)   Housing: 34.4% of $1950

               $= 0.344 \times \$1950 = \$670.80$

       Transportation: 16% of $1950

               $= 0.16 \times \$1750 = \$312.00$

        $\$670.80 - \$312.00 = \$358.80$

9.   a)   20.7% of $200,000 $= 0.207 \times 200,000$

                         $= \$41,400$

      $\$200,000 + \$41,400 = \$241,400$

     b)   1.4% of $220,000 $= 0.014 \times 220,000$

                         $= \$3080$

      $\$220,000 + \$3080 = \$223,080$

     c)   Bremerton, WA: -30.9% of $200,000

                         $= -0.309 \times \$200,000$

                         $= -\$61,800$

      $\$200,000 - \$61,800 = \$138,200$

      Seattle, WA: -23.5% of $200,000

                         $= -0.235 \times \$200,000$

                         $= -\$47,000$

      $\$200,000 - \$47,000 = \$153,000$

      $\$153,000 - \$138,200 = \$14,800$

11. $\$250 + \$130(18) = \$250 + \$2340 = \$2590$

      Savings: $\$2590 - \$2500 = \$90$

13.   15 year mortgage: $\$887.63(12)(15)$

                   $= \$159,773.40$

     30 year mortgage: $\$572.90(12)(30)$

                   $= \$206,244.00$

     Savings: $\$206,244.00 - \$159,773.40$

                   $= \$46,470.60$

15.   a) $10 \times 10 \times 10 \times 10 = 10,000$

     b) 1 in 10,000

17. $38,687.0 \text{ mi} - 38,451.4 \text{ mi} = 235.6 \text{ mi}$

     $\dfrac{235.6 \text{ mi}}{12.6 \text{ gal}} \approx 18.698 \approx 18.7 \text{ mpg}$

19.   By mail: $(\$52.80 + \$5.60 + \$8.56) \times 4$

      $= \$66.96 \times 4 = \$267.84$

      Tire store: $\$324 + 0.08 \times \$324$

      $= \$324 + \$25.92 = \$349.92$

      Savings: $\$349.92 - \$267.84 = \$82.08$

21. $15,000 \text{ ft} - 3000 \text{ ft} = 12,000 \text{ ft}$ decrease in elevation. Temperature increases $2.4°\text{F}$ for every 1000 ft decrease in elevation.

     $2.4°\text{F} \times 12 = 28.8°\text{F}$

     $-6°\text{F} + 28.8°\text{F} = 22.8°\text{F}$

     The precipitation at the airport will be snow.

23.   The family paid more than $10,162.50 but less than $28,925, so they paid $10,162.50 plus 25% of the amount over $73,800.

     If x = adjusted gross income, taxes are

     $12,715 = 10,162.50 + .25(x - 73,800)$

     $12,715 = 10,162.50 + .25x - 18,450$

     $21,002.5 = .25x$

     $x = \$84,010$

25.   a) $1 \times 60 \times 24 \times 365$

        $\dfrac{525,600}{128} = 4106.25 \text{ gal}$

     b) $\dfrac{4106.25}{1000} \times \$11.20 = 4.10625 \times \$11.20$

        $= \$45.99$

27. a) $\dfrac{20,000}{20.8} - \dfrac{20,000}{21.6} \approx 961.538 - 925.926$

$= 35.612 \approx 35.61$ gal

b) $35.61 \times \$3.00 = \$106.83$

c) $140,000,000 \times 35.61 = 4,985,400,000$ gal

29. Cost after 1 year: $\$999 + 0.02(\$999)$

$= \$999 + \$19.98 = \$1018.98$

Cost after 2 years: $\$1018.98 + 0.02(\$1018.98)$

$= \$1018.98 + \$20.38 = \$1039.36$

31. After paying the $100 deductible, Yungchen must pay 20% of the cost of x-rays.

First x-ray:

$\$100 + 0.20(\$620) = \$100 + \$124 = \$224$

Second x-ray: $0.20(\$980) = \$196$

Total: $\$224 + \$196 = \$420$

33. a) water/milk: $3(1) = 3$ cups   salt: $3\left(\dfrac{1}{8}\right) = \dfrac{3}{8}$ tsp

Cream of wheat: $3(3) = 9$ tbsp $= \dfrac{9}{16}$ cup (because 16 tbsp = 1 cup)

b) water/milk: $\dfrac{2 + 3\frac{3}{4}}{2} = \dfrac{\frac{23}{4}}{2} = \dfrac{23}{8} = 2\dfrac{7}{8}$ cups

salt: $\dfrac{\frac{1}{4} + \frac{1}{2}}{2} = \dfrac{\frac{3}{4}}{2} = \dfrac{3}{8}$ tsp

cream of wheat: $\dfrac{\frac{1}{2} + \frac{3}{4}}{2} = \dfrac{\frac{5}{4}}{2} = \dfrac{5}{8}$ cups

c) water/milk: $3\dfrac{3}{4} - 1 = \dfrac{15}{4} - \dfrac{4}{4} = \dfrac{11}{4} = 2\dfrac{3}{4}$ cups

salt: $\dfrac{1}{2} - \dfrac{1}{8} = \dfrac{4}{8} - \dfrac{1}{8} = \dfrac{3}{8}$ tsp   cream of wheat: $\dfrac{3}{4} - \dfrac{3}{16} = \dfrac{12}{16} - \dfrac{3}{16} = \dfrac{9}{16}$ cup = 9 tbsp

d) Differences exist in water/milk because the amount for 4 servings is not twice that for 2 servings.

Differences also exist in Cream of Wheat because $\dfrac{1}{2}$ cup is not twice 3 tbsp.

35. a) $\$425 - \$240$ (one box of 20 DVDs) $= \$185$

$\$185 - \$180$ (one box of 12 DVDs) $= \$5$

One box of 20 DVDs and one box of 12 DVDs are the maximum number of DVDs that can be purchased.

b) $\$240 + \$180 = \$420$

37. 1 ft^2 would be 12 in. by 12 in.

Thus, 1 ft$^2 = 12$ in. $\times 12$ in. $= 144$ in.^{2}

39. Area of original rectangle $= lw$

Area of new rectangle $= (2l)(2w) = 4lw$

Thus, if the length and width of a rectangle are doubled, the area is 4 times as large.

41. Volume of original cube $= lwh$

Volume of new cube $= (2l)(2w)(2h) = 8lwh$ Thus, if the length, width, and height of a cube are doubled, the volume is 8 times as large.

43. $\dfrac{10 \text{ pieces}}{\$x} = \dfrac{1000 \text{ pieces}}{\$10}$

$1000x = 10(10)$

$\dfrac{1000x}{1000} = \dfrac{100}{1000}$

$x = \dfrac{100}{1000} = \$0.10 = 10¢$

45. 3

47. a) refresh
    b) workout

49.

51.

8	6	16
18	10	2
4	14	12

53. $8+6+2+4 = 20; 3+7+5+1 = 16;$

$10+14+12+8 = 44$

The sum of the four corner entries is
4 times the number in the center
of the middle row.

55. 45, 36, 99

Multiply the number in the center of the
middle row by 9.

57. $3 \times 2 \times 1 = 6$ ways

59.

	7	
3	1	4
5	8	6
	2	

Other answers are possible, but 1 and
8 must appear in the center.

61.

1	2	3	4	5
2	3	4	5	1
3	4	5	1	2
4	5	1	2	3
5	1	2	3	4

Other answers are possible.

63. Mark plays the drums.

65. Areas of the colored regions are:
    $1\times1,\ 1\times1,\ 2\times2,\ 3\times3,\ 5\times5,\ 8\times8,\ 13\times13,$
    $21\times21;\ 1+1+4+9+25+64+169+441$
    $\qquad\qquad = 714$ square units

67. Thomas would have opened the box labeled *grapes and cherries*. Because all the boxes are labeled incorrectly, whichever fruit he pulls from the box of grapes and cherries, will be the only fruit in that box. If he pulled a grape, he labeled the box *grape*. If he pulled a cherry he labeled the box *cherries*. That left two boxes whose original labels were incorrect. Because all labels must be changed, there was only one way for Thomas to assign the two remaining labels.

Review Exercises

1. 23, 28, 33 (Add 5 to previous number.)

2. 16, 13, 10 (Subtract 3 from the previous number)

3. 64, -128, 256 (Multiply previous number by –2.)

4. 25, 32, 40 (19 + 6 = 25, 25 + 7 = 32, 32 + 8 = 40)

5. 10, 4, -3 (subtract 1, then 2, then 3, ...)

6. $\dfrac{3}{8},\dfrac{3}{16},\dfrac{3}{32}$ (Multiply previous number by $\dfrac{1}{2}$.)

7.

8.

9. c

10. a) The final number is twice the original number.
    b) The final number is twice the original number.
    c) Conjecture: The final number is twice the original number.
    d) $n, 10n, 10n + 5, \dfrac{10n + 5}{5} = \dfrac{10n}{5} + \dfrac{5}{5} = 2n + 1, 2n + 1 - 1 = 2n$

11. This process will always result in an answer of 3. $n, n + 5, 6(n + 5) = 6n + 30, 6n + 30 - 12$
    $= 6n + 18, \dfrac{6n + 18}{2} = \dfrac{6n}{2} + \dfrac{18}{2} = 3n + 9, \dfrac{3n + 9}{3} = \dfrac{3n}{3} + \dfrac{9}{3} = n + 3, n + 3 - n = 3$

12. $1^2 + 2^2 = 5, 5$ is an odd number.

(Note: Answers for Ex. 13 - 25 will vary depending on how you round your numbers. The answers may differ from the answers in the back of the textbook. However, your answers should be something near the answers given. All answers are approximate.)

13. $205,123 \times 4002 \approx 200,000 \times 4000 = 800,000,000$

14. $215.9 + 128.752 + 3.6 + 861 + 792$
    $\approx 250 + 150 + 0 + 850 + 750 = 2000$

15. 21% of $2095 \approx 20\%$ of 2000
    $= 0.20 \times 2000 = 400$

16. Answers will vary.

17. 48 bricks $\times \$3.97 \approx 50 \times 4 = \$200$

18. 8% of $\$20,000 \approx 7\%$ of 2000
    $= 0.08 \times 20,000 = \$1600$

19. $\dfrac{1.1 \text{ mi}}{22 \text{ min}} \approx \dfrac{1 \text{ mi}}{20 \text{ min}} = \dfrac{3 \text{ mi}}{60 \text{ min}} = 3 \text{ mph}$

20. $\$2.49 + \$0.79 + \$1.89 +$
    $\$0.10 + \$2.19 + \$6.75$
    $\approx \$2 + \$1 + \$2 + \$0 + \$2 + \$7 = \$14.00$

21. $5 \text{ in.} = \dfrac{20}{4} \text{ in.} = 20\left(\dfrac{1}{4}\right) \text{ in.} = 20(0.1) \text{ mi} = 2 \text{ mi}$

22. $2.35 \text{ million} - 1.95 \text{ million} = 0.4 \text{ million}$

23. $2.8 \text{ million} - 1.8 \text{ million} = 1.0 \text{ million}$

24. 13 square units

25. $\text{Length} = 1.75 \text{ in.}, 1.75(12.5) = 21.875 \approx 22 \text{ ft}$

    $\text{Height} = 0.625 \text{ in.}, 0.625(12.5) = 7.8125 \approx 8 \text{ ft}$

26. $\$50 + \$40(12) = \$530$

    Savings: $\$530 - \$500 = \$30$

27. $4(\$1.99) = \$7.96$ for four six-packs

    Savings: $\$7.96 - \$4.99 = \$2.97$

28. Freemac: $\$15 \times 4 \times 2 = \$120$

    Sylvan: $\$25 \times 4 \text{ hours} = \$100$

    $\$120 - \$100 = \$20$

    Sylvan Rental is less expensive by \$20.

29. Cost per person with 5 people: $\dfrac{\$445}{5} = \$89$

    Cost per person with 6 people: $\dfrac{\$510}{6} = \$85$

    $\$89 - \$85 = \$4$ savings

30. a) $\dfrac{30 \text{ lb}}{2500 \text{ ft}^2} = \dfrac{x}{24{,}000 \text{ ft}^2}$;

    $x = \dfrac{30 \times 24{,}000}{2500} = 288 \text{ lb}$

    b) $\dfrac{150 \text{ lb}}{30 \text{ lb/bag}} = 5 \text{ bags, and } 5 \times 2500 = 12{,}500 \text{ ft}^2$

31. 10% of $\$1030 = 0.10 \times \$1030 = \$103$

    $\$103 \times 7 = \$721$

    Savings: $\$721 - \$60 = \$661$

32. $\dfrac{1.5 \text{ mg}}{10 \text{ lb}} = \dfrac{x \text{ mg}}{52 \text{ lb}}$

    $10x = 52(1.5)$

    $\dfrac{10x}{10} = \dfrac{78}{10}$

    $x = 7.8 \text{ mg}$

33. $\$5500 - 0.30(\$5500) = \$5500 - \$1650$

    $= \$3850$ take-home

    28% of $\$3850 = 0.28 \times \$3850 = \$1078$

34. 9 A.M. Eastern is 6 A.M. Pacific,
    from 6 A.M. Pacific to 1:35 P.M. Pacific
    is 7 hr 35 min , 7 hr 35 min $-$ 50 min stop
    = 6 hr 45 min

35. 3 P.M. $- 4$ hr = 11 A.M.
    July 26, 11:00 A.M.

36. a) $\dfrac{65 \text{ mi}}{1 \text{ hr}} \times \dfrac{1.6 \text{ km}}{\text{mi}} = \dfrac{104 \text{ km}}{1 \text{ hr}} \approx 104 \text{ km/hr}$

    b) $\dfrac{90 \text{ km}}{1 \text{ hr}} \times \dfrac{1 \text{ mi}}{1.6 \text{ km}} = \dfrac{90 \text{ mi}}{1.6 \text{ km}}$

    $\approx 56.25 \text{ mi/hr}$

37. Each figure has an additional two dots. To get the hundredth figure, 97 more figures must be drawn, $97(2) = 194$ dots added to the third figure. Thus, $194 + 7 = 201$.

38.

21	7	8	18
10	16	15	13
14	12	11	17
9	19	20	6

39.

23	25	15
13	21	29
27	17	19

40. 59 min 59 sec Since it doubles every second, the jar was half full 1 second earlier than 1 hour.

41. 6

42. Nothing. Each friend paid $9 for a total of $27; $25 to the hotel, $2 to the clerk.
$25 for the room + $3 for each friend + $2 for the clerk = $30

43. Let $x =$ the total score on the fifth exam
$$\frac{93 + 88 + 81 + 86 + x}{5} = 80, \quad \frac{348 + x}{5} = 80, \quad 348 + x = 400, \quad x = 52$$

44. Yes; 3 quarters and 4 dimes, or 1 half dollar, 1 quarter and 4 dimes, or 1 quarter and 9 dimes.
Other answers are possible.

45. $6 \text{ cm} \times 6 \text{ cm} \times 6 \text{ cm} = 216 \text{ cm}^3$

46. Place six coins in each pan with one coin off to the side. If it balances, the heavier coin is the one on the side. If the pan does not balance, take the six coins on the heavier side and split them into two groups of three. Select the three heavier coins and weigh two coins. If the pan balances, it is the third coin. If the pan does not balance, you can identify the heavier coin.

47. $1 + 500 = 501, 2 + 499 = 501, \ldots$
There are 250 such pairs and $250(501) = 125{,}250$.

48. 16 blue: 4 green $\rightarrow$ 8 blue, 2 yellow $\rightarrow$ 5 blue, 2 white $\rightarrow$ 3 blue

49. 90: 101, 111, 121, 131, 141, 151, 161, 171, 181, 191, …

50. The fifth figure will be an octagon with sides of equal length. Inside the octagon will be a seven sided figure with sides of equal length. The figure will have one antenna.

51. 61: The sixth figure will have 6 rows of 6 tiles and 5 rows of 5 tiles ($6 \times 6 + 5 \times 5 = 36 + 25 = 61$).

52. Some possible answers are given below. There are other possibilities.

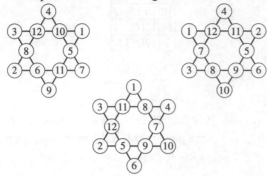

53. a) 2

b) There are 3 choices for the first spot. Once that person is standing, there are 2 choices for the second spot and 1 for the third. Thus, $3 \times 2 \times 1 = 6$.

c) $4 \times 3 \times 2 \times 1 = 24$

d) $5 \times 4 \times 3 \times 2 \times 1 = 120$

e) $n(n-1)(n-2) \cdots 1$, (or $n!$), where $n =$ the number of people in line

Chapter Test

1. 26, 32, 38   (Add 6 to previous number.)

2. $\dfrac{1}{5}, \dfrac{1}{6}, \dfrac{1}{7}$ (Add 1 to the denominator of the previous number.)

3. a) The result is the original number plus 1.

b) The result is the original number plus 1.

c) Conjecture: The result will always be the original number plus 1.

d) $n, 5n, 5n+10, \dfrac{5n+10}{5} = \dfrac{5n}{5} + \dfrac{10}{5} = n+2, n+2-1 = n+1$

(Note: Answers for #4 - #6 will vary depending on how you round your numbers. The answers may differ from the answers in the back of the textbook. However, your answers should be something near the answers given. All answers are approximate.)

4. $0.51 \times 96,000 \approx 0.5 \times 100,000 = 50,000$

5. $\dfrac{188,000}{0.11} \approx \dfrac{200,000}{0.1} \approx 2,000,000$

6.   9 square units

7.   a) $\dfrac{130 \text{ lb}}{63 \text{ in.}} \approx 2.0635$

$\dfrac{2.0635}{63 \text{ in.}} = 0.032754$

$0.032754 \times 703 \approx 23.03$

b) He is in the at risk range.

8.   a) 7.9 billion

b) 2.8 billion

9. $\$50.40 - \$35.00 = \$15.40$

$\dfrac{\$15.40}{\$0.20} = 77 \text{ miles}$

10. $\dfrac{\$15}{\$2.59} \approx 5.79$

The maximum number of 6 packs is 5.

$\$15.00 - (5 \times \$2.59) = \$15.00 - \$12.95 = \$2.05$

$\dfrac{\$2.05}{\$0.80} = 2.5625$

Thus, two individual cans can be purchased.

6 packs	Indiv. cans	Number of cans
5	2	32
4	5	29
3	9	27
2	12	24
1	15	21
0	18	18

The maximum number of cans is 32.

11. 1 cut yields 2 equal pieces.
Cut each of these 2 equal pieces to get
4 equal pieces. 3 cuts → 3(2.5 min) = 7.5 min

12. 2.5 in. by 1.875 in.

$\approx 2.5 \times 15.8$ by $1.875 \times 15.8 = 39.5$ in. by 29.625 in.

≈ 39.5 in. by 29.6 in.

(The actual dimensions are 100.5 cm by 76.5 cm.)

13. $\$12.75 \times 40 = \$510$
$\$12.75 \times 1.5 \times 10 = \$191.25$
$\$510 + \$191.25 = \$701.25$
$\$701.25 - \$652.25 = \$49.00$

14.

40	15	20
5	25	45
30	35	10

15. Mary drove the first 15 miles at 60 mph which took $\dfrac{15}{60} = \dfrac{1}{4}$ hr, and the second 15 miles at 30 mph which

took $\dfrac{15}{30} = \dfrac{1}{2}$ hr for a total time of $\dfrac{3}{4}$ hr. If she drove the entire 30 miles at 45 mph, the trip would take

$\dfrac{30}{45} = \dfrac{2}{3}$ hr (40 min) which is less than $\dfrac{3}{4}$ hr (45 min).

16. $3\,\text{tsp} = 1\,\text{tbsp}$

$\dfrac{6\,\text{lb}}{2\,\text{lb}} = 3;\ 3 \times \dfrac{1}{2}\,\text{tsp} = \dfrac{3}{2}\,\text{tsp}$ or $1\dfrac{1}{2}\,\text{tsp}$

$1\dfrac{1}{2}\,\text{tsp} = \dfrac{1}{2}\,\text{tbsp}$

17. Area of lawn including walkway: $(10+2) \times (12+2) = 12 \times 14 = 168\,\text{m}^2$

Area of lawn only: $10 \times 12 = 120\,\text{m}^2$

Area of walkway: $168 - 120 = 48\,\text{m}^2$

18. 243 jelly beans; $260 - 17 = 243, 234 + 9 = 243, 274 - 31 = 243$

19. a) $3 \times \$3.99 = \$11.97$

  b) $9(\$1.75 \times 0.75) = 11.8125 \approx \$11.81$

  c) $\$11.97 - \$11.81 = \$0.16$ Using the coupon is least expensive by $0.16.

20. 24 (The first position can hold any of four letters, the second any of the three remaining letters, and so on. $4 \times 3 \times 2 \times 1 = 24$

CHAPTER TWO

SETS

Exercise Set 2.1

1. Set
3. Description, Roster form, Set-builder notation
5. Infinite
7. Equivalent
9. Empty or null
11. Not well defined, "best" is interpreted differently by different people.
13. Well defined, the contents can be clearly determined.
15. Well defined, the contents can be clearly determined.
17. Infinite, the number of elements in the set is not a natural number.
19. Infinite, the number of elements in the set is not a natural number.
21. Infinite, the number of elements in the set is not a natural number.
23. $\{$ Hawaii $\}$

25. $\{11,12,13,14,...,177\}$ 27. $B = \{2, 4, 6, 8, ...\}$

29. $\{\ \}$ or $\varnothing$ 31. $E = \{14, 15, 16, 17, ..., 84\}$

33. $\{$ Bangkok, London, Paris, Singapore, New York $\}$

35. $\{\ \}$ or $\varnothing$

37. $\{2012, 2013, 2014\}$

39. $\{2011\}$ 41. $B = \{x | x \in N \text{ and } 6 < x < 15\}$ or
$B = \{x | x \in N \text{ and } 7 \leq x \leq 14\}$

43. $C = \{x | x \in N \text{ and } x \text{ is a multiple of } 3\}$ 45. $E = \{x | x \in N \text{ and } x \text{ is odd}\}$

47. $C = \{x | x \text{ is February}\}$

49. Set A is the set of natural numbers less than or equal to 7.
51. Set V is the set of vowels in the English alphabet
53. Set T is the set of species of trees.
55. Set S is the set of seasons.
57. $\{$ Facebook, Twitter, LinkedIn, Pinterest $\}$

59. $\{$ Twitter, LinkedIn, Pinterest $\}$ 61. $\{2011, 2012, 2013, 2014\}$

13

63. $\{\,2007,\ 2008,\ 2009,\ 2010\,\}$

65. False; $\{e\}$ is a set, and not an element of the set.

67. False; h is not an element of the set.

69. False; 3 is an element of the set.

71. True; 9 is an odd natural number.

73. $n(A) = 4$

75. $n(C) = 0$

77. Both; A and B contain exactly the same elements

79. Neither; the sets have a different number of elements.

81. Equivalent; both sets contain the same number of elements, 4.

83. a) Set A is the set of natural numbers greater than 2. Set B is the set of all numbers greater than 2.

    b) Set A contains only natural numbers. Set B contains other types of numbers, including fractions and decimal numbers.

    c) $A = \{\,3, 4, 5, 6, \ldots\}$

    d) No; because there are an infinite number of elements between any two elements in set B, we cannot write set B in roster form.

85. Cardinal; 7 tells how many.

87. Ordinal; 16th tells Lincoln's relative position.

89. Answers will vary

91. Answers will vary

Exercise Set 2.2

1. Subset

3. 2^n, where n is the number of elements in the set.

5. True; $\{$circle$\}$ is a subset of $\{$ square, circle, triangle $\}$.

7. False; potato is not in the second set.

9. True; $\{$ cheesecake, pie $\}$ is a subset of $\{$ pie, cookie, cheesecake, brownie $\}$.

11. False; no subset is a proper subset of itself.

13. True; spade is an element of $\{$ diamond, heart, spade, club $\}$.

15. False; $\{$quarter$\}$ is a set, not an element.

17. True; tiger is not an element of $\{$zebra, giraffe, polar bear$\}$

19. True; $\{$chair$\}$ is a proper subset of $\{$sofa, table, chair$\}$.

21. False; the set $\{\varnothing\}$ contains the element $\varnothing$.

23. False; the set $\{0\}$ contains the element 0.

25. False; 0 is a number and $\{\ \}$ is a set.

27. $B \subseteq A,\ B \subset A$

29. $A \subseteq B,\ A \subset B$

31. $B \subseteq A,\ B \subset A$

33. $A = B,\ A \subseteq B,\ B \subseteq A$

35. $\{\ \}$ is the only subset.

37. $\{\ \}, \{$cow$\}, \{$horse$\}, \{$cow, horse$\}$

39. a) $\{\ \},\{a\},\{b\},\{c\},\{d\},\{a,b\},\{a,c\},\{a,d\},$

   $\{b,c\},\{b,d\},\{c,d\},\{a,b,c\},\{a,b,d\},$

   $\{a,c,d\},\{b,c,d\},\{a,b,c,d\}$

   b) All the sets in part (a) are proper subsets of
   A except $\{a,b,c,d\}$.

41. False; A could be equal to B.

43. True; every set is a subset of itself.

45. True; $\varnothing$ is a proper subset of every set except itself.

47. True; every set is a subset of the universal set.

49. True; $\varnothing$ is a proper subset of every set except itself and $U \neq \varnothing$.

51. True; $\varnothing$ is a subset of every set.

53. The number of different variations is equal to the number of subsets of
   $\begin{Bmatrix} \text{cell phone holder, bluetooth radio, anchor lights, cupholders, depth guage, navigation lights,} \\ \text{dinette table, battery charging system, ski mirror} \end{Bmatrix}$,

   which is $2^9 = 2\times2\times2\times2\times2\times2\times2\times2\times2 = 512$.

55. The number of options is equal to the number of subsets of
   $\{$ cucumbers, onions, tomatoes, carrots, green peppers, olives, mushrooms$\}$, which is

   $2^7 = 2\times2\times2\times2\times2\times2\times2 = 128$.

57. $E = F$ since they are both subsets of each other.

59. a) Yes.

   b) No, c is an element of set D.

   c) Yes, each element of $\{a,b\}$ is an element of set D.

61. A one element set has one proper subset, namely the empty set. A one element set has two subsets, namely itself and the empty set. One is one-half of two. Thus, the set must have one element.

63. Yes

Section 2.3

1. Complement

3. Intersection

5. Cartesian

7. $m \times n$

9.

11.

13.

15.

17.

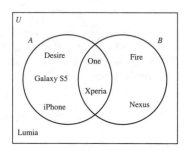

19.

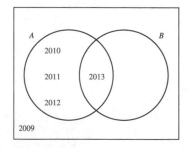

21.   The set of retail stores in the United States that do not sell children's clothing

23.   The set of cities in the United States that do not have a professional sports team

25.   The set of cities in the United States that have a professional sports team or a symphony

27.   The set of cities in the United States that have a professional sports team and do not have a symphony

29.   The set of furniture stores in the U.S. that sell mattresses or leather furniture

31. The set of furniture stores in the U.S. that do not sell outdoor furniture and sell leather furniture

33. The set of furniture stores in the U.S. that sell mattresses or outdoor furniture or leather furniture

35. $A = \{a, b, c, h, t, w\}$

37. $A \cap B = \{a, b, c, h, t, w,\} \cap \{a, d, f, g, h, r, v\} = \{a, h\}$

39. $A \cup B = \{a, b, c, h, t, w\} \cup \{a, d, f, g, h, r, v\} = \{a, b, c, d, f, g, h, r, t, v, w\}$

41. $A' \cap B' = \{d, f, g, m, p, r, v, z\} \cap \{b, c, m, p, t, w, z\} = \{m, p, z\}$

43. $A - B = \{b, c, t, w\}$

45. $A = \{1, 2, 7, 8, 9\}$

47. $U = \{1, 2, 3, 4, 5, 6, 7, 8, 9, 10, 11\}$

49. $A' \cup B = \{3, 4, 5, 6, 10, 11\} \cup \{2, 3, 5, 6, 9\} = \{2, 3, 4, 5, 6, 9, 10, 11\}$

51. $A' \cap B = \{3, 4, 5, 6, 10, 11\} \cup \{2, 3, 5, 6, 9\} = \{3, 5, 6\}$

53. $A' - B = \{4, 10, 11\}$

55. $A \cup B = \{1, 2, 4, 5, 7\} \cup \{2, 3, 5, 6\} = \{1, 2, 3, 4, 5, 6, 7\}$

57. $(A \cup B)'$ From #55, $A \cup B = \{1, 2, 3, 4, 5, 6, 7\}$. $(A \cup B)' = \{1, 2, 3, 4, 5, 6, 7\}' = \{8\}$

59. $(A \cup B)' \cap B$: From #57, $(A \cup B)' = \{8\}$. $(A \cup B)' \cap B = \{8\} \cap \{2, 3, 5, 6\} = \{\ \}$

61. $(B \cup A)' \cap (B' \cup A')$: From #57, $(A \cup B)' = (B \cup A)' = \{8\}$.

$(B \cup A)' \cap (B' \cup A') = \{8\} \cap \left[\{2, 3, 5, 6\}' \cup \{1, 2, 4, 5, 7\}'\right] = \{8\} \cap (\{1, 4, 7, 8\} \cup \{3, 6, 8\})$

$= \{8\} \cap \{1, 3, 4, 6, 7, 8\} = \{8\}$

63. $(A - B)' = \{2, 3, 5, 6, 8\}$

65. $B \cup C = \{b, c, d, f, g\} \cup \{a, b, f, i, j\} = \{a, b, c, d, f, g, i, j\}$

67. $A' = \{b, e, h, j, k\}$, $B' = \{a, e, h, i, j, k\}$.

$A' \cup B' = \{b, e, h, j, k\} \cup \{a, e, h, i, j, k\} = \{a, b, e, h, i, j, k\}$

69. $(A \cap B) \cup C = (\{a, c, d, f, g, i\} \cap \{b, c, d, f, g\}) \cup \{a, b, f, i, j\}$

$= \{c, d, f, g\} \cup \{a, b, f, i, j\}$

$= \{a, b, c, d, f, g, i, j\}$

71. $(A' \cup C) \cup (A \cap B) = \left[\{a, c, d, f, g, i\}' \cup \{a, b, f, i, j\}\right] \cup (\{a, c, d, f, g, i\} \cap \{b, c, d, f, g\})$

$= (\{b, e, h, j, k\} \cup \{a, b, f, i, j\}) \cup \{c, d, f, g\} = \{a, b, e, f, h, i, j, k\} \cup \{c, d, f, g\}$

$= \{a, b, c, d, e, f, g, h, i, j, k\}$, or U

73. $A - B = \{a, i\}$, $(A - B)' = \{b, c, d, e, f, g, h, j, k\}$, $C = \{a, b, f, i, j\}$

$(A - B)' - C = \{c, d, e, g, h, k\}$

75. $\{(1, a), (1, b), (2, a), (2, b), (3, a), (3, b)\}$

77. No. The ordered pairs are not the same. For example $(1, a) \neq (a, 1)$

79. 6

81. $A \cap B = \{1,3,5,7,9\} \cap \{2,4,6,8\} = \{\ \}$

83. $(B \cup C)' = (\{2,4,6,8\} \cup \{1,2,3,4,5\})' = \{1,2,3,4,5,6,8\}' = \{7,9\}$

85. $A \cap B' = \{1,3,5,7,9\} \cap \{2,4,6,8\}' = \{1,3,5,7,9\} \cap \{1,3,5,7,9\} = \{1,3,5,7,9\}$, or A

87. $(A \cup C) \cap B = (\{1,3,5,7,9\} \cup \{1,2,3,4,5\}) \cap \{2,4,6,8\} = \{1,2,3,4,5,7,9\} \cap \{2,4,6,8\} = \{2,4\}$

89. $(A' \cup B') \cap C = \left[\{1,3,5,7,9\}' \cup \{2,4,6,8\}'\right] \cap \{1,2,3,4,5\}$

$= (\{2,4,6,8\} \cup \{1,3,5,7,9\}) \cap \{1,2,3,4,5\} = \{1,2,3,4,5,6,7,8,9\} \cap \{1,2,3,4,5\} = \{1,2,3,4,5\}$, or C

91. A set and its complement will always be disjoint since the complement of a set is all of the elements in the universal set that are not in the set. Therefore, a set and its complement will have no elements in common.

For example, if $U = \{1,2,3\}, A = \{1,2\}$, and $A' = \{3\}$, then $A \cap A' = \{\ \}$.

93. Let $A = \{$ customers who owned dogs $\}$ and $B = \{$ customers who owned cats$\}$.

$n(A \cup B) = n(A) + n(B) - n(A \cap B) = 27 + 38 - 16 = 49$

95. a) $A \cup B = \{a,b,c,d\} \cup \{b,d,e,f,g,h\} = \{a,b,c,d,e,f,g,h\}, n(A \cup B) = 8,$

$A \cap B = \{a,b,c,d\} \cap \{b,d,e,f,g,h\} = \{b,d\}, n(A \cap B) = 2.$

$n(A) + n(B) - n(A \cap B) = 4 + 6 - 2 = 8$

Therefore, $n(A \cup B) = n(A) + n(B) - n(A \cap B)$.

b) Answers will vary.

c) Elements in the intersection of A and B are counted twice in $n(A) + n(B)$.

97. $A \cup B = \{1,2,3,4,\ldots\} \cup \{4,8,12,16,\ldots\} = \{1,2,3,4,\ldots\}$, or A

99. $B \cup C = \{4,8,12,16,\ldots\} \cup \{2,4,6,8,\ldots\} = \{2,4,6,8,\ldots\}$, or C

101. $A \cap C = \{1,2,3,4,\ldots\} \cap \{2,4,6,8,\ldots\} = \{2,4,6,8,\ldots\}$, or C

103. $(B \cup C)' \cup C$: From #99, $B \cup C = C$. $(B \cup C)' \cup C = C' \cup C = \{2,4,6,8,\ldots\}' \cup \{2,4,6,8,\ldots\}$

$= \{0,1,2,3,4,\ldots\}$, or U

105. $A \cap A' = \{\ \}$

107. $A \cup \varnothing = A$

109. $A' \cup U = U$

111. If $A \cap B = B$, then $B \subseteq A$.

113. If $A \cap B = \varnothing$, then A and B are disjoint sets.

Exercise Set 2.4

1. 8

3. a) $A' \cap B'$

b) $A' \cup B'$

5. $A' \cap B'$ is represented by regions V and VI. If $B \cap C$ contains 12 elements and region V contains 4 elements, then region VI contains $12 - 4 = 8$ elements

7.

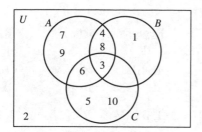

9.

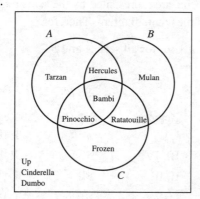

11.

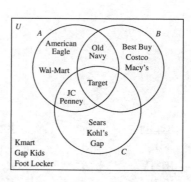

13.

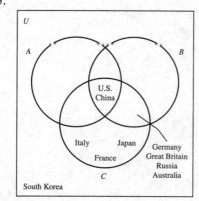

15. Canada, VI

17. Greece, V

19. Spain, VIII

21. VI

23. III

25. III

27. V

29. II

31. VII

33. I

35. VIII

37. VI

39. $A = \{1, 3, 4, 5, 9, 10\}$

41. $C = \{4, 5, 6, 8, 9, 11\}$

43. $A \cap B = \{3, 4, 5\}$

45. $(B \cap C)' = \{1, 2, 3, 7, 9, 10, 11, 12, 13, 14\}$

47. $(A \cup C)' = \{2, 7, 12, 13, 14\}$

49. $(A - B)' = \{2, 3, 4, 5, 6, 7, 8, 11, 12, 13, 14\}$

51. $(A \cap B)'$ $A' \cup B'$

Set	Regions	Set	Regions
A	I, II	A	I, II
B	II, III	A'	III, IV
$A \cap B$	II	B	II, III
$(A \cap B)'$	I, III, IV	B'	I, IV
		$A' \cup B'$	I, III, IV

Both statements are represented by the same regions, I, III, IV, of the Venn diagram. Therefore,

$(A \cap B)' = A' \cup B'$ for all sets A and B.

53. $A' \cup B'$ $A \cap B$

Set	Regions	Set	Regions
A	I, II	A	I, II
A'	III, IV	B	II, III
B	II, III	$A \cap B$	II
B'	I, IV		
$A' \cup B'$	I, III, IV		

Since the two statements are not represented by the same regions, it is not true that $A' \cup B' = A \cap B$ for all sets A and B.

55. $A' \cap B'$ $A \cup B'$

Set	Regions	Set	Regions
A	I, II	A	I, II
A'	III, IV	B	II, III
B	II, III	B'	I, IV
B'	I, IV	$A \cup B'$	I, II, IV
$A' \cap B'$	IV		

Since the two statements are not represented by the Same regions, it is not true that $A' \cap B' = A \cup B'$ for all sets A and B.

57. $A \cap (B \cup C)$ $(A \cap B) \cup C$

Set	Regions	Set	Regions
B	II, III, V, VI	A	I, II, IV, V
C	IV , V, VI, VII	B	II, III, V, VI
$B \cup C$	II, III, IV, V, VI, VII	$A \cap B$	II, V
A	I, II, IV, V	C	IV, V, VI, VII
$A \cap (B \cup C)$	II, IV, V	$(A \cap B) \cup C$	II, IV, V, VI, VII

Since the two statements are not represented by the sameregions, it is not true that $A \cap (B \cup C) = (A \cap B) \cup C$ for all sets A and B and C..

59. $A \cap (B \cup C)$ $(B \cup C) \cap A$

Set	Regions	Set	Regions
B	II, III, V, VI	B	II, III, V, VI
C	IV , V, VI, VII	C	IV, V, VI, VII
$B \cup C$	II, III, IV, V, VI, VII	$B \cup C$	II, III, IV, V, VI, VII
A	I, II, IV, V	A	I, II, IV, V
$A \cap (B \cup C)$	II, IV, V	$(B \cup C) \cap A$	II, IV, V

Both statements are represented by the same regions, II, IV, V, of the Venn diagram.

Therefore, $A \cap (B \cup C) = (B \cup C) \cap A$ for all sets $A, B,$ and C.

61. $A \cap (B \cup C)$ $(A \cap B) \cup (A \cap C)$

Set	Regions	Set	Regions
B	II, III, V, VI	A	I, II, IV, V
C	IV, V, VI, VII	B	II, III, V, VI
$B \cup C$	II, III, IV, V, VI, VII	$A \cap B$	II, V
A	I, II, IV, V	C	IV, V, VI, VII
$A \cap (B \cup C)$	II, IV, V	$A \cap C$	IV, V
		$(A \cap B) \cup (A \cap C)$	II, IV, V

Both statements are represented by the same regions, II, IV, V, of the Venn diagram.

Therefore, $A \cap (B \cup C) = (A \cap B) \cup (A \cap C)$ for all sets $A, B,$ and C.

63. $(A \cup B) \cap (B \cup C)$ $B \cup (A \cap C)$

Set	Regions	Set	Regions
A	I, II, IV, V	A	I, II, IV, V
B	II, III, V, VI	C	IV, V, VI, VII
$A \cup B$	I, II, III, IV, V, VI	$A \cap C$	IV, V
C	IV, V, VI, VII	B	II, III, V, VI
$B \cup C$	II, III, IV, V, VI, VII	$B \cup (A \cap C)$	II, III, IV, V, VI
$(A \cup B) \cap (B \cup C)$	II, III, IV, V, VI		

Both statements are represented by the same regions, II, III, IV, V, VI, of the Venn diagram.

Therefore, $(A \cup B) \cap (B \cup C) = B \cup (A \cap C)$ for all sets $A, B,$ and C.

65. $(A \cup B)'$

67. $(A \cup B) \cap C'$

69. a) $A \cap B = \{3, 6\}, (A \cap B)' = \{1, 2, 4, 5, 7, 8\}$

   $A' = \{4, 5, 7, 8\}, B' = \{1, 2, 4, 5, 8\}$

   $A' \cup B' = \{1, 2, 4, 5, 7, 8\}$

   Both equal $\{1, 2, 4, 5, 7, 8\}$

   b) Answers will vary.

71.

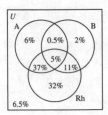

73. a) A: Office Building Construction Projects, B: Plumbing Projects, C: Budget Greater Than $300,000

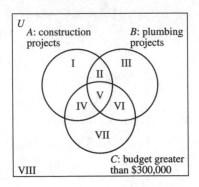

b) Region V; $A \cap B \cap C$

c) Region VI; $A' \cap B \cap C$

d) Region I; $A \cap B' \cap C'$

75. a)

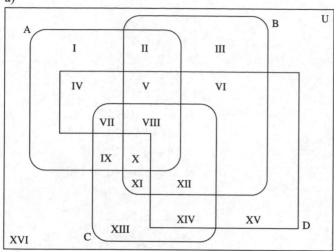

b)

Region	Set	Region	Set
I	$A \cap B' \cap C' \cap D'$	IX	$A \cap B' \cap C \cap D'$
II	$A \cap B \cap C' \cap D'$	X	$A \cap B \cap C \cap D'$
III	$A' \cap B \cap C' \cap D'$	XI	$A' \cap B \cap C \cap D'$
IV	$A \cap B' \cap C' \cap D$	XII	$A' \cap B \cap C \cap D$
V	$A \cap B \cap C' \cap D$	XIII	$A' \cap B' \cap C \cap D'$
VI	$A' \cap B \cap C' \cap D$	XIV	$A' \cap B' \cap C \cap D$
VII	$A \cap B' \cap C \cap D$	XV	$A' \cap B' \cap C' \cap D$
VIII	$A \cap B \cap C \cap D$	XVI	$A' \cap B' \cap C' \cap D'$

Exercise Set 2.5

1.

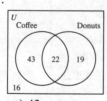

a) 43
b) 19
c) $100 - (43 + 22 + 19)$, or 16

3.

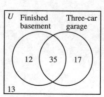

a) 12   b) 17
c) 64, the sum of the numbers in Regions I, II, III

5.

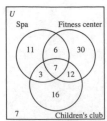

a) 11   b) 11+30+16, or 57

c) 11+6+30+3+7+12+16, or 85

d) 3+6+12, or 21

e) 7

7.

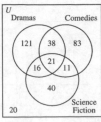

a) 20   b) 121

c) 121+83+40, or 244

d) 16+38+11, or 65

e) 350−20−40, or 290

9.

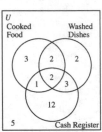

a) 3   b) 12

c) 3

d) 12+3+2, or 17

e) 1+2+3+2, or 8

11.

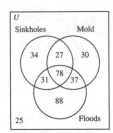

a) 30+37, or 67

b) 350−25−88, or 237

c) 37   d) 25

13. The Venn diagram shows the number of cars driven by women is 37, the sum of the numbers in Regions II, IV, V. This exceeds the 35 women the agent claims to have surveyed.

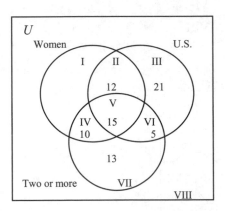

15. First fill in 125, 100, and 90 on the Venn diagram. Referring to the labels in the Venn diagram and the given information, we see that

$a+c = 60$

$b+c = 50$

$a+b+c = 200-125 = 75$

Adding the first two equations and subtracting the third from this sum gives $c = 60+50-75 = 35$.

Then $a = 25$ and $b = 15$. Then $d = 180-110-25-35 = 10$. We now have labeled all the regions except the region outside the three circles, so the number of farmers growing at least one of the crops is $125+25+110+15+35+10+90$, or 410. Thus the number growing none of the crops is $500-410$, or 90.

a) 410

b) 35

c) 90

d) $15+25+10$, or 50

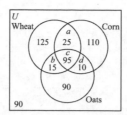

17. From the given information we can generate the Venn diagram. First fill in 4 for Region V. Then since the intersections in pairs all have 6 elements, we can fill in 2 for each of Regions II, IV, and VI. This already accounts for the 10 elements $A \cup B \cup C$, so the remaining 2 elements in U must be in Region VIII.

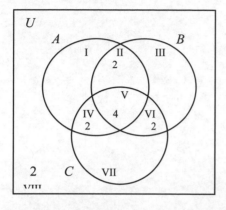

a) 10, the sum of the numbers in Regions I, II, III, IV, V, VI

b) 10, the sum of the numbers in Regions III, IV, V, VI, VIII

c) 6, the sum of the numbers in Regions I, III, IV, VI, VIII

Exercise Set 2.6

1. Infinite

3. $\{2, 3, 4, 5, 6, \ldots, n+1, \ldots\}$
   $\downarrow\ \downarrow\downarrow\ \downarrow\downarrow\qquad\ \downarrow$
   $\{3, 4, 5, 6, 7, \ldots, n+2, \ldots\}$

5. $\{3, 5, 7, 9, 11, \ldots, 2n+1, \ldots\}$
   $\downarrow\downarrow\ \downarrow\ \downarrow\ \downarrow\qquad\ \downarrow$
   $\{5, 7, 9, 11, 13, \ldots, 2n+3, \ldots\}$

7. $\{5, 7, 9, 11, 13\ldots, 2n+3, \ldots\}$
   $\downarrow\downarrow\ \downarrow\ \downarrow\ \downarrow\quad\ \downarrow$
   $\{9, 11, 13, 15, \ldots, 2n+5, \ldots\}$

9. $\left\{\dfrac{1}{2}, \dfrac{1}{4}, \dfrac{1}{6}, \dfrac{1}{8}, \dfrac{1}{10}, \ldots, \dfrac{1}{2n}, \ldots\right\}$
   $\downarrow\downarrow\downarrow\downarrow\ \downarrow\qquad\ \downarrow$
   $\left\{\dfrac{1}{4}, \dfrac{1}{6}, \dfrac{1}{8}, \dfrac{1}{10}, \ldots, \dfrac{1}{2n+2}, \ldots\right\}$

11. $\left\{ \dfrac{4}{11}, \dfrac{5}{11}, \dfrac{6}{11}, \dfrac{7}{11}, \dots, \dfrac{n+3}{11}, \dots \right\}$

$\quad\quad \downarrow \ \downarrow \ \downarrow \ \downarrow \quad\quad \downarrow$

$\left\{ \dfrac{5}{11}, \dfrac{6}{11}, \dfrac{7}{11}, \dfrac{8}{11}, \dots, \dfrac{n+4}{11}, \dots \right\}$

13. $\{1, 2, 3, 4, 5, \dots, n, \dots\}$

$\quad\ \downarrow \downarrow \downarrow \downarrow \downarrow \quad\ \downarrow$

$\{2, 4, 6, 8, 10, \dots, 2n, \dots\}$

15. $\{1, 2, 3, 4, 5, \dots, n, \dots\}$

$\quad\ \downarrow \downarrow \downarrow \downarrow \downarrow \quad\ \downarrow$

$\{3, 5, 7, 9, 11, \dots, 2n+1, \dots\}$

17. $\{1, 2, 3, 4, 5, \dots, n, \dots\}$

$\quad\ \downarrow \downarrow \downarrow \downarrow \downarrow \quad\ \downarrow$

$\{2, 5, 8, 11, 14, \dots, 3n-1, \dots\}$

19. $\{1, 2, 3, 4, 5, \dots, n, \dots\}$

$\quad\ \downarrow \downarrow \downarrow \downarrow \downarrow \quad\ \downarrow$

$\left\{ \dfrac{1}{3}, \dfrac{1}{6}, \dfrac{1}{9}, \dfrac{1}{12}, \dfrac{1}{15}, \dots, \dfrac{1}{3n}, \dots \right\}$

21. $\{1, 2, 3, 4, 7, \dots, n, \dots\}$

$\quad\ \downarrow \downarrow \downarrow \downarrow \quad\ \downarrow$

$\left\{ \dfrac{1}{3}, \dfrac{1}{4}, \dfrac{1}{5}, \dfrac{1}{6}, \dfrac{1}{7}, \dots, \dfrac{1}{n+2}, \dots \right\}$

23. $\{1, 2, 3, 4, 5, \dots, n, \dots\}$

$\quad\ \downarrow \downarrow \downarrow \downarrow \downarrow \quad\ \downarrow$

$\{1, 4, 9, 16, 25, \dots, n^2, \dots\}$

25. $\{1, 2, 3, 4, 5, \dots, n, \dots\}$

$\quad\ \downarrow \downarrow \downarrow \downarrow \downarrow \quad\ \downarrow$

$\{3, 9, 27, 81, 243, \dots, 3^n, \dots\}$

27. $=$

29. $=$

31. $=$

Review Exercises

1. True

2. False; the word *best* makes the statement not well defined.

3. True

4. False; no set is a proper subset of itself.

5. False; the elements 6, 12, 18, 24, … are members of both sets.

6. True

7. False; the two sets do not contain exactly the same elements.

8. True

9. True

10. True

11. True

12. True

13. True

14. True

15. $A = \{7, 9, 11, 13, 15\}$

16. $B = \{$ Colorado, Nebraska, Missouri, Oklahoma $\}$

17. $C = \{1, 2, 3, 4, ..., 174\}$

18. $D = \{9, 10, 11, 12, ..., 80\}$

19. $A = \{x \mid x \in N \text{ and } 50 < x < 150\}$ or $A = \{x \mid x \text{ in } N \text{ and } 51 \leq x \leq 149\}$

20. $B = \{x \mid x \in N \text{ and } x > 42\}$

21. $C = \{x \mid x \in N \text{ and } x < 7\}$

22. $D = \{x \mid x \in N \text{ and } 27 \leq x \leq 51\}$

23. A is the set of capital letters in the English alphabet from E through M, inclusive.

24. B is the set of U.S. coins with a value of less than one dollar.

25. C is the set of the first three lowercase letters in the English alphabet.

26. D is the set of numbers greater than or equal to 3 and less than 9.

27. $A \cap B = \{1, 3, 5, 7\} \cap \{3, 7, 9, 10\} = \{3, 7\}$

28. $A \cup B' = \{1, 3, 5, 7\} \cup \{3, 7, 9, 10\}' = \{1, 3, 5, 7\} \cup \{1, 2, 4, 5, 6, 8\} = \{1, 2, 3, 4, 5, 6, 7, 8\}$

29. $A' \cap B = \{1, 3, 5, 7\}' \cap \{3, 7, 9, 10\} = \{2, 4, 6, 8, 9, 10\} \cap \{5, 7, 9, 10\} = \{9, 10\}$

30. $(A \cup B)' \cup C = (\{1, 3, 5, 7\} \cup \{3, 7, 9, 10\})' \cup \{1, 7, 10\} = \{1, 3, 5, 7, 9, 10\}' \cup \{1, 7, 10\}$

    $= \{2, 4, 6, 8\} \cup \{1, 7, 10\} = \{1, 2, 4, 6, 7, 8, 10\}$

31. $A - B = \{1, 3, 5, 7\} - \{3, 7, 9, 10\} = \{1, 5\}$

32. $A - C' = \{1, 3, 5, 7\} - \{1, 7, 10\}' = \{1, 3, 5, 7\} - \{2, 3, 4, 5, 6, 8, 9\} = \{1, 7\}$

33. $\{(1, 1), (1, 7), (1, 10), (3, 1), (3, 7), (3, 10), (5, 1), (5, 7), (5, 10), (7, 1), (7, 7), (7, 10)\}$

34. $\{(3, 1), (3, 3), (3, 5), (3, 7), (7, 1), (7, 3), (7, 5), (7, 7), (9, 1), (9, 3), (9, 5), (9, 7), (10, 1), (10, 3), (10, 5), (10, 7)\}$

35. $2^4 = 2 \times 2 \times 2 \times 2 = 16$

36. $2^4 - 1 = (2 \times 2 \times 2 \times 2) - 1 = 16 - 1 = 15$

37.

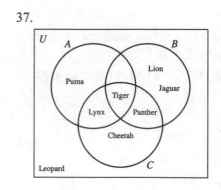

38. $A \cup C = \{\, b, c, d, e, f, h, k, l \,\}$

39. $A \cap B' = \{\, e, k \,\}$

40. $A \cup B \cup C = \{\, a, b, c, d, e, f, g, h, k, l \,\}$

41. $A \cap B \cap C = \{\, f \,\}$

42. $(A \cup B) \cap C = \{\, c, e, f \,\}$

43. $A - B' = \{\, d, f, l \,\}$

44. $\left(A' \cup B'\right)'$ $\qquad\qquad A \cap B$

Set	Regions
A	I, II
A'	III, IV
B	II, III
B'	I, IV
$A' \cup B'$	I, III, IV
$\left(A' \cup B'\right)'$	II

Set	Regions
A	I, II
B	II, III
$A \cap B$	II

Both statements are represented by the same region, II, of the Venn diagram. Therefore, $\left(A' \cup B'\right)' = A \cap B$ for all sets A and B.

45. $\left(A \cup B'\right) \cup \left(A \cup C'\right)$ $\qquad\qquad\qquad\qquad A \cup \left(B \cap C\right)'$

Set	Regions
A	I, II, IV, V
B	II, III, V, VI
B'	I, IV, VII, VIII
$A \cup B'$	I, II, IV, V, VII, VIII
C	IV, V, VI, VII
C'	I, II, III, VIII
$A \cup C'$	I, II, III, IV, V, VIII
$\left(A \cup B'\right) \cup \left(A \cup C'\right)$	I, II, III, IV, V, VII, VIII

Set	Regions
B	II, III, V, VI
C	IV, V, VI, VII
$B \cap C$	V, VI
$\left(B \cap C\right)'$	I, II, III, IV, VII, VIII
A	I, II, IV, V
$A \cup \left(B \cap C\right)'$	I, II, III, IV, V, VII, VIII

Both statements are represented by the same regions, I, II, III, IV, V, VII, VIII, of the Venn diagram.

Therefore, $\left(A \cup B'\right) \cup \left(A \cup C'\right) = A \cup \left(B \cap C\right)'$ for all sets $A, B,$ and C.

46. II

47. III

48. I

49. IV

50. IV

51. II

52. II

53. The company paid $450 since the sum of the numbers in Regions I through IV is 450.

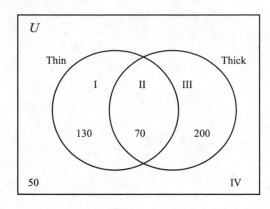

54. a) 131, the sum of the numbers in Regions I through VII
    b) 32, Region I
    c) 10, Region II
    d) 65, the sum of the numbers in Regions I, IV, VII

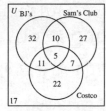

55. a) 128, Region I
    b) 264, the sum of the numbers in Regions I, III, VII
    c) 40, Region II
    d) 168, the sum of the numbers in Regions III, VI
    e) 99, the sum of the numbers in Regions II, IV, VI

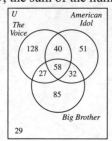

56. {2, 4, 6, 8, 10, ..., 2n, ...}
    ↓ ↓ ↓ ↓ ↓   ↓
    {4, 6, 8, 10, 12, ..., 2n + 2, ...}

57. {3, 5, 7, 9, 11, ..., 2n + 1, ...}
    ↓ ↓ ↓ ↓ ↓   ↓
    {5, 7, 9, 11, 13, ..., 2n + 3, ...}

58. {1, 2, 3, 4, 5, ...,   n, ...}
    ↓ ↓ ↓ ↓ ↓   ↓
    {5, 8, 11, 14, 17, ..., 3n + 2, ...}

59. {1, 2, 3, 4,   5, ...,   n, ...}
    ↓ ↓ ↓ ↓ ↓   ↓
    {4, 9, 14, 19, 24, ..., 5n − 1, ...}

Chapter Test

1. True

2. False; the sets do not contain exactly the same elements.

3. True

4. False; the second set does not contain the element 7.

5. False; the set has $2^4 = 2 \times 2 \times 2 \times 2 = 16$ subsets.

6. True

7. False; for any set A, $A \cup A' = U$, not $\{ \ \}$.

8. True

9. $A = \{1, 2, 3, 4, 5, 6, 7, 8, 9, 10, 11\}$

10. Set A is the set of natural numbers less than 12.

11. $A \cap B = \{3, 5, 7, 9\} \cap \{7, 9, 11, 15\} = \{7, 9\}$

12. $A \cup C' = \{3, 5, 7, 9\} \cup \{3, 11, 15\}'$
    $= \{3, 5, 7, 9\} \cup \{1, 5, 7, 9, 13\} = \{1, 3, 5, 7, 9, 13\}$

13. $A \cap \left(B \cap C'\right) = \{3, 5, 7, 9\} \cap \left(\{7, 9, 11, 15\} \cap \{1, 5, 7, 9, 13\}\right) = \{3, 5, 7, 9\} \cap \{7, 9\} = \{7, 9\}$

14. $n\left(A \cap B'\right) = n\left(\{3, 5, 7, 9\} \cap \{7, 9, 11, 15\}'\right) = n\left(\{3, 5, 7, 9\} \cap \{1, 3, 5, 15\}\right) = n\left(\{3, 5\}\right) = 2$

15. $A - B = \{3, 5, 7, 9\} - \{7, 9, 11, 15\} = \{3, 5\}$

16. $A \times C = \{(3, 3), (3, 11), (3, 15), (5, 3), (5, 11), (5, 15), (7, 3), (7, 11), (7, 15), (9, 3), (9, 11), (9, 15)\}$

17.
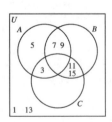

18. $A \cap (B \cup C')$ $(A \cap B) \cup (A \cap C')$

Set	Regions		Set	Regions
B	II, III, V, VI		A	I, II, IV, V
C	IV, V, VI, VII		B	II, III, V, VI
C'	I, II, III, VIII		$A \cap B$	II, V
$B \cup C'$	I, II, III, V, VI, VIII		C	IV, V, VI, VII
A	I, II, IV, V		C'	I, II, III, VIII
$A \cap (B \cup C')$	I, II, V		$A \cap C'$	I, II
			$(A \cap B) \cup (A \cap C')$	I, II, V

Both statements are represented by the same regions, I, II, V, of the Venn diagram.

Therefore, $A \cap (B \cup C') = (A \cap B) \cup (A \cap C')$ for all sets $A, B,$ and C.

19.

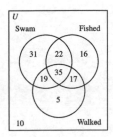

a) 58, the sum of the numbers in Regions II, IV, VI
b) 10, Region VIII
c) 145, the sum of the numbers in all regions except VIII
d) 22, Region II
e) 69, the sum of the numbers in Regions I, II, III
f) 16, Region III

20. $\{7, 8, \ 9, \ 10, 11, ..., n + 6, ...\}$
      $\ \downarrow \downarrow \ \downarrow \ \downarrow \ \downarrow \qquad \downarrow$
      $\{8, 9, \ 10, 11, 12, ..., n + 7, ...\}$

CHAPTER THREE

LOGIC

Exercise Set 3.1

1. Statement

3. Compound

5. a) Not   b) And
   c) Or

7. Some

9. Compound; biconditional ↔

11. Compound; conjunction, ∧

13. Compound; conditional, →

15. Compound; negation, ~

17. Some Mustangs are not Fords

19. All turtles have claws.

21. Some bicycles have three wheels.

23. No pedestrians are in the crosswalk.

25. Some mountain climbers are teachers.

27. $\sim p$

29. $\sim q \vee \sim p$

31. $\sim p \rightarrow \sim q$

33. $\sim q \leftrightarrow p$

35. $\sim p \wedge q$

37. $\sim (p \vee q)$

39. Brie does not have a MacBook.

41. Joe has an iPad and Brie has a Macbook.

43. If Joe does not have an iPad then Brie has a MacBook.

45. Joe does not have an iPad or Brie does not have a MacBook.

47. It is false that Joe has an iPad and Brie has a MacBook.

49. $(p \wedge \sim q) \wedge r$

51. $(p \wedge q) \vee r$

53. $(r \wedge q) \rightarrow p$

55. $(r \leftrightarrow q) \wedge p$

57. The water is 70º or the sun is shining, and we do not go swimming.

59. The water is not 70º , and the sun is shining or we go swimming.

61. If we do not go swimming, then the sun is shining and the water is 70º.

63. If the sun is shining then we go swimming, and the water is 70º.

65. The sun is shining if and only if the water is 70º, and we go swimming.

67. Not permissible. In the list of choices, the connective "or" is the exclusive or, thus one can order either the soup or the salad but not both items.

69. Not permissible. Potatoes and pasta cannot be ordered together.

71. a) b: started bonfire; m: forgot marshmallows; $b \wedge \sim m$
    b) Conjunction

73. a) w: work out; g: gain weight; $\sim (w \rightarrow \sim g)$
    b) Negation

75. a) f: food has fiber; v: food has vitamins; h: be healthy; $(f \vee v) \rightarrow h$
    b) Conditional

77. a) c: may take course; f: fail previous exam; p: passed placement test; $c \leftrightarrow (\sim f \vee p)$
    b) Biconditional

79. a) *c*: classroom is empty; *w*: is the
    weekend; *s*: is 7:00 a.m.; $(c \leftrightarrow w) \vee s$

    b) Disjunction

81. Answers will vary.

Exercise Set 3.2

1. Opposite
3. False

5.

p	p	∧	~p
T	T	F	F
F	F	F	T
	1	3	2

7.

p	q	q	∨	~p
T	T	T	T	T
T	F	F	F	T
F	T	T	T	T
F	F	F	T	T
		1	3	2

9.

p	q	~p	∨	~q
T	T	F	F	F
T	F	F	T	T
F	T	T	T	F
F	F	T	T	T
		1	3	2

11.

p	q	~(	p	∧	~q)
T	T	T	T	F	F
T	F	F	T	T	T
F	T	T	F	F	F
F	F	T	F	F	T
		4	1	3	2

13.

p	q	r	p ∨ (~q ∨ r)				
T	T	T	T	T	F	T	T
T	T	F	T	T	F	F	F
T	F	T	T	T	T	T	T
T	F	F	T	T	T	T	F
F	T	T	F	T	F	T	T
F	T	F	F	F	F	F	F
F	F	T	F	T	T	T	T
F	F	F	F	T	T	T	F
			1	5	2	4	3

15.

p	q	r	(p ∧ ~q) ∨ r				
T	T	T	T	F	F	T	T
T	T	F	T	F	F	F	F
T	F	T	T	T	T	T	T
T	F	F	T	T	T	T	F
F	T	T	F	F	F	T	T
F	T	F	F	F	F	F	F
F	F	T	F	F	T	T	T
F	F	F	F	F	T	F	F
			1	4	2	5	3

17.

p	q	r	(r ∨ ~p) ∧ ~q				
T	T	T	T	T	F	F	F
T	T	F	F	F	F	F	F
T	F	T	T	T	F	T	F
T	F	F	T	F	F	F	T
F	T	T	T	F	T	F	F
F	T	F	F	T	T	F	F
F	F	T	T	T	T	T	T
F	F	F	F	T	T	T	T
			1	4	2	5	3

19. p: Apples are a good source of fiber.

q: Oranges are a good source of vitamin C.

In symbolic form the statement is p ∧ q.

p	q	p ∧ q
T	T	T
T	F	F
F	T	F
F	F	F
		1

21. p: I have worked all week.

    q: I have been paid.

    In symbolic form the statement is $p \wedge \sim q$.

p	q	p	$\wedge$	$\sim q$
T	T	T	F	F
T	F	T	T	T
F	T	F	F	F
F	F	F	F	T
		1	3	2

23. p: Jasper is a tutor.

    q: Mark is a secretary.

    In symbolic form the statement is $\sim (p \wedge q)$.

p	q	$\sim$	(p	$\wedge$	q)
T	T	F	T	T	T
T	F	T	T	F	F
F	T	T	F	F	T
F	F	T	F	F	F
		4	1	3	2

25. p: The copier is out of toner.

    q: The lens is dirty.

    r : The corona wires are broken.

    The statement is $p \vee (q \vee r)$.

p	q	r	p	$\vee$	(q $\vee$ r)
T	T	T	T	T	T
T	T	F	T	T	T
T	F	T	T	T	T
T	F	F	T	T	F
F	T	T	F	T	T
F	T	F	F	T	T
F	F	T	F	T	T
F	F	F	F	F	F
			2	3	1

27. a) $(p \wedge \sim q) \vee r$

     (T $\wedge$ T) $\vee$ T

       T $\vee$ T

         T

    Therefore the statement is true.

    b) $(p \wedge \sim q) \vee r$

     (F $\wedge$ F) $\vee$ T

       F $\vee$ T

         T

    Therefore the statement is true.

29. (a) $(\sim p \wedge \sim q) \vee \sim r$

     (F $\wedge$ T) $\vee$ F

       F $\vee$ F

        F

    Therefore the statement is false.

    (b) $(\sim p \wedge \sim q) \vee \sim r$

     (T $\wedge$ F) $\vee$ F

       F $\vee$ F

        F

    Therefore the statement is false.

31. (a) $(p \vee \sim q) \wedge \sim (p \wedge \sim r)$

     (T $\vee$ T) $\wedge \sim$(T $\wedge$ F)

       T $\wedge$ $\sim$F

        T

    Therefore the statement is true.

    (b) $(p \vee \sim q) \wedge \sim (p \wedge \sim r)$

     (F $\vee$ F) $\wedge \sim$ (F $\wedge$ F)

       F $\wedge$ $\sim$F

        F

33. (a) $(\sim r \wedge p) \vee q$

     (F $\wedge$ T) $\vee$ F

       F $\vee$ F

        F

    Therefore the statement is false.

    (b) $(\sim r \wedge p) \vee q$

     (F $\wedge$ F) $\vee$ T

       F $\vee$ T

        T

35. (a) $(\sim p \vee \sim q) \vee (\sim r \vee q)$

     (F $\vee$ T) $\vee$ (F $\vee$ F)

       T $\vee$ F

        T

    Therefore the statement is true.

    (b) $(\sim p \vee \sim q) \vee (\sim r \vee q)$

     (T $\vee$ F) $\vee$ (F $\vee$ T)

       T $\vee$ T

        T

37. $8 + 7 = 20 - 5$ and $63 \div 7 = 3 \cdot 3$

    T   ∧   T

          T

Therefore the statement is true.

39. C: Chevrolet makes trucks.

T: Toyota makes shoes.

  C ∨ T

  T ∨ F

     T

Therefore the statement is true.

41. Y: The New York Yankees are a baseball team.

G: The New York Giants are a football team.

N: The Brooklyn Nets are a basketball team.

  (Y ∧ G) ∧ ~N

 (T ∧ T) ∧ F

    T   ∧ F

       F

Therefore the statement is false.

43. C: Chicago is in Mexico.

L: Los Angeles is in California.

D: Dallas is in Canada.

 (C ∨ L) ∨ D

 (F ∨ T) ∨ F

   T   ∨ F

     T

Therefore the statement is true.

45. p: India produced more feature films than the United States.

q: China produced more feature films than Japan.

  p ∧ q

  T ∧ T

  True

47. p: The United States produced more feature films than Japan.

q: Japan produced more feature films than China.

  p ∧ q

  T ∧ F

  False

49. p: 30% of Americans get 6 hours of sleep.

q: 9% get 5 hours of sleep.

  ~ (p ∧ q)

  ~ (F ∧ T)

    ~F

   True

51. p: 13% of Americans get ≤ 5 hrs. of sleep.

q: 32% of Americans get ≥ 6 hrs. of sleep.

r: 30% of Americans get ≥ 8 hrs. of sleep.

  (p ∨ q) ∧ r

  (T ∨ F) ∧ F

   T   ∧ F

   False

53.

p ∧ ~q
F
T
F
F

True in case 2.

55.

p ∨ ~q
T
T
F
T

True in cases 1, 2, and 4.

57.

$(r \lor q) \land p$
T
T
T
F
F
F
F
F

True in cases 1, 2, and 3.

59.

$q \lor (p \land \sim r)$
T
T
F
T
T
T
F
F

True in cases 1, 2, 4, 5, and 6.

61. (a) Mr. Duncan qualifies for the loan.
Mrs. Tuttle qualifies for the loan.

(b) Mrs. Rusinek does not qualify,
since their combined income is less
than $46,000.

63. (a) Xavier qualifies for the special
fare.

(b) The other 4 do not qualify:
Gina returns after 04/01;
Kara returns on Monday;
Christos does not stay
at least one Saturday; and
Chang returns on Monday.

65.

p	q	r					
			$[(q \land \sim r)$	$\land$	$(\sim p \lor \sim q)]$	$\lor$	$(p \lor \sim r\square)$
T	T	T	F	F	F	T	T
T	T	F	T	F	F	T	T
T	F	T	F	F	T	T	T
T	F	F	F	F	T	T	T
F	T	T	F	F	T	F	F
F	T	F	T	T	T	T	T
F	F	T	F	F	T	F	F
F	F	F	F	F	T	T	T
			1	3	2	5	4

67. Yes

p	q	r	$(p \land \sim q) \lor r$			$(q \land \sim r) \lor p$		
T	T	T	F	T	T	F	T	T
T	T	F	F	F	F	T	T	T
T	F	T	T	T	T	F	T	T
T	F	F	T	T	F	F	T	T
F	T	T	F	T	T	F	F	F
F	T	F	F	F	F	T	T	F
F	F	T	F	T	T	F	F	F
F	F	F	F	F	F	F	F	F

Exercise Set 3.3

1. False
3. True
5. Self-contradiction

7.

p	q	~(p → q)	
T	T	F	T
T	F	T	F
F	T	F	T
F	F	F	T
		2	1

9.

p	q	~p	↔	~q
T	T	F	T	F
T	F	F	F	T
F	T	T	F	F
F	F	T	T	T
		1	1	2

11.

p	q	(p →q)	↔	p
T	T	T	T	T
T	F	F	F	T
F	T	T	F	F
F	F	T	F	F
		1	3	2

13.

p	q	p	↔	(q ∨ p)
T	T	T	T	T
T	F	T	T	T
F	T	F	F	T
F	F	F	T	F
		1	3	2

15.

p	q	q → (p → ~ q)				
T	T	T	F	T	F	F
T	F	F	T	T	T	T
F	T	T	T	F	T	F
F	F	F	T	F	T	T
		4	5	1	3	2

17.

p	q	r	~p	→	(q	∧	r)
T	T	T	F	T	T	T	T
T	T	F	F	T	T	F	F
T	F	T	F	T	F	F	T
T	F	F	F	T	F	F	F
F	T	T	T	T	T	T	T
F	T	F	T	F	T	F	F
F	F	T	T	F	F	F	T
F	F	F	T	F	F	F	F
			4	5	1	3	2

19.

p	q	r	~p	↔	(q	∨	~r)
T	T	T	F	F	T	T	F
T	T	F	F	F	T	T	T
T	F	T	T	T	F	F	F
T	F	F	T	F	F	T	T
F	T	T	F	T	T	T	F
F	T	F	F	T	T	T	T
F	F	T	T	F	F	F	F
F	F	F	T	T	F	T	T
			1	5	2	4	3

21.

p	q	r	(~r	∨	~q)	→	p
T	T	T	F	F	F	T	T
T	T	F	T	T	F	T	T
T	F	T	F	T	T	T	T
T	F	F	T	T	T	T	T
F	T	T	F	F	F	T	F
F	T	F	T	T	F	F	F
F	F	T	F	T	T	F	F
F	F	F	T	T	T	F	F
			1	3	2	5	4

23.

p	q	r	(p	→	q)	↔	(~q	→	~r)
T	T	T		T		T	F	T	F
T	T	F		T		T	F	T	T
T	F	T		F		T	T	F	F
T	F	F		F		F	T	T	T
F	T	T		T		T	F	T	F
F	T	F		T		T	F	T	T
F	F	T		T		F	T	F	F
F	F	F		T		T	T	T	T
				1		5	2	4	3

25.   p: p: It is Monday; q: The library is open; r: We can eat study together.

p	q	r	p	→	(q	∧	r)
T	T	T	T	T			T
T	T	F	T	F			F
T	F	T	T	F			F
T	F	F	T	F			F
F	T	T	F	T			T
F	T	F	F	T			F
F	F	T	F	T			F
F	F	F	F	T			F
			1	3			2

27.   p: The election was fair; q: The polling station stayed open until 8 P.M.; r: We will request a recount.

p	q	r	(p	↔	q)	∨	r
T	T	T		T		T	T
T	T	F		T		T	F
T	F	T		F		T	T
T	F	F		F		F	F
F	T	T		F		T	T
F	T	F		F		F	F
F	F	T		T		T	T
F	F	F		T		T	F
				1		3	2

29.   p: It is too cold; q: We can take a walk; r: We can go to the gym.

p	q	r	(~ p	→	q)	∨	r
T	T	T		T		T	T
T	T	F		T		T	F
T	F	T		T		T	T
T	F	F		T		T	F
F	T	T		T		T	T
F	T	F		T		T	F
F	F	T		F		T	T
F	F	F		F		F	F
				1		3	2

31.

p	q	(p ∧ q)	∨	~ q
T	T	T	F	F
T	F	F	T	T
F	T	F	F	F
F	F	F	T	T
		1	3	2

Neither

33.

p	q	~ p	∧	(q ↔ ~ q)
T	T	F	F	F
T	F	F	F	F
F	T	T	F	F
F	F	T	F	F
		1	3	2

Self-contradiction

35.

p	q	(~ q	→	p)	∨	~ q
T	T		T		T	F
T	F		T		T	T
F	T		T		T	F
F	F		F		T	T
			1		3	2

Tautology

37.

p	~ p	→	p
T	F	T	T
F	T	F	F
	1	3	2

Not an implication

39.

p	q	~p	→	~(p ∧ q)
T	T	F	T	F
T	F	F	T	T
F	T	T	T	T
F	F	T	T	T
		1	3	2

An implication

41.

p	q	[(p → q)	∧	(q → p)]	→	(p ↔ q)
T	T	T	T	T	T	T
T	F	F	F	T	T	F
T	T	T	F	F	T	F
T	F	T	T	T	T	T
		1	3	2	5	4

Implication

43. $p \to (q \to r)$
   $T \to (F \to F)$
   $T \to \quad T$
   $\quad\quad T$

45. $(p \land q) \leftrightarrow (q \lor \sim r)$
   $(T \land F) \leftrightarrow (F \lor T)$
   $\quad F \quad \leftrightarrow \quad T$
   $\quad\quad\quad F$

47. $(\sim p \land \sim q) \lor \sim r$
   $(F \land T) \lor T$
   $\quad F \quad \lor T$
   $\quad\quad\quad T$

49. $(\sim p \leftrightarrow r) \lor (\sim q \leftrightarrow r)$
   $(F \leftrightarrow T) \lor (T \leftrightarrow F)$
   $\quad F \quad \lor \quad T$
   $\quad\quad\quad T$

51. If $9^2 = 81$, then $\sqrt{81} = 9$.

   $T \to T$
   $\quad T$

53. If a cat has whiskers or a fish
   can swim, then a dog
   lays eggs.
   $(T \lor T) \to F$
   $\quad T \quad \to \quad F$
   $\quad\quad\quad F$

55. Apple makes computers, if and only if Nike makes
   sports shoes or Rolex makes watches.
   $T \leftrightarrow (T \lor T)$
   $T \leftrightarrow \quad T$
   $\quad\quad T$

57. Independance Day is in July and Labor Day is in
   September, if and only if Thanksgiving is in April.
   $(T \land T) \leftrightarrow F$
   $\quad T \quad \leftrightarrow F$
   $\quad\quad\quad F$

59. Io has a diameter of 1000–3161 miles, or Thebe may
   have water, and Io may have atmosphere.
   $(T \lor F) \land T$
   $\quad T \land T$
   $\quad\quad T$

61. Phoebe has a larger diameter than Rhea if and only
   if Callisto may have water ice, and Calypso has a
   diameter of 6–49 miles.
   $(F \leftrightarrow T) \land T$
   $\quad F \quad \land T$
   $\quad\quad\quad F$

63. The number of communications credits needed is
    more than the number of mathematics credits needed
    and the number of cultural issues credits needed is
    equal to the number of humanities credits needed, if
    and only if the number of social sciences credits
    needed is more than the numbe of natural sciences
    credits needed..

$$(T \wedge T) \leftrightarrow F$$
$$T \quad \leftrightarrow F$$
$$F$$

For 65 – 67 *p*: Muhundan spoke at the teachers' conference.
          q: Muhundan received the outstanding teacher award
          Assume *p* and *q* are false.

65. $q \to p$ 67. $\sim q \leftrightarrow p$
     $F \to F$ $T \leftrightarrow F$
       T F

69. No, the statement only states what will occur if your sister gets straight A's. If your sister does not get straight
    A's, your parents may still get her a computer.

71.

p	q	r	[p	∨	(q	→	~ r)]	↔	(p ∧ ~ q)
T	T	T		T		F	F	F	F
T	T	F		T		T	T	F	F
T	F	T		T		T	F	T	T
T	F	F		T		T	T	T	T
F	T	T		F		F	F	T	F
F	T	F		T		T	T	F	F
F	F	T		T		T	F	F	F
F	F	F		T		T	T	F	F
				3		2	1	5	4

73.

p	q	r	s	(p ∨ q)	→	(r ∧ s)	(q → ~p)	∨	(r ↔ s)
					(a)			(b)	
T	T	T	T	T	T	T	F	T	T
T	T	T	F	T	F	F	F	F	F
T	T	F	T	T	F	F	F	F	F
T	T	F	F	T	F	F	F	T	T
T	F	T	T	T	T	T	T	T	T
T	F	T	F	T	F	F	T	T	F
T	F	F	T	T	F	F	T	T	F
T	F	F	F	T	F	F	T	T	T
F	T	T	T	T	T	T	T	T	T
F	T	T	F	T	F	F	T	T	F
F	T	F	T	T	F	F	T	T	F
F	T	F	F	T	F	F	T	T	T
F	F	T	T	F	T	T	T	T	T
F	F	T	F	F	T	F	T	T	F
F	F	F	T	F	T	F	T	T	F
F	F	F	F	F	T	F	T	T	T
				1	3	2	1	3	2

75.

Tiger	Boots	Sam	Sue
Blue	Yellow	Red	Green
Nine Lives	Whiskas	Friskies	Meow Mix

Exercise Set 3.4

1. Equivalent

3. $p \rightarrow q$ is equivalent to $\sim p \vee q$.

5. $q \rightarrow p$

7. $\sim q \rightarrow \sim p$

9.

p	q	p → q	~p ∨ q
T	T	T	F T T
T	F	F	F F F
F	T	T	T T T
F	F	T	T T F
		1	1 3 2

The statements are equivalent.

11.

p	q	~q → ~p	p → q
T	T	F T F	T
T	F	T F F	F
F	T	F T T	T
F	F	T T T	T
		1 3 2	1

The statements are equivalent.

13.

p	q	r	(p ∨ q) ∧ r	p ∨ (q ∧ r)
T	T	T	T T T	T T T
T	T	F	T F F	T T F
T	F	T	T T T	T T F
T	F	F	T F F	T T F
F	T	T	T T T	F T T
F	T	F	T F F	F F F
F	F	T	F F T	F F F
F	F	F	F F F	F F F
			1 3 2	2 3 1

The statements are not equivalent.

15. Yes, equivalent.

p	q	(p→q) ∧ (q→p)	p ↔ q
T	T	T T T	T
T	F	F F T	F
F	T	T F F	F
F	F	T T T	T
		1 3 2	

17. $\sim(p \wedge q) \Leftrightarrow \sim p \vee q$ (by law 1) and this is not equivalent to $\sim p \wedge \sim q$.

19. Equivalent by law 2.

21. Not equivalent

23. Yes, equivalent

25. p: Taylor Swift is a country singer.
    q: Wiz Khalifa sings opera.
    In symbolic form, the statement is $\sim(p \wedge q)$.
    Applying DeMorgan's Laws we get: $\sim p \vee \sim q$.
    Taylor Swift is not a country singer or Wiz Khalifa does not sing opera.

27. p: The dog was a bulldog.
    q: The dog was a boxer.
    In symbolic form, the statement is $\sim p \wedge \sim q$.
    Applying DeMorgan's Laws we get: $\sim(p \vee q)$.
    It is false that the dog was a bulldog or the dog was a boxer.

29. p: Ashley takes the new job.
    q: Ashley will move.
    r: Ashley will buy a new house in town.
    In symbolic form, the statement is
    $p \to (\sim q \vee r)$. Applying DeMorgan's Laws
    we get: $p \to \sim(q \wedge \sim r)$. If Ashley takes the
    new job, it is false that she will move and will
    not buy a new house in town.

31. p: Janette buys a new car.
    q: Janette sells her old car.
    In symbolic form, the statement is $p \to q$.
    Since $p \to q \Leftrightarrow \sim p \vee q$, an equivalent
    statement is: Janette does not buy a new car or
    she sells her old car.

33. p: Chase is hiding.
    q: The pitcher is broken.
    In symbolic form, the statement is $\sim p \vee q$.
    $\sim p \vee q \Leftrightarrow p \to q$. If Chase is hiding, then the
    pitcher is broken.

35. p: The Foo Fighters will practice
    q: The Foo Fighters will sound good.
    In symbolic form, the statement is $p \vee \sim q$.
    Since $p \vee \sim q \Leftrightarrow \sim p \to \sim q$, an equivalent
    statement is: If the Foo Fighters will not
    practice then they will not sound good.

37. p: We go to Chicago.
    q: We go to Navy Pier.
    In symbolic form, the statement is $\sim(p \to q)$.
    $\sim(p \to q) \Leftrightarrow p \wedge \sim q$.
    We go to Chicago and we do not go to Navy Pier.

39. p: I am cold.
    q: The heater is working.
    In symbolic form, the statement is $p \wedge \sim q$.
    $p \wedge \sim q \Leftrightarrow \sim(p \to q)$.
    It is false that if I am cold then the heater is working.

41. p: Amazon has a sale.
    q: We will buy $100 worth of books.
    In symbolic form, the statement is $\sim(p \to q)$.
    $\sim(p \to q) \Leftrightarrow p \wedge \sim q$.
    Amazon has a sale and we will not buy $100 worth
    of books.

43. Converse: If Nanette needs extra yarn, then she
    teaches macramé.
    Inverse: If Nanette does not teach macramé, then
    she does not need extra yarn.
    Contrapositive: If Nanette does not need
    extra yarn, then she does not teach macramé.

45. Converse: If I buy silver jewelry, then I go to
    Mexico.
    Inverse: If I do not go to Mexico, then I do not
    buy silver jewelry.
    Contrapositive: If I do not buy silver jewelry
    then I do not go to Mexico.

47. Converse: If I do not stay on my diet, then the menu
    includes calzones.
    Inverse: If the menu does not include calzones, then I
    do stay on my diet.
    Contrapositive: If I stay on my diet,
    then the menu does not include calzones.

49. If a natural number is not divisible by 7, then it is not divisible by 14. True

51. If a natural number is not divisible by 6, then it is not divisible by 3. False

53. If two lines are not parallel, then the two lines intersect in at least one point. True

55. p: The player gets a red card.

q: The player sits out.

In symbolic form, the statements are:

a) $p \rightarrow q$, b) $q \rightarrow p$, c) $\sim q \vee p$.

p	q	a) $p \rightarrow q$	b) $q \rightarrow p$	c) $\sim q \vee p$
T	T	T T T	T T T	F T T
T	F	T F F	F T T	T T T
F	T	F T F	T F F	F F F
F	F	F T F	F T F	T T F
		1 3 2	1 3 2	1 3 2

Since the truth tables for (a) and (b) are different we conclude that only statements (b) and (c) are equivalent.

57. p: The shoes are on sale.

q: The purse is on sale.

In symbolic form, the statements are: a) $\sim p \wedge q$, b) $\sim p \rightarrow \sim q$, c) $\sim (p \vee \sim q)$. If we use DeMorgan's Laws on statement (a), we get statement (c). Therefore, statements (a) and (c) are equivalent. If we look at the truth tables for statements (a), (b), and (c), we see that only statements (a) and (c) are equivalent.

p	q	a) $\sim p \wedge q$	b) $\sim p \rightarrow \sim q$	c) $\sim (p \vee \sim q)$
T	T	F F T	F T F	F T T F
T	F	F F F	F T T	F T T T
F	T	T T T	T F F	T F F F
F	F	T F F	T T T	F F T T
		1 3 2	1 3 2	4 1 3 2

59. p: We go hiking.

q: We go fishig.

In symbolic form, the statements are: a) $\sim p \vee q$, b) $p \rightarrow \sim q$, c) $q \rightarrow \sim p$. Looking at the truth table for all three statements, we can determine that only statements (b) and (c) are equivalent.

p	q	a) $\sim p \vee q$	b) $p \rightarrow \sim q$	c) $q \rightarrow \sim p$
T	T	F T T	T F F	T F F
T	F	F F F	T T T	F T F
F	T	T T T	F T F	T T T
F	F	T T F	F T T	F T T
		1 3 2	1 3 2	1 3 2

61. p: The grass grows.

q: The trees are blooming.

In symbolic form, the statements are: a) $p \wedge q$,

b) $q \to \sim p$, c) $\sim q \vee p$. Using the fact that $p \to q$

$\Leftrightarrow \sim p \vee q$, on statement (b) we get $\sim q \vee \sim p$.

Therefore, statements (b) and (c) are equivalent.

Looking at the truth table for statements (a) and (b)

we can conclude that only statements (b) and (c) are

equivalent.

p	q	$p \wedge q$	$q \to \sim p$		
T	T	T	T	F	F
T	F	F	F	T	F
F	T	F	T	T	T
F	F	F	F	T	T
		1	1	3	2

63. p: The corn bag goes in the hole.

q: you are awarded three points.

In symbolic form, the statements are:

a) $p \to q$, b) $\sim (p \wedge q)$, and c) $\sim p \wedge \sim q$.

p	q	$p \to q$			$\sim (p \wedge q)$,				$\sim p \wedge \sim q$		
T	T	T	T	T	F	T	T	T	F	F	F
T	F	T	F	F	T	T	F	F	F	F	T
F	T	F	T	T	T	F	F	T	T	F	F
F	F	F	T	F	T	F	F	F	T	T	T
		1	3	2	4	1	3	2	1	3	2

Therefore, none of the statements are equivalent.

65. p: The pay is good.

q: Today is Monday.

r : I will take the job.

Looking at the truth tables for statements (a), (b), and

(c), we see that none of the statements are equivalent.

			a)			b)				c)		
p	q	r	$(p \wedge q) \to r$			$\sim r \to \sim (p \vee q)$				$(p \wedge q) \vee r$		
T	T	T	T	T	T	F	T	F	T	T	T	T
T	T	F	T	F	F	T	F	F	T	T	T	F
T	F	T	F	T	T	F	T	F	T	F	T	T
T	F	F	F	T	F	T	F	F	T	F	F	F
F	T	T	F	T	T	F	T	F	T	F	T	T
F	T	F	F	T	F	T	F	F	T	F	F	F
F	F	T	F	T	T	F	T	T	F	F	T	T
F	F	F	F	T	F	T	T	T	F	F	F	F
			1	3	2	1	4	3	2	1	3	2

67. p: The package was sent by Federal Express.

q: The package was sent by United Parcel Service.

r : The package arrived on time.

Using the fact that p → q ⟺ ~ p ∨ q to rewrite statement (c), we get p ∨ (~ q ∧ r). Therefore, statements (a) and (c) are equivalent. Looking at the truth table for statements (a) and (b), we can conclude that only statements (a) and (c) are equivalent.

			a)	b)	c)
p	q	r	p ∨ (~q ∧ r)	r ↔ (p ∨ ~q)	~p → (~q ∧ r)
T	T	T	T T F F T	T T T T F	F T F F T
T	T	F	T T F F F	F F T T F	T T F F F
T	F	T	T T T T T	T T T T T	F T T T T
T	F	F	T T T F F	F F T T T	T T T F F
F	T	T	F F F F T	T F F F F	F F F F T
F	T	F	F F F F F	F T F F F	T F F F F
F	F	T	F T T T T	T T F T T	F T T T T
F	F	F	F F T F F	F F F T T	T F T F F
			1 5 2 4 3	1 5 2 4 3	1 5 2 4 3

69. True. If $p \rightarrow q$ is false, it must be of the form T → F . Therefore, the converse must be of the form F → T , which is true.

71. False. A conditional statement and its contrapositive always have the same truth values.

73. If we use DeMorgan's Laws to rewrite ~ p ∨ q, we get ~ (p ∧ ~ q). Since ~ p ∨ q ⟺ ~ (p ∧ ~ q) and p → q ⟺ ~ p ∨ q, we can conclude that p → q ⟺ ~ (p ∧ ~ q). Other answers are possible.

75. Answers will vary.

77.

Exercise Set 3.5
1. Valid
3. Fallacy
5. Valid
7. Inverse
9. Syllogism
11. Syllogism

13. This argument is the fallacy of the inverse, therefore it is invalid.

15. This is the law of detachment, so it is a valid argument.

17. This argument is a disjunctive syllogism and therefore is valid.

19. This argument is the fallacy of the converse. Therefore it is invalid.

21. This argument is the fallacy of the inverse, Therefore it is invalid.

23. This argument is the law of syllogism and therefore it is valid.

25.

p	q	r	[(p ↔ q) ∧ (q ∧ r)] → (p ∨ r)				
T	T	T	T	T	T	T	T
T	T	F	T	F	F	T	T
T	F	T	F	F	F	T	T
T	F	F	F	F	F	T	T
F	T	T	F	F	T	T	T
F	T	F	F	F	F	T	F
F	F	T	T	F	F	T	T
F	F	F	T	F	F	T	F
			1	3	2	5	4

The argument is valid.

27.

p	q	r	[(r ↔ p) ∧ (~p ∧ q)] → (p ∧ r)						
T	T	T	T	F	F F T	T	T
T	T	F	F	F	F F T	T	F
T	F	T	T	F	F F F	T	T
T	F	F	F	F	F F F	T	F
F	T	T	F	F	T T T	T	F
F	T	F	T	T	T T T	F	F
F	F	T	F	F	T F F	T	F
F	F	F	T	F	T F F	T	F
			1	5	2 4 3	7	6

The argument is invalid.

29.

p	q	r	[(p → q) ∧ (q ∨ r) ∧ (r ∨ p)] → p						
T	T	T	T	T	T	T	T	T T
T	T	F	T	T	T	T	T	T T
T	F	T	F	F	T	F	T	T T
T	F	F	F	F	F	F	T	T T
F	T	T	T	T	T	T	T	F F
F	T	F	T	T	T	F	F	T F
F	F	T	T	T	T	T	T	F F
F	F	F	T	F	F	F	F	T F
			1	3	2	5	4	7 6

The argument is invalid.

31.

p	q	r	[(p → q) ∧ (r → ~ p) ∧ (p ∨ r)] → (q ∨ ~ p)
T	T	T	T F TF F F T T T T F
T	T	F	T T FT F T T T T T F
T	F	T	F F TF F F T T F F F
T	F	F	F F FT F F T T F F F
F	T	T	T T TT T T T T T T T
F	T	F	T T FT T F F T T T T
F	F	T	T T TT T T T T F T T
F	F	F	T T FT T F F T F T T
			1 5 24 3 7 6 11 9 10 8

The argument is valid.

33. a) p: Your cell phone rings during class.
    q: The teacher gets angry.

    p → q
    ~q
    ∴ ~p

   b) The argument is valid by the law of contraposition.

35. a) p: We visit the zoo.
    q: We see a zebra.

    p → q
    p
    ∴ q

   b) The argument is valid (law of detachment).

37. a) p: The guitar is a Les Paul model.
    q: The guitar is made by Gibson.

    p → q
    ~q
    ∴ ~p

   b) This argument is valid by the law of contraposition.

39. a) p: We take kayaks on the river.
    q: We see alligators.

    p → q
    q
    ∴ p

   b) This is the fallacy of the converse; thus the argument is invalid.

41. a) p: Erica Kane will marry Samuel Woods.
    q: Erica Kane will marry David Hayward.

    p ∨ q
    ~ p
    ∴ q

   b) The argument is a disjunctive syllogism and is therefore valid.

43. a) p: It is cold.
    q: The graduation will be held indoors.
    r: The fireworks will be postponed.

    p → q
    q → r
    ∴ p → r

   b) This argument is valid because of the law of syllogism.

45. p: Jevon is a bowler.
    q: Michael is a golfer.
    r: Alisha is a curler.

    p ∧ q
    q → r
    ∴ r → p

    The argument is valid.

47. s: It is snowing.
    g: I am going skiing.
    c: I will wear a coat.

    s ∧ g
    g → c
    ∴ s → c

    The argument is valid.

49. p: You read *Riptide Ultra-Glide*.
    q: You can understand *Tiger Shrimp Tango*.

    p → q
    ~q
    ∴ ~p

    The argument is the law of contraposition and is valid.

51. p: Max is playing Game Boy with the sound off.
    q: Max is wearing headphones.

    p ∨ q
    ~p
    ∴ q

    This argument is an example of disjunctive syllogism and is therefore valid.

53. t: A tiger is a cat.
    g: A giraffe is a dog.

    t ∧ g
    ~t ∨ ~g
    ∴ ~t

    The argument is valid.

55. c: The baby is crying.
    h: The baby is hungry.

    c ∧ ~h
    h → c
    ∴ h

    The argument is invalid.

57. p: You liked *This Is Spinal Tap*.   p → q
    q: You liked *Best In Show*.   q → ~r
    r: You liked *A Mighty Wind*.   ∴ p → r

    b) Using the law of syllogism p → ~r, so this argument is invalid.

59. Therefore, the radio is made by RCA.
    (law of detachment)

61. Therefore, I am stressed out.
    (disjunctive syllogism)

63. Therefore, you did not close the deal.
    (law of contraposition)

65. Yes. It is not necessary for the premises or the conclusion to be true statements for the argument to be valid.

67. Yes. If the conclusion does not follow from the set of premises, then the argument is invalid.

69. p: Lynn wins the contest. $p \lor q \to r$
    q: Lynn strikes oil. $\underline{r \to s}$
    r : Lynn will be rich. $\therefore \sim s \to \sim p$
    s: Lynn will stop working.

p	q	r	s	[((p ∨ q)	→	r)	∧	(r → s)]	→	(~ s →	~ p)
T	T	T	T		T	T	T	T	T	T	T
T	T	T	F		T	T	T	F	F	T	F
T	T	F	T		T	F	F	T	T	T	T
T	T	F	F		T	F	F	T	T	T	F
T	F	T	T		T	T	T	T	T	T	T
T	F	T	F		T	T	T	F	F	T	F
T	F	F	T		T	F	F	T	T	T	T
T	F	F	F		T	F	F	T	T	T	F
F	T	T	T		T	T	T	T	T	T	T
F	T	T	F		T	T	T	F	F	T	T
F	T	F	T		T	F	F	T	T	T	T
F	T	F	F		T	F	F	T	T	T	T
F	F	T	T		F	T	T	T	T	T	T
F	F	T	F		F	T	T	F	F	T	T
F	F	F	T		F	T	F	T	T	T	T
F	F	F	F		F	T	F	T	T	T	T
				1	3	2	5	4	7	6	

The argument is valid.

71. a) p: I think.
       q: I am.
       $p \to q$
       $\underline{\sim p}$
       $\therefore \sim q$

    b) No, the argument is invalid.

    c) This argument is the fallacy of the inverse.

Exercise Set 3.6
1. Euler
3. Invalid
5. No

7. Valid

9. Invalid

11. Valid

13. Invalid

15. Invalid

17. Invalid

19. Valid

21. Invalid

23. Invalid

25. Valid

27. Yes. If the conjunction of the premises is false in all cases, then the argument is valid regardless of the truth value of the conclusion.

29. $[(p \rightarrow q) \wedge (p \vee q)] \rightarrow \sim p$ can be expressed as a set statement by $[(P' \cup Q) \cap (P \cup Q)] \subseteq P'$. If this statement is true, then the argument is valid; otherwise, the argument is invalid.

Set	Regions
$P' \cup Q$	II, III, IV
$P \cup Q$	I, II, III
$(P' \cup Q) \cap (P \cup Q)$	II, III
P'	III, IV

Since $(P' \cup Q) \cap (P \cup Q)$ is not a subset of P', the argument is invalid.

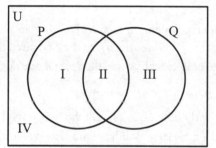

Exercise Set 3.7

1. Series

3. Closed

5. a) $p \wedge q$

b)

		Light
p	q	$p \wedge q$
T	T	T
T	F	F
F	T	F
F	F	F

7. a) $(p \vee q) \wedge \overline{q}$

b)

			Light	
p	q	$(p \vee q)$	$\wedge$	$\sim q$
T	T	T	F	F
T	F	T	T	T
F	T	T	F	F
F	F	F	F	T

9. a) $(p \wedge q) \wedge [(p \wedge \sim q) \vee r]$

   b) It is clear from the $(p \wedge q)$ condition that both p and q must be closed for the bulb to light. In this case the upper branch of the parallel portion of the circuit is open since it includes $\bar{q}$, so for the bulb to light, r must be closed. Thus the bulb lights only if p, q and r are all T.

11. a) $p \vee q \vee (r \wedge \sim p)$

    b) Reading from the circuit, we can see that the only case in which the bulb will *not* light is: p is F, q is F, and r is F.

13.

15.

17.

19.

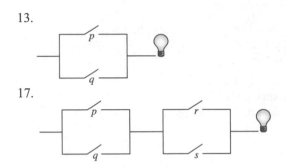

21. $p \vee \sim q; \ \sim p \wedge q$

    Not equivalent; in fact, by De Morgan's laws, $p \vee \sim q \Leftrightarrow \sim (\sim p \wedge q)$ so the first circuit will light the bulb exactly when the second one does not.

23. $[(p \wedge q) \vee r] \wedge p; \ (q \vee r) \wedge p$

    Clearly both circuits light the bulb only if p is T. In this case $p \wedge q$ has the same value as q alone, so $(p \wedge q) \vee r$ will have the same value as $q \vee r$. Thus the two circuits are equivalent.

p	q	r	$(p \wedge q) \vee r \ \wedge p$	$(q \vee r) \wedge p$
T	T	T	T T T T T	T T T
T	T	F	T T F T T	T T T
T	F	T	F T T T T	T T T
T	F	F	F F F F T	F F T
F	T	T	F T T F F	T F F
F	T	F	F F F F F	T F F
F	F	T	F T T F F	T F F
F	F	F	F F F F F	F F F
			1 2 3	1 2

25. $(p \vee \sim p) \wedge q \wedge r; \ p \wedge q \wedge r$

    The second circuit will only light the bulb when p, q, and r are all true. In the first circuit $(p \vee \sim p)$ is always true. Therefore, if p is false and q and r are true, the first circuit will light the bulb, and the second circuit will not. The circuits are not equivalent.

27. One of the two switches will always be open.

29. a) $p \rightarrow q \Leftrightarrow \sim p \vee q$

    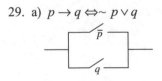

    b) $\sim (p \rightarrow q) \Leftrightarrow \sim p \wedge q$

Review Exercises

1. Some Acuras are not Hondas.

2. Some pets are allowed in this park.

3. No women are presidents.

4. All pine trees are green.

5. The coffee is hot or the coffee is strong.

6. The coffee is not hot and the coffee is strong.

7. If the coffee is hot, then the coffee is strong and it is not Maxwell House.

8. The coffee is Maxwell House if and only if the coffee is not strong.

9. The coffee is not Maxwell House, if and only if the coffee is strong and the coffee is not hot.

10. The coffee is Maxwell House or the coffee is not hot, and the coffee is not strong.

11. $q \wedge \sim r$

12. $r \rightarrow \sim p$

13. $(r \rightarrow q) \vee \sim p$

14. $(q \leftrightarrow p) \wedge \sim r$

15. $(r \wedge q) \vee \sim p$

16. $\sim (r \wedge q)$

17.

p	q	$(p \vee q) \wedge \sim p$		
T	T	T	F	F
T	F	T	F	F
F	T	T	T	T
F	F	F	F	T
		1	3	2

18.

p	q	$q \leftrightarrow (p \vee \sim q)$				
T	T	T	T	T	T	F
T	F	F	F	T	T	T
F	T	T	F	F	F	F
F	F	F	F	F	T	T
		1	5	2	4	3

19.

p	q	r	$(p \vee q)$	$\leftrightarrow$	$(p \vee r)$
T	T	T	T	T	T
T	T	F	T	T	T
T	F	T	T	T	T
T	F	F	T	T	T
F	T	T	T	T	T
F	T	F	T	F	F
F	F	T	F	F	T
F	F	F	F	T	F
			1	3	2

20.

p	q	r	p	$\wedge$	$(\sim q$	$\vee$	$r)$
T	T	T	T	T	F	T	T
T	T	F	T	F	F	F	F
T	F	T	T	T	T	T	T
T	F	F	T	T	T	T	F
F	T	T	F	F	F	T	T
F	T	F	F	F	F	F	F
F	F	T	F	F	T	T	T
F	F	F	F	F	T	T	F
			4	5	1	3	2

21.

p	q	r	p	→	(q	∧	~ r)
T	T	T	T	F	T	F	F
T	T	F	T	T	T	T	T
T	F	T	T	F	F	F	F
T	F	F	T	F	F	F	T
F	T	T	F	T	T	F	F
F	T	F	F	T	T	T	T
F	F	T	F	T	F	F	F
F	F	F	F	T	F	F	T
			4	5	1	3	2

22.

p	q	r	(p	∧ q)	→	~ r
T	T	T	T		F	F
T	T	F	T		T	T
T	F	T	F		T	F
T	F	F	F		T	T
F	T	T	F		T	F
F	T	F	F		T	T
F	F	T	F		T	F
F	F	F	F		T	T
			1		3	2

23. p: Apple makes iPhones.
    q: Dell makes canoes.
    r: Hewlett Packard makes laser printers.

    (p ∧ q) ∨ r
    (T ∧ F) ∨ T
      F   ∨ T
          T

24. p: ESPN is a sports network.
    q: CNN is a news network.
    r: Nickelodeon is a cooking network.

    (p → q) ↔ r
    (T → T) ↔ F
      T   ↔ F
         F

25. p: Oregon borders the Pacific Ocean.
    q: California borders the Atlantic Ocean.
    r: Minnesota is south of Texas.

    (p ∨ q) → r
    (T ∨ F) → F
      T → F
        F

26. p: President's Day is in February.
    q: Memorial Day is in May.
    r: Labor Day is in December.

    p ∨ (q ∧ r)
    T ∨ (T ∧ F)
    T ∨   F
      T

27. (~ p ∨ q) → ~ (p ∧ ~ q)
    (F ∨ F) → ~ (T ∨ T)
      F   → ~ T
          T

28. (p ↔ q) → (~ p ∨ r)
    (T ↔ F) → (F ∨ F)
      F   → F
        T

29. ~ r ↔ [(p ∨ q) ↔ ~ p]
    T ↔ [(T ∨ F) ↔ F]
    T ↔ [T ↔ F]
    T ↔   F
      F

30. ~ [(q ∧ r) → (~ p ∨ r)]
    ~ [(F ∧ F) → (F ∨ F)]
    ~ [F →   F]
        ~ T
        F

31.

p	q	~(p ∧ ~q)	~p ∧ q
T	T	T T F F	F F T
T	F	F T T T	F F F
F	T	T F F F	T T T
F	F	T F F T	T F F
		4 1 3 2	1 3 2

The statements are not equivalent.

32.

p	q	p ∨ q	~p → q
T	T	T T T	F T T
T	F	T T F	F T F
F	T	F T T	T T T
F	F	F F F	T F F
		1 3 2	1 3 2

33.

p	q	r	~p ∨ (q ∧ r)	(~p ∨ q) ∧ (~p ∨ r)
T	T	T	F T T T	F T T T F T T
T	T	F	F F F F	F T T F F F F
T	F	T	F F F F	F F F F F T T
T	F	F	F F F F	F F F F F F F
F	T	T	T T T T	T T T T T T T
F	T	F	T T T F	T T T T T T F
F	F	T	T T F T	T T F T T T T
F	F	F	T T F T	T T F T T T F
			2 3 1	1 3 2 7 4 6 5

The statements are equivalent.

34.

p	q	(~q → p) ∧ p	~(~p ↔ q) ∨ p
T	T	F T T T T	T F F T T T
T	F	T T T T T	F F T F T T
F	T	F T F F F	F T T T F F
F	F	T F F F F	T T F F T F
		1 3 2 5 4	4 1 3 2 6 5

The statements are not equivalent.

35. p: A grasshopper is an insect.

q: A spider is an insect.
In symbolic form, the statement is p ∧ ~q. We are given that p ∧ ~q ⇔ ~(p → q). So an equivalent statement is: It is false that if a grasshopper is an insect, then a spider is an insect.

36. p: Lynn Swann played for the Steelers.

q: Jack Tatum played for the Raiders.
In symbolic form, the statement is p ∨ q. We are given that ~p ∨ q ⇔ (p → q), so p ∨ q ⇔ (~p → q). Thus an equivalent statement is: If Lynn Swann did not play for the Steelers then Jack Tatum played for the Raiders.

37. p: Altec Lansing only produces speakers.

q: Harmon Kardon only produces stereo receivers.
The symbolic form is ~ (p ∨ q).
Using De Morgan's Laws, we get
~ (p ∨ q) ⇔ ~ p ∧ ~ q.
Altec Lansing does not produce only speakers and
Harmon Kardon does not produce only stereo receivers.

38. p: Travis Tritt won an Academy Award.

q: Randy Jackson does commercials for
Milk Bone Dog Biscuits.
The symbolic form is ~ p ∧ ~ q.
Using De Morgan's Laws, we get
~ p ∧ ~ q ⇔ ~ (p ∨ q). It is false
that Travis Tritt won an Academy
Award or Randy Jackson does
commercials for Milk Bone Dog
Biscuits.

39. p: The temperature is above 32 degrees Fahrenheit.

q: We will go ice fishing at O'Leary's Lake.
The symbolic form is $\sim p \rightarrow q \Leftrightarrow p \vee q$

The temperature is above 32 degrees Fahrenheit or
we will go ice fishing at O'Leary's Lake.

40. a) Converse: If we get Evan Longoria's autograph,
then we go to the Rays game.

b) Inverse: If we do not go to the Rays game, then
we do not get Evan Longoria's autograph.

c) Contrapositive: If we do not get Evan
Longoria's autograph, then we do not go to the
Rays game.

41. a) Converse: If we are going to learn the table's
value, then we take the table to *Antiques
Roadshow*.

b) Inverse: If we do not take the table to *Antiques
Roadshow*, then we will not learn the table's
value.

c) Contrapositive: If we are not going to learn the
table's value then we do not take the table to
Antiques Roadshow.

42. a) Converse: If you do not sell more doughnuts,
then you do not advertise.

b) Inverse: If If you advertise, then you sell more
doughnuts.

c) Contrapositive: If you sell more doughnuts,
then you advertise.

43. a) Converse: If we will not buy a desk at Miller's
Furniture, then the desk is made by Winner's Only
and is in the Rose catalog.

b) Inverse: If the desk is not made by
Winner's Only or is not in the Rose catalog,
then we will buy a desk at Miller's Furniture.

c) Contrapositive: If we will buy a desk at Miller's
Furniture, then the desk is not made by Winner's Only
or it is not in the Rose catalogue.

44. a) Converse: If I let you attend the prom,
then you get straight A's on your report
card.

b) Inverse: If you do not get straight A's on
your report card, then I will not let you
attend the prom.

c) Contrapositive: If I do not let you attend
the prom, then you do not get straight
A's on your report card.

45. p: Cheap Trick plays at the White House.

q: Jacque is the president.
In symbolic form, the statements are: a) p → q,
b) ~ p ∨ q, and c) ~ (p ∧ ~ q). Using the fact that
p → q is equivalent to ~ p ∨ q, statements (a) and (b)
are equivalent. Using DeMorgan's Laws on
statement (b) we get ~ (p ∧ ~ q).
Therefore all 3 statements are equivalent.

46. p: The screwdriver is on the workbench.

q: The screwdriver is on the counter.

In symbolic form, the statements are: a) $p \leftrightarrow \sim q$,
b) $\sim q \rightarrow \sim p$, and c) $\sim (q \wedge \sim p)$. Looking at the truth tables for statements (a), (b), and (c) we can conclude that none of the statements are equivalent.

		a)			b)			c)			
p	q	\multicolumn{3}{	c	}{$p \leftrightarrow \sim q$}	\multicolumn{3}{	c	}{$\sim q \rightarrow \sim p$}	\multicolumn{4}{	c	}{$\sim (q \wedge \sim p)$}

p	q	p	$\leftrightarrow$	$\sim q$	$\sim q$	$\rightarrow$	$\sim p$	$\sim$	$(q$	$\wedge$	$\sim p)$
T	T	T	F	F	F	T	F	T	T	F	F
T	F	T	T	T	T	F	F	T	F	F	F
F	T	F	T	F	F	T	T	F	T	T	T
F	F	F	F	T	T	T	T	T	F	F	T
		1	3	2	1	3	2	4	1	3	2

47. p: $2 + 3 = 6$.

q: $3 + 1 = 5$.

In symbolic form, the statements are: a) $p \rightarrow q$,
b) $p \leftrightarrow \sim q$, and c) $\sim q \rightarrow \sim p$.

Statement (c) is the contrapositive of statement (a). Therefore statements (a) and (c) are equivalent. For p and q false, (a) and (c) are true but (b) is false, so (b) is not equivalent to (a) and (c).

48. p: The sale is on Tuesday.

q: I have money.

r : I will go to the sale.

In symbolic form the statements are: a) $(p \wedge q) \rightarrow r$,
b) $r \rightarrow (p \wedge q)$, and c) $r \vee (p \wedge q)$. The truth table for statements (a), (b), and (c) shows that none of the statements are equivalent.

p	q	r	\multicolumn{3}{c}{$(p \wedge q) \rightarrow r$}	\multicolumn{3}{c}{$r \rightarrow (p \wedge q)$}	\multicolumn{3}{c}{$r \vee (p \wedge q)$}
T	T	T	T		T T	T T	T	T T	T
T	T	F	T		F F	F T	T	F T	T
T	F	T	F		T T	T F	F	T T	F
T	F	F	F		T F	F T	F	F F	F
F	T	T	F		T T	T F	F	T T	F
F	T	F	F		T F	F T	F	F F	F
F	F	T	F		T T	T F	F	T T	F
F	F	F	F		T F	F T	F	F F	F
			1		3 2	1 3	2	1 3	2

49.

p	q	\multicolumn{6}{c}{$[(p \rightarrow q) \wedge \sim p] \rightarrow \sim q$}
T	T	T	F F		T	F
T	F	F	F F		T	T
F	T	T	T T		F	F
F	F	T	T T		T	T
		1	3 2		5	4

The argument is invalid. This is the fallacy of the inverse.

50.

p	q	r	[(p ∧ q) ∧ (q → r)] → (p → r)
T	T	T	T T T T T
T	T	F	T F F T F
T	F	T	F F T T T
T	F	F	F F T T F
F	T	T	F F T T T
F	T	F	F F F T T
F	F	T	F F T T T
F	F	F	F F T T T
			1 3 2 5 4

The argument is valid.

51. p: Jose is the manager.

    q: Kevin is the coach.

    r: Tim is the umpire

      p → q

      q → r

   ∴ p → r

This argument is in the form of the law of syllogism so it is valid.

52. p: The truck is a diesel.

    q: The truck is too cold to start.

    r : The car has a flat tire.

      p → q

      q ∨ r

   ∴ ~ p

If p is F, q is T, and r is either T or F, the premises are both true but the conclusion is false, so the argument is invalid.

53. Invalid

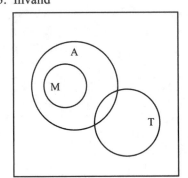

54. Valid

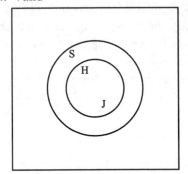

55.

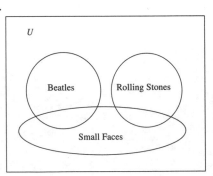

Invalid

56.

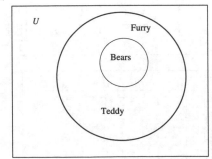

Invalid

57. a) $p \wedge [(q \wedge r) \vee \sim p]$

b) For the bulb to be on, the first switch on the left must be closed, so p is T. The eliminates the bottom branch of the parallel portion, so both switches on the top branch must be closed. Thus the bulb lights exactly when p, q, and r are all closed.

58.

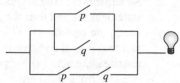

59. Symbolically the two circuits are $(p \vee q) \wedge (\sim q \vee \sim p)$ and $(p \wedge \sim q) \vee (q \wedge \sim p)$.

p	q	$\overline{p}$	$\overline{q}$	$(p \vee q)$	$\wedge$	$(\sim q \vee \sim p)$
T	T	F	F	T	F	F
T	F	F	T	T	T	T
F	T	T	F	T	T	T
F	F	T	T	F	F	T

p	q	$\overline{p}$	$\overline{q}$	$(p \wedge \sim q)$	$\vee$	$(q \wedge \sim p)$
T	T	F	F	F	F	F
T	F	F	T	T	T	F
F	T	T	F	F	T	T
F	F	T	T	F	F	F

The next-to last columns of these truth tables are identical, so the circuits are equivalent.

Chapter Test

1. $(\sim p \vee q) \wedge \sim r$

2. $(r \to q) \vee \sim p$

3. $\sim (r \leftrightarrow \sim q)$

4. Phobos is not a moon of Mars and Rosalind is a moon of Uranus, if and only if Callisto is not a moon of Jupiter.

5. If Phobos is a moon of Mars or Callisto is not a Moon of Jupiter, then Rosalind is a moon of Uranus.

6.

p	q	r	[~ (p → r)]	∧	q
T	T	T	F T	F	T
T	T	F	T F	T	T
T	F	T	F T	F	F
T	F	F	T F	F	F
F	T	T	F T	F	T
F	T	F	F T	F	T
F	F	T	F T	F	F
F	F	F	F T	F	F
			2 1	4	3

7.

p	q	r	(q ↔ ~r)	∨	p
T	T	T	T F F	T	T
T	T	F	T T T	T	T
T	F	T	F T F	T	T
T	F	F	F F T	T	T
F	T	T	T F F	F	F
F	T	F	T T T	T	F
F	F	T	F T F	T	F
F	F	F	F F T	F	F
			1 3 2	5	4

8. p: $2 + 6 = 8$
   q: $7 - 12 = 5$
   $p \vee q$
   $T \vee F$
   T

9. p: Harrison Ford is an actor.
   q: Gerald Ford was a president.
   r : The Mississippi is a river.
   $(p \vee q) \leftrightarrow r$
   $(T \vee T) \leftrightarrow T$
   $T \quad \leftrightarrow T$
   T

10. $(\sim p \wedge q) \leftrightarrow (q \vee \sim r)$
    $(F \wedge F) \leftrightarrow (F \wedge F)$
    $F \quad \leftrightarrow \quad F$
    T

11. $[\sim(r \to \sim p)] \wedge (q \to p)$
    $[\sim(T \to F)] \wedge (F \to T)$
    $[\quad \sim(F) \quad] \wedge \quad T$
    $\quad\quad T \quad\quad \wedge \quad T$
    $\quad\quad\quad\quad T$

12. By DeMorgan's Laws ,
    $\sim p \vee q \Leftrightarrow \sim (\sim (\sim p) \wedge \sim q).$
    and this is equivalent to
    $\sim(p \wedge \sim q).$

13. p: The bird is red.
    q: It is a cardinal.
    In symbolic form the statements
    are: a) $p \to q$, b) $\sim p \vee q$,
    and c) $\sim p \to \sim q$.
    Statement (c) is the inverse of
    statement (a) and thus they
    cannot be equivalent. Using the
    fact that $p \to q \Leftrightarrow \sim p \vee q$, to
    rewrite statement (a) we get
    $\sim p \vee q$. (a) and (b) are
    equivalent.

14. p: The test is today.   q: The concert is tonight. In symbolic form the
    statements are: a) $\sim (p \vee q)$, b) $\sim p \wedge \sim q$, and $\sim p \to \sim q$.
    Applying DeMorgan's Law to statement (a) we get: $\sim p \wedge \sim q$.
    Therefore statements (a) and (b) are equivalent. When we compare
    the truth tables for statements (a), (b), and (c) we see that only
    statements (a) and (b) are equivalent.

p	q	$\sim (p \vee q)$		$\sim p \wedge \sim q$			$\sim p \to \sim q$		
T	T	F	T	F	F	F	F	T	F
T	F	F	T	F	F	T	F	T	T
F	T	F	T	T	F	F	T	F	F
F	F	T	F	T	T	T	T	T	T
		2	1	1	3	2	1	3	2

15. s: The soccer team won the
    game.
    f: Sue played fullback.
    p: The team is in second place.

    $$s \to f$$
    $$\underline{f \to p}$$
    $$s \to p$$
    This argument is the law of
    syllogism and therefore it is valid.

16.

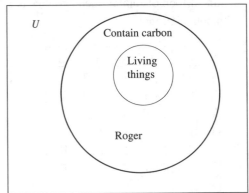

Invalid

17. Some Gryffindors are not brave.

18. Nick did not play football or
    Max did not play baseball.

19. Converse: If today is Saturday,
    then the garbage truck comes.
    Inverse: If the garbage truck
    does not come today, then today
    is not Saturday.
    Contrapositive: If today is not
    Saturday, then the garbage truck
    does not come.

20.

CHAPTER FOUR

SYSTEMS OF NUMERATION

Exercise Set 4.1

1. How many

3. Hindu-Arabic

5. Subtract

7. Multiplicative

9. $100+100+100+10+10$
$+1+1+1+1+1=325$

11. $10,000+1000+1000+100+100+100+10+10$
$+10+10+1=12341$

13. $100,000+100,000+100,000+10,000$
$+10,000+10,000+1000+1000+1000+1000$
$+100+100+10+1+1+1+1=334,214$

15.

17.

19.

21. $10+10+5+1+1=27$

23. $100+(100-10)+(5-1)=194$

25. $1000+1000+500+100+(50-10)+1+1=2642$

27. $1000+1000+(1000-100)+(50-10)+5+1$
$=2946$

29. $4(1000)+(500-100)+(100-10)+(10-1)$
$=4499$

31. $10(1000)+1000+1000+500+100+50+10$
$+5+1=12,666$

33. XLIII

35. CCXCIV

37. MCMXIV

39. $\overline{\text{IV}}$DCCXCIII

41. $\overline{\text{IX}}$CMXCIX

43. $\overline{\text{XLVI}}$CCLXXXI

45. $7(10)+4=74$

47. $4(1000)+8(10)+1=4081$

49. $8(1000)+5(100)+5(10)=8550$

51. $4(1000)+3=4003$

53.

55.

Copyright © 2017 Pearson Education, Inc.

57.

四千二百六十

59.

七千零五十六

61. $30 + 4 = 34$

63. $800 + 70 + 8 = 878$

65. $2 \times 1000 + 800 + 80 + 3 = 2883$

67. $\sigma\alpha$

69. $\psi\,\pi\,\delta$

71. $'\varepsilon\varepsilon$

73.

1021, MXXI, 一, $'\alpha\kappa\alpha$

一千零二十一

75.

527, 𐤙𐤙𐤙𐤙𐤙∩∩||||||, DXXVII, $\phi\kappa\zeta$

77. A number is a quantity, and it answers the question "How many?" A numeral is a symbol used to represent the number.

79. $\overline{\text{CMXCIX}}$CMXCIX

81. Turn the book upside down.

83. 1888, MDCCCLXXXVIII

Exercise Set 4.2

1. 10

3. Hundreds

5. Units

7. a) 1
    b) 10
    c) Subtraction

9. a) 0
    b) 1
    c) 5

11. $(6 \times 10) + (3 \times 1)$

13. $(7 \times 100) + (1 \times 10) + (2 \times 1)$

15. $(4 \times 1000) + (3 \times 100) + (8 \times 10) + (7 \times 1)$

17. $(1 \times 10,000) + (6 \times 1000) + (4 \times 100) + (0 \times 10) + (2 \times 1)$

19. $(3 \times 100,000) + (4 \times 10,000) + (6 \times 1000) + (8 \times 100) + (6 \times 10) + (1 \times 1)$

21. $10 + 10 + 1 + 1 + 1 + 1 = 24$

23. $(10 + 1 + 1 + 1)(60) + (1 + 1 + 1 + 1)(1) = 13(60) + 4(1) = 780 + 4 = 784$

25. $1(60^2) + (10 + 10 + 1)(60) + (10 - (1 + 1))(1) = 3600 + 21(60) + (10 - 2)(1) = 3600 + 1260 + 8 = 4868$

27. 32 is 32 units. ⟨⟨⟨❚❚

29. 471 is 7 groups of 60 and 51 units remaining. ⟨𝖳𝖨𝖨𝖨 ⟨⟨⟨⟨⟨𝖨

31. 3605 is 1 group of 3600 and 5 units remaining. 𝖨   𝖨𝖨𝖨𝖨𝖨

33. $6(20)+17(1)=137$

35. $13(18\times20)+0(20)+2(1)=4680+0+2=4682$

37. $11(18\times20)+2(20)+0(1)=3960+40+0=4000$

39. •••

41.

$$
\begin{array}{r}
14 \\
20\overline{\smash)297} \\
280 \\
\hline
17
\end{array}
$$

$297=14(20)+17(1)$

43.

$$
\begin{array}{r}
6 \\
360\overline{\smash)2163} \\
2160 \\
\hline
3
\end{array}
$$

$2163=6(360)+0(20)+3(1)$

45. Hindu-Arabic:

$5(18\times20)+7(20)+4(1)=1800+140+4=1944$

Babylonian: $1944=32(60)+24(1)$

⟨⟨⟨𝖨𝖨 ⟨⟨𝖨𝖨𝖨𝖨

47.

$$\left(\triangle\times\varobar^{2}\right)+\left(\square\times\varobar\right)+\left(\lozenge\times1\right)$$

49. The Mayan system has a different base and the numbers are written vertically.

51. a) $999,999=6(18\times20^{3})+18(18\times20^{2})+17(18\times20)+13(20)+19(1)$

•

•••

••

•••

••••

b) $999,999=4(60)^{3}+37(60)^{2}+46(60)+39(1)$

𝖨𝖨𝖨𝖨 ⟨⟨⟨⟨𝖳𝖨𝖨𝖨 ⟨⟨⟨⟨𝖨𝖨𝖨𝖨𝖨𝖨 ⟨⟨⟨⟨𝖳𝖨

53. $3(60)+33(1)=180+33=213$

32

$213-32=181$

$181=3(60)+1(1)$ ❚❚❚ ❚

55. $6(18\times20)+7(20)+13(1)=2160+140+13=2313$

$2655-2313=342$

$342=17(20)+2(1)$

$\overset{\bullet\bullet}{\underset{\bullet\bullet}{\equiv}}$

57. Answers will vary.

Exercise Set 4.3

1. Base

3. 5

5. $12_3=(1\times3)+(2\times1)=5$

7. $23_4=(2\times4)+(3\times1)=11$

9. $241_8=(2\times64)+(4\times8)+(1\times1)=161$

11. $309_{12}=(3\times144)+(0\times12)+(9\times1)=441$

13. $573_{16}=(5\times256)+(7\times16)+(3\times1)=1395$

15. $110101_2=32+16+4+1=53$

17. $7654_8=(7\times512)+(6\times64)+(5\times8)+(4\times1)=4012$

19. $A91_{12}=(10\times144)+(9\times12)+1=1549$

21. $C679_{16}=(12\times4096)+(6\times256)+(7\times16)+9=50,809$

23. To convert 7 to base 2   ... 16 8 4 2 1

$$
\begin{array}{ccc}
\underset{\dfrac{4}{3}}{4\overline{)\overset{1}{7}}} &
\underset{\dfrac{2}{1}}{2\overline{)\overset{1}{3}}} &
\underset{\dfrac{1}{0}}{1\overline{)\overset{1}{1}}}
\end{array}
$$

$7=111_2$

25. To convert 23 to base 3   ... 81 27 9 3 1

$$
\begin{array}{ccc}
\underset{\dfrac{18}{5}}{9\overline{)\overset{2}{23}}} &
\underset{\dfrac{3}{2}}{3\overline{)\overset{1}{5}}} &
\underset{\dfrac{2}{0}}{1\overline{)\overset{2}{2}}}
\end{array}
$$

$23=212_3$

27. To convert 190 to base 4   ... 64 16 4 1

$$\begin{array}{c} 2 \\ 64\overline{)190} \\ \underline{128} \\ 62 \end{array} \quad \begin{array}{c} 3 \\ 16\overline{)62} \\ \underline{48} \\ 14 \end{array} \quad \begin{array}{c} 3 \\ 4\overline{)14} \\ \underline{12} \\ 2 \end{array} \quad \begin{array}{c} 2 \\ 1\overline{)2} \\ \underline{2} \\ 0 \end{array}$$

$190 = 2332_4$

29. To convert 102 to base 5   ... 625 125 25 5 1

$$\begin{array}{c} 4 \\ 25\overline{)102} \\ \underline{100} \\ 2 \end{array} \quad \begin{array}{c} 0 \\ 5\overline{)2} \\ \underline{0} \\ 2 \end{array} \quad \begin{array}{c} 2 \\ 1\overline{)2} \\ \underline{2} \\ 0 \end{array}$$

$102 = 402_5$

31. To convert 1098 to base 8   ... 4096 512 64 8 1

$$\begin{array}{c} 2 \\ 512\overline{)1098} \\ \underline{1024} \\ 74 \end{array} \quad \begin{array}{c} 1 \\ 64\overline{)74} \\ \underline{64} \\ 10 \end{array} \quad \begin{array}{c} 1 \\ 8\overline{)10} \\ \underline{8} \\ 2 \end{array} \quad \begin{array}{c} 2 \\ 1\overline{)2} \\ \underline{2} \\ 0 \end{array}$$

$1098 = 2112_8$

33. To convert 9004 to base 12   ... 20,736 1728 144 12 1

$$\begin{array}{c} 5 \\ 1728\overline{)9004} \\ \underline{8640} \\ 364 \end{array} \quad \begin{array}{c} 2 \\ 144\overline{)364} \\ \underline{288} \\ 76 \end{array} \quad \begin{array}{c} 6 \\ 12\overline{)76} \\ \underline{72} \\ 4 \end{array} \quad \begin{array}{c} 4 \\ 1\overline{)4} \\ \underline{4} \\ 0 \end{array}$$

$9004 = 5264_{12}$

35. To convert 9455 to base 16   ... 65,536 4096 256 16 1

$$\begin{array}{c} 2 \\ 4096\overline{)9455} \\ \underline{8192} \\ 1263 \end{array} \quad \begin{array}{c} 4 \\ 256\overline{)1263} \\ \underline{1024} \\ 239 \end{array} \quad \begin{array}{c} E = 14 \\ 16\overline{)239} \\ \underline{224} \\ 15 \end{array} \quad \begin{array}{c} 15 \quad = F \\ 1\overline{)15} \\ \underline{15} \\ 0 \end{array}$$

$9455 = 24EF_{16}$

37. To convert 2017 to base 3   ... 2187 729 243 81 27 9 3 1

$$\begin{array}{c} 2 \\ 729\overline{)2017} \\ \underline{1458} \\ 559 \end{array} \, \begin{array}{c} 2 \\ 243\overline{)559} \\ \underline{486} \\ 73 \end{array} \, \begin{array}{c} 0 \\ 81\overline{)73} \\ \underline{0} \\ 73 \end{array} \, \begin{array}{c} 2 \\ 27\overline{)73} \\ \underline{54} \\ 19 \end{array} \, \begin{array}{c} 2 \\ 9\overline{)19} \\ \underline{18} \\ 1 \end{array} \, \begin{array}{c} 0 \\ 3\overline{)1} \\ \underline{0} \\ 1 \end{array} \, \begin{array}{c} 1 \\ 1\overline{)1} \\ \underline{1} \\ 0 \end{array}$$

$2017 = 2202201_3$

39. To convert 2017 to base 5   ... 3125 625 125 25 5 1

$$\begin{array}{c} 3 \\ 625\overline{)2017} \\ \underline{1875} \\ 142 \end{array} \quad \begin{array}{c} 1 \\ 125\overline{)142} \\ \underline{125} \\ 17 \end{array} \quad \begin{array}{c} 0 \\ 25\overline{)17} \\ \underline{0} \\ 17 \end{array} \quad \begin{array}{c} 3 \\ 5\overline{)17} \\ \underline{15} \\ 2 \end{array} \quad \begin{array}{c} 2 \\ 1\overline{)2} \\ \underline{2} \\ 0 \end{array}$$

$2017 = 31032_5$

41. To convert 2017 to base 7   ... 2401 343 49 7 1

$$\begin{array}{c} 5 \\ 343\overline{)2017} \\ \underline{1715} \\ 302 \end{array} \quad \begin{array}{c} 6 \\ 49\overline{)302} \\ \underline{294} \\ 8 \end{array} \quad \begin{array}{c} 1 \\ 7\overline{)8} \\ \underline{7} \\ 1 \end{array} \quad \begin{array}{c} 1 \\ 1\overline{)1} \\ \underline{1} \\ 0 \end{array}$$

$2017 = 5611_7$

43. To convert 2017 to base 12   ... 1728 144 12 1

$$\begin{array}{r} 1 \\ 1728 \overline{)\ 2017} \\ \underline{1728} \\ 289 \end{array} \quad \begin{array}{r} 2 \\ 144 \overline{)\ 289} \\ \underline{288} \\ 1 \end{array} \quad \begin{array}{r} 0 \\ 12 \overline{)\ 1} \\ \underline{0} \\ 1 \end{array} \quad \begin{array}{r} 1 \\ 1 \overline{)\ 1} \\ \underline{1} \\ 0 \end{array}$$

$$2017 = 1201_{12}$$

45. $2(5) + 3(1) = 10 + 3 = 13$

47. $3(5^2) + 0(5) + 3(1) = 3(25) + 0 + 3$

   $= 75 + 0 + 3 = 78$

49. To convert   ... 25 5 1

$$\begin{array}{r} 3 = \ominus \\ 5 \overline{)\ 19} \\ \underline{15} \\ 4 \end{array} \quad \begin{array}{r} 4 = \oslash \\ 1 \overline{)\ 4} \\ \underline{4} \\ 0 \end{array}$$

$$19 = {\ominus\oslash}_5$$

51. To convert   ... 125 25 5 1

$$\begin{array}{r} 2 = \oslash \\ 25 \overline{)\ 74} \\ \underline{50} \\ 24 \end{array} \quad \begin{array}{r} 4 = \oslash \\ 5 \overline{)\ 24} \\ \underline{20} \\ 4 \end{array} \quad \begin{array}{r} 4 = \oslash \\ 1 \overline{)\ 4} \\ \underline{4} \\ 0 \end{array}$$

$$74 = {\oslash\oslash\oslash}_5$$

53. $1(4) + 3(1) = 4 + 3 = 7$

55. $2(4^2) + 1(4) + 0(1) = 2(16) + 4 + 0 = 32 + 4 + 0 = 36$

57. To convert   ... 16 4 1

$$\begin{array}{r} 2 = \text{(go)} \\ 4 \overline{)\ 10} \\ \underline{8} \\ 2 \end{array} \quad \begin{array}{r} 2 = \text{(go)} \\ 1 \overline{)\ 2} \\ \underline{2} \\ 0 \end{array}$$

$$10 = \text{(go)} \ \text{(go)} \ _4$$

59. To convert   ... 64 16 4 1

$$\begin{array}{r} 3 = \text{(gr)} \\ 16 \overline{)\ 60} \\ \underline{48} \\ 12 \end{array} \quad \begin{array}{r} 3 = \text{(gr)} \\ 4 \overline{)\ 12} \\ \underline{12} \\ 0 \end{array} \quad \begin{array}{r} 0 = \text{(b)} \\ 1 \overline{)\ 0} \\ \underline{0} \\ 0 \end{array}$$

$$60 = \text{(gr)} \ \text{(gr)} \ \text{(b)} \ _4$$

61.

a)

```
5 | 683
5 | 136    3 ↑
5 | 27     1 ↑
5 | 5      2 ↑
5 | 1      0 ↑
    0      1 ↑
```

$683 = 10213_5$

b)

```
8 | 763
8 | 95     3 ↑
8 | 11     7 ↑
8 | 1      3 ↑
    0      1 ↑
```

$763 = 1373_8$

63.   a) $10_2 = (1 \times 2) + (0 \times 1) = 2$

b) $10_8 = (1 \times 8) + (0 \times 1) = 8$

c) $10_{16} = (1 \times 16) + (0 \times 1) = 16$

d) $10_{32} = (1 \times 32) + (0 \times 1) = 32$

e) In general, for any base b, $10_b = (1 \times b) + (0 \times 1) = b$

65. $1(b^2) + 1(b) + 1 = 43$

$b^2 + b + 1 = 43$

$b^2 + b - 42 = 0$

$(b+7)(b-6) = 0$

$b + 7 = 0$ or $b - 6 = 0$

$b = -7$ or $b = 6$

Since the base cannot be negative, $b = 6$.

67. a) 1_3, 2_3, 10_3, 11_3, 12_3, 20_3, 21_3, 22_3, 100_3, 101_3, 102_3, 110_3, 111_3, 112_3, 120_3, 121_3, 122_3, 200_3, 201_3, 202_3
b) 1000_3

69. Answers will vary.

For #71, blue = 0 = b, red = 1 = r, gold = 2 = go, green = 3 = gr

71. a) $3(4^4) + 1(4^3) + 2(4^2) + 3(4) + 0(1) = 3(256) + 64 + 2(16) + 12 + 0 = 768 + 64 + 32 + 12 + 0 = 876$

b) To convert   … 256   64   16   4   1

```
       2   = (go)         3   = (gr)        0   = (b)          1   = (r)
64 | 177              16 | 49            4 | 1              1 | 1
   128                    48                0                  1
    49                     1                1                  0
```

$177 = $ (go) (gr) (b) (r) $_4$

73. Answers will vary.

Exercise Set 4.4

1. 12_5

3. 5

5. 22_5

7. 12_5
        21_5
       110_3

9. 132_5
        34_5
       221_5

11. 654_7
        463_7
       1450_7

13. 1011_2
       1110_2
      11001_2

15. $A734_{12}$
       $128B_{12}$
       $BA03_{12}$

17. $A734_{16}$
       $128B_{16}$
       $B9BF_{16}$

19. 32_5
        -24_5
          3_5

21. 338_9
       -274_9
         54_9

23. 1101_2
       -111_2
        110_2

25. 4223_7
       -304_7
       3616_7

27. 4232_5
       -2341_5
       1341_5

29. $A3B3_{12}$
       $-21B4_{12}$
       $81BB_{12}$

31. 42_5
       $\times\ 2_5$
        134_5

33. 212_3
       $\times\ \ 2_3$
       1201_3

35. 37_8
       $\times 21_8$
         37
         76
       1017_8

37. $B12_{12}$
       $\times\ 83_{12}$
         2936
         7494
       77676_{12}

39. 110_2
       $\times\ 11_2$
         110
         110
       10010_2

41. 316_7
       $\times\ 16_7$
        2541
         316
       6031_7

43. $2_4 \times 1_4 = 2_4$
     $2_4 \times 2_4 = 10_4$
     $2_4 \times 3_4 = 12_4$

$$2_4 \overline{)\, 312_4} \qquad 123_4$$

```
      123₄
 2₄ ) 312₄
      2
      11
      10
      12
      12
       0
```

45. $3_7 \times 1_7 = 3_7$
     $3_7 \times 2_7 = 6_7$
     $3_7 \times 3_7 = 12_7$
     $3_7 \times 4_7 = 15_7$
     $3_7 \times 5_7 = 21_7$
     $3_7 \times 6_7 = 24_7$

```
      146₇   R2₇
 3₇ ) 506₇
      3
      20
      15
      26
      24
       2
```

47. $3_8 \times 1_8 = 3_8$
     $3_8 \times 2_8 = 6_8$
     $3_8 \times 4_8 = 14_8$
     $3_8 \times 5_8 = 17_8$

```
      52₈    R2₈
 3₈ ) 200₈
      17
      10
       6
       2
```

49. $2_4 \times 1_4 = 2_4$
     $2_4 \times 2_4 = 10_4$
     $2_4 \times 3_4 = 12_4$

```
      103₄   R1₄
 2₄ ) 213₄
      2
      01
      00
      13
      12
       1
```

51. $3_5 \times 1_5 = 3_5$
     $3_5 \times 2_5 = 11_5$
     $3_5 \times 3_5 = 14_5$
     $3_5 \times 4_5 = 22_5$

```
      41₅    R1₅
 3₅ ) 224₅
      22
      04
       3
       1
```

53. $6_7 \times 1_7 = 6_7$
     $6_7 \times 2_7 = 15_7$
     $6_7 \times 3_7 = 24_7$
     $6_7 \times 4_7 = 33_7$
     $6_7 \times 5_7 = 42_7$
     $6_7 \times 6_7 = 51_7$

```
      45₇    R2₇
 6₇ ) 404₇
      33
      44
      42
       2
```

55. $\begin{array}{r} 3_5 \\ + 3_5 \\ \hline 11_5 = \end{array}$ 🌓🌗$_5$

57. $\begin{array}{r} 23_5 \\ + 13_5 \\ \hline 41_5 = \end{array}$ 🌕🌗$_5$

For #59-65, blue = 0 = b, red = 1 = r, gold = 2 = go, green = 3 = gr

59. $\begin{array}{r} 3_4 \\ + 2_4 \\ \hline 11_4 = \end{array}$ (r)(r)$_4$

61. $\begin{array}{r} 32_4 \\ + 11_4 \\ \hline 103_4 = \end{array}$ (r)(b)(gr)$_4$

63. $\begin{array}{r} 31_4 \\ - 13_4 \\ \hline 12_4 = \end{array}$ (r)(go)$_4$

65. $\begin{array}{r} 231_4 \\ - 103_4 \\ \hline 122_4 = \end{array}$ (r)(go)(go)$_4$

67. $\begin{array}{r} FAB_{16} \\ \times \ \ 4_{16} \\ \hline 2C \\ 28 \ \ \\ 3C \ \ \ \\ \hline 3EAC_{16} \end{array}$

69. a) $\begin{array}{r} 462_8 \\ \times \ 35_8 \\ \hline 2772 \\ 1626 \ \ \\ \hline 21252_8 \end{array}$

b) $462_8 = 4(8^2) + 6(8) + 2(1) = 4(64) + 48 + 2$
$= 256 + 48 + 2 = 306$
$35_8 = 3(8) + 5(1) = 24 + 5 = 29$

c) $306 \times 29 = 8874$

d) $21252_8 = 2(8^4) + 1(8^3) + 2(8^2) + 5(8) + 2(1)$
$= 2(4096) + 512 + 2(64) + 40 + 2$
$= 8192 + 512 + 128 + 40 + 2 = 8874$

e) Yes, in part a), the numbers were multiplied in base 8 and then converted to base 10 in part d). In part b), the numbers were converted to base 10 first, then multiplied in part c).

71. ⬤ = 0, ⬤ = 1, ⬤ = 2, ⬤ = 3 (Gold = 0, Purple = 1, Blue = 2, Red =3)

73.   Answers will vary.

Exercise Set 4.5
1.   a) Divided
     b) Doubled
3.   Three, two

5.

```
15 –  17
 7 –  34
 3 –  68
 1 – 136
      255
```

7.

```
27 –  31
13 –  62
 6 – 124
 3 – 248
 1 – 496
      837
```

9.

```
35 –  236
17 –  472
 8 –  944
 4 – 1888
 2 – 3776
 1 – 7552
      8260
```

11.

```
93 –   93
46 –  186
23 –  372
11 –  744
 5 – 1488
 2 – 2976
 1 – 5952
      8649
```

13.

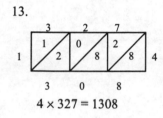

$4 \times 327 = 1308$

15.

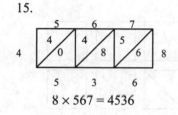

$8 \times 567 = 4536$

17.

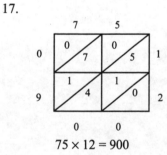

$75 \times 12 = 900$

19.

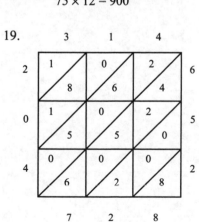

$$314 \times 652 = 204{,}728$$

21.

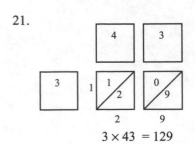

$$3 \times 43 = 129$$

23.

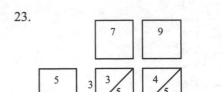

$$5 \times 79 = 395$$

25.

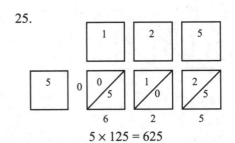

$$5 \times 125 = 625$$

27.

6 7 4 2

9   6 | 5/4 | 6/3 | 3/6 | 1/8
       0   6   7   8

$$9 \times 6742 = 60{,}678$$

29.   a) 253×46; Place the factors of 8 until the
          correct factors and placements are found
          so the rest of the rectangle can be completed.

   b)

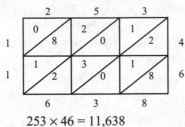

$253 \times 46 = 11{,}638$

31.   a) 4×382; Place the factors of 12 until the correct
          factors and placements are found so the rest
          can be completed.

   b)

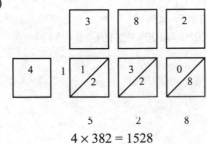

$4 \times 382 = 1528$

33.   13 – 22
      6 – 44
      3 – 88
      1 – <u>176</u>
      286 =

35.

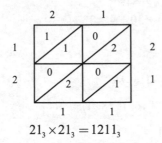

$21_3 \times 21_3 = 1211_3$

37. Answers will vary.

Review Exercises

1. $100 + 100 + 10 + 10 + 10 + 1 + 1 = 232$

2. $10{,}000 + 10{,}000 + 10{,}000 + 10{,}000 + 1000$
$+ 1000 + 10 + 10 + 10 + 10 + 10 + 1 + 1 + 1$
$= 42{,}053$

3. $100{,}000 + 100{,}000 + 100 + 100 + 100 + 100 + 10$
$+ 1000 + 10 + 1 = 200{,}421$

4. $1{,}000{,}000 + 100{,}000 + 100{,}000 + 10{,}000 + 1000$
$+ 1000 + 1000 + 100 + 100 + 100 + 10 + 10 + 10$
$= 1{,}214{,}330$

5.

6.

7.

8.

9. $10 + (10 - 1) = 19$

10. $100 + 100 + (50 - 10) + 5 + 1 + 1 = 247$

11. $1000 + 1000 + (500 - 100) + 10 + 10 + 10$
$+ 5 + 1 + 1 + 1 = 2437$

12. $5(1000) + (500 + 200) + 50 + (10 - 1) = 5759$

13. XXIV

14. CDLIX

15. MCMLXIV

16. $\overline{\text{V}}$ICDXCI

17. $3(10) + 2 = 32$

18. $4(10) + 5 = 45$

19. $2(100) + 6(10) + 7 = 267$

20. $3(1000) + 4(100) + 2(10) + 9 = 3429$

21.

22.

23.

24.

25. $80 + 1 = 81$

26. $500 + 40 + 8 = 548$

27. $600 + 5 = 605$

28. $3000 + 300 + 30 + 4 = 3334$

29. $\kappa\beta$

30. $\psi o\gamma$

31. $\omega\xi\zeta$

32. $'\delta\sigma\iota\theta$

33. $(3 \times 60) + (11 \times 1) = 191$

34. $(3 \times 60) + (9 \times 1) = 189$

35. $(1 \times 60^2) + (1 \times 60) + (13 \times 1) = 3673$

36. $(2 \times 60^2) + (3 \times 60) + (8 \times 1) = 7388$

37.

38.

39. ⅼⅼ ⅼ ≪

40. ≪ ⅼⅼⅼ ⅼⅼⅼⅼ

41. $2(20) + 8 = 48$

42. $3(18 \times 20) + 14(20) + 7 = 1367$

43. $7(18 \times 20) + 0(20) + 1 = 2521$

44. $2(18 \times 20^2) + 4(18 \times 20) + 4(20) + 3 = 15{,}923$

45.

$$20 \overline{\smash{\big)}\, \begin{array}{r} 3 \\ 69 \end{array}}$$
$$\underline{60}$$
$$9$$

$$69 = 3(3 \times 20) + 9(1)$$

46.

$$360 \overline{\smash{\big)}\, \begin{array}{r} 2 \\ 812 \end{array}} \qquad 20 \overline{\smash{\big)}\, \begin{array}{r} 4 \\ 92 \end{array}}$$
$$\underline{720} \qquad\qquad \underline{80}$$
$$92 \qquad\qquad\quad 12$$

$$812 = 2(18 \times 20) + 4(20) + 12(1)$$

47.

$$360 \overline{\smash{\big)}\, \begin{array}{r} 4 \\ 1571 \end{array}} \qquad 20 \overline{\smash{\big)}\, \begin{array}{r} 6 \\ 131 \end{array}}$$
$$\underline{1440} \qquad\qquad \underline{120}$$
$$131 \qquad\qquad\quad 11$$

$$1571 = 4(18 \times 20) + 6(20) + 11(1)$$

48.

$$7200 \overline{\smash{\big)}\, \begin{array}{r} 2 \\ 17{,}913 \\ \underline{14{,}400} \\ 3513 \end{array}} \qquad 3600 \overline{\smash{\big)}\, \begin{array}{r} 9 \\ 3513 \\ \underline{3240} \\ 273 \end{array}} \qquad 20 \overline{\smash{\big)}\, \begin{array}{r} 13 \\ 273 \\ \underline{260} \\ 13 \end{array}}$$

$$17{,}913 = 2\left(18 \times 20^2\right) + 9(18 \times 20) + 13(20) + 13(1)$$

$$\cdot\cdot$$
$$\cdot\cdot\cdot\cdot$$
$$\underset{=}{\cdot\cdot\cdot}$$
$$\underset{=}{\cdot\cdot\cdot}$$

49. $12_3 = 1(3) + 2(1) = 5$

50. $111_2 = 1\left(2^2\right) + 1(2) + 1(1) = 4 + 2 + 1 = 7$

51. $130_4 = 1\left(4^2\right) + 3(4) + 0(1) = 16 + 12 + 0 = 28$

52. $3425_7 = 3\left(7^3\right) + 4\left(7^2\right) + 2(7) + 5(1) = 3(343) + 4(49) + 14 + 5 = 1029 + 196 + 14 + 5 = 1244$

53. $A94_{12} = 10\left(12^2\right) + 9(12) + 4(1) = 1440 + 108 + 4 = 1552$

54. $20220_3 = 2\left(3^4\right) + 0\left(3^3\right) + 2\left(3^2\right) + 2(3) + 0(1) = 2(81) + 0 + 2(9) + 6 + 0 = 162 + 0 + 18 + 6 + 0 = 186$

55. To convert 463 to base 2   ... 512 256 128 64 32 16 8 4 2 1

$$256 \overline{\smash{\big)}\, \begin{array}{r} 1 \\ 463 \\ \underline{256} \\ 207 \end{array}} \quad 128 \overline{\smash{\big)}\, \begin{array}{r} 1 \\ 207 \\ \underline{128} \\ 79 \end{array}} \quad 64 \overline{\smash{\big)}\, \begin{array}{r} 1 \\ 79 \\ \underline{64} \\ 15 \end{array}} \quad 32 \overline{\smash{\big)}\, \begin{array}{r} 0 \\ 15 \\ \underline{0} \\ 15 \end{array}} \quad 16 \overline{\smash{\big)}\, \begin{array}{r} 0 \\ 15 \\ \underline{0} \\ 15 \end{array}} \quad 8 \overline{\smash{\big)}\, \begin{array}{r} 1 \\ 15 \\ \underline{8} \\ 7 \end{array}} \quad 4 \overline{\smash{\big)}\, \begin{array}{r} 1 \\ 7 \\ \underline{4} \\ 3 \end{array}} \quad 2 \overline{\smash{\big)}\, \begin{array}{r} 1 \\ 3 \\ \underline{2} \\ 1 \end{array}} \quad 1 \overline{\smash{\big)}\, \begin{array}{r} 1 \\ 1 \\ \underline{1} \\ 0 \end{array}}$$

$463 = 111001111_2$

56. To convert 463 to base 4   ... 1024 256 64 16 4 1

$$256 \overline{\smash{\big)}\, \begin{array}{r} 1 \\ 463 \\ \underline{256} \\ 207 \end{array}} \quad 64 \overline{\smash{\big)}\, \begin{array}{r} 3 \\ 207 \\ \underline{192} \\ 15 \end{array}} \quad 16 \overline{\smash{\big)}\, \begin{array}{r} 0 \\ 15 \\ \underline{0} \\ 15 \end{array}} \quad 4 \overline{\smash{\big)}\, \begin{array}{r} 3 \\ 15 \\ \underline{12} \\ 3 \end{array}} \quad 1 \overline{\smash{\big)}\, \begin{array}{r} 3 \\ 3 \\ \underline{3} \\ 0 \end{array}}$$

$463 = 13033_4$

57. To convert 463 to base 5 ...   625 125 25 5 1

$$125 \overline{\smash{\big)}\, \begin{array}{r} 3 \\ 463 \\ \underline{375} \\ 88 \end{array}} \quad 25 \overline{\smash{\big)}\, \begin{array}{r} 3 \\ 88 \\ \underline{75} \\ 13 \end{array}} \quad 5 \overline{\smash{\big)}\, \begin{array}{r} 2 \\ 13 \\ \underline{10} \\ 3 \end{array}} \quad 3 \overline{\smash{\big)}\, \begin{array}{r} 1 \\ 3 \\ \underline{3} \\ 0 \end{array}}$$

$463 = 3323_5$

58. To convert 463 to base 8   ... 512 64 8 1

$$64 \overline{\smash{\big)}\, \begin{array}{r} 7 \\ 463 \\ \underline{448} \\ 15 \end{array}} \quad 8 \overline{\smash{\big)}\, \begin{array}{r} 1 \\ 15 \\ \underline{8} \\ 7 \end{array}} \quad 1 \overline{\smash{\big)}\, \begin{array}{r} 7 \\ 7 \\ \underline{7} \\ 0 \end{array}}$$

$463 = 717_8$

59. To convert 463 to base 12   … 1728   144   12   1

$$\begin{array}{r} 3 \\ 144 \overline{)463} \\ \underline{432} \\ 31 \end{array} \qquad \begin{array}{r} 2 \\ 12 \overline{)31} \\ \underline{24} \\ 7 \end{array} \qquad \begin{array}{r} 7 \\ 1 \overline{)7} \\ \underline{7} \\ 0 \end{array} \qquad 463 = 327_{12}$$

60. To convert 463 to base 16   … 65,536   4096   256   16   1

$$\begin{array}{r} 1 \\ 256 \overline{)463} \\ \underline{256} \\ 207 \end{array} \qquad \begin{array}{r} 12 = C \\ 16 \overline{)207} \\ \underline{192} \\ 15 \end{array} \qquad \begin{array}{r} 15 = F \\ 1 \overline{)15} \\ \underline{15} \\ 0 \end{array} \qquad 493 = 1CF_{16}$$

61. $\begin{array}{r} 121_4 \\ +322_4 \\ \hline 1103_4 \end{array}$

62. $\begin{array}{r} 10110_2 \\ +11001_2 \\ \hline 101111_2 \end{array}$

63. $\begin{array}{r} 9B_{12} \\ +87_{12} \\ \hline 166_{12} \end{array}$

64. $\begin{array}{r} 2B9_{16} \\ +456_{16} \\ \hline 70F_{16} \end{array}$

65. $\begin{array}{r} 3024_5 \\ +4023_5 \\ \hline 12102_5 \end{array}$

66. $\begin{array}{r} 3407_8 \\ +7014_8 \\ \hline 12423_8 \end{array}$

67. $\begin{array}{r} 321_4 \\ -133_4 \\ \hline 122_4 \end{array}$

68. $\begin{array}{r} 1001_2 \\ -101_2 \\ \hline 100_2 \end{array}$

69. $\begin{array}{r} A7B_{12} \\ -95_{12} \\ \hline 9A6_{12} \end{array}$

70. $\begin{array}{r} 4321_5 \\ -442_5 \\ \hline 3324_5 \end{array}$

71. $\begin{array}{r} 1713_8 \\ -1243_8 \\ \hline 450_8 \end{array}$

72. $\begin{array}{r} F64_{16} \\ -2A3_{16} \\ \hline CC1_{16} \end{array}$

73. $\begin{array}{r} 431_6 \\ \times\ 3_6 \\ \hline 2133_6 \end{array}$

74. $\begin{array}{r} 2321_4 \\ \times\ 3_4 \\ \hline 20223_4 \end{array}$

75. $\begin{array}{r} 34_5 \\ \times 21_5 \\ \hline 34 \\ \underline{123} \\ 1314_5 \end{array}$

76. $\begin{array}{r} 476_8 \\ \times 23_8 \\ \hline 1672 \\ \underline{1174} \\ 13632_8 \end{array}$

77. $\begin{array}{r} 126_{12} \\ \times\ 47_{12} \\ \hline 856 \\ \underline{4A0} \\ 5656_{12} \end{array}$

78. $\begin{array}{r} 1A3_{16} \\ \times\ 12_{16} \\ \hline 346 \\ \underline{1A3} \\ 1D76_{16} \end{array}$

79.
$$\begin{array}{l} 2_3 \times 1_3 = 2_3 \\ 2_3 \times 2_3 = 11_3 \\ 2_3 \times 3_3 = 20_3 \\ 2_3 \times 4_3 = 22_3 \end{array}$$

$$\begin{array}{r} 21_3 \quad \text{R } 1_3 \\ 2_3 \overline{)120_3} \\ \underline{11} \\ 10 \\ \underline{2} \\ 1 \end{array}$$

80.
$$\begin{array}{l} 2_4 \times 1_4 = 2_4 \\ 2_4 \times 2_4 = 10_4 \\ 2_4 \times 3_4 = 12_4 \end{array}$$

$$\begin{array}{r} 130_4 \\ 2_4 \overline{)320_4} \\ \underline{2} \\ 12 \\ \underline{12} \\ 0 \\ \underline{0} \\ 0 \end{array}$$

81. $3_5 \times 1_5 = 3_5$

$3_5 \times 2_5 = 11_5$

$3_5 \times 3_5 = 14_5$

$3_5 \times 4_5 = 22_5$

$$
\begin{array}{r}
23_5 \ \ \text{R}1_5 \\
3_5\,\overline{)\,130_5} \\
\underline{11} \\
20 \\
\underline{14} \\
1
\end{array}
$$

82. $4_6 \times 1_6 = 4_6$

$4_6 \times 2_6 = 12_6$

$4_6 \times 3_6 = 20_6$

$4_6 \times 4_6 = 24_6$

$4_6 \times 5_6 = 32_6$

$$
\begin{array}{r}
433_6 \\
4_6\,\overline{)\,3020_6} \\
\underline{24} \\
22 \\
\underline{20} \\
20 \\
\underline{20} \\
0
\end{array}
$$

83. $3_6 \times 1_6 = 3_6$

$3_6 \times 2_6 = 10_6$

$3_6 \times 3_6 = 13_6$

$3_6 \times 4_6 = 20_6$

$3_6 \times 5_6 = 23_6$

$$
\begin{array}{r}
411_6 \ \ \text{R}1_6 \\
3_6\,\overline{)\,2034_6} \\
\underline{20} \\
03 \\
\underline{3} \\
04 \\
\underline{3} \\
1
\end{array}
$$

84. $6_8 \times 1_8 = 6_8$

$6_8 \times 2_8 = 14_8$

$6_8 \times 3_8 = 22_8$

$6_8 \times 4_8 = 30_8$

$6_8 \times 5_8 = 36_8$

$6_8 \times 6_8 = 44_8$

$6_8 \times 7_8 = 52_8$

$$
\begin{array}{r}
664_8 \ \ \text{R}2_8 \\
6_8\,\overline{)\,5072_8} \\
\underline{44} \\
47 \\
\underline{44} \\
32 \\
\underline{30} \\
2
\end{array}
$$

85. $125 \ - \ 23$

$\ \ \cancel{62} \ \ \ - \ \ \ 46$

$\ \ \ 31 \ - \ \ \ 92$

$\ \ \ 15 \ - \ 184$

$\ \ \ \ 7 \ - \ 368$

$\ \ \ \ 3 \ - \ 736$

$\ \ \ \ 1 \ - \ \underline{1472}$

$\ \ \ \ \ \ \ \ \ \ \ \ \ \ 2875$

86.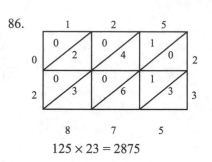

$125 \times 23 = 2875$

87.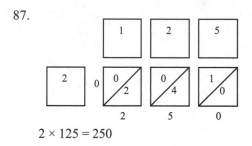

$2 \times 125 = 250$

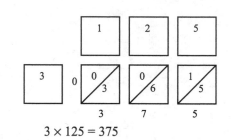

$3 \times 125 = 375$

$2 \times 252 = 250$, therefore $20 \times 125 = 2500$

Therefore, $125 \times 23 = 2500 + 375 = 2875$.

Chapter Test

1. $(5000 - 1000) + (500 - 100) + 50 + 10 + 5 + 1 + 1 = 4467$

2. $31(60) + 13(1) = 1860 + 13 = 1873$

3. $8(1000) + 0 + 9(10) = 8000 + 0 + 90 = 8090$

4. $2(18 \times 20) + 12(20) + 9(1) = 2(360) + 240 + 9$
$= 720 + 240 + 9 = 969$

5. $100{,}000 + 10{,}000 + 10{,}000 + 1000 + 1000$
$+ 100 + 10 + 10 + 10 + 10 + 1 + 1 = 122{,}142$

6. $2(1000) + 700 + 40 + 5 = 2000 + 700 + 40 + 5$
$= 2745$

7. ꝏ9∩∩IIII

8. $'\beta\upsilon o f$

9.

$$\begin{array}{r} 3 \\ 360\overline{\smash)1434} \\ \underline{1080} \\ 354 \end{array} \qquad \begin{array}{r} 17 \\ 20\overline{\smash)354} \\ \underline{340} \\ 14 \end{array}$$

$$1434 = 3(18 \times 20) + 17(20) + 14(1)$$

10.

$$\begin{array}{r} 26 \\ 60\overline{\smash)1596} \\ \underline{1560} \\ 36 \end{array} \quad \text{<<𝕀𝕀𝕀𝕀𝕀𝕀 <<<𝕀𝕀𝕀𝕀𝕀𝕀}$$

$$1596 = 26(60) + 36(1)$$

11. MMDCCXLIX

12. $706_8 = 7(8^2) + 0(8) + 6(1) = 7(64) + 0 + 6$

 $= 448 + \quad 0 + 6 = 454$

13. $B92_{12} = 11(12^2) + 9(12) + 2(1) = 11(144) + 108 + 2 = 1584 + 108 + 2 = 1694$

14. To convert 49 to base 2   … 64 32 16 8 4 2 1

$$\begin{array}{ccccccc} 1 & 0 & 0 & 0 & 0 & 1 \\ 32\overline{\smash)49} & 16\overline{\smash)17} & 8\overline{\smash)1} & 4\overline{\smash)1} & 2\overline{\smash)1} & 1\overline{\smash)1} \\ \underline{32} & \underline{16} & \underline{0} & \underline{0} & \underline{0} & \underline{1} \\ 17 & 1 & 1 & 1 & 1 & 0 \end{array}$$

$$49 = 110001_2$$

15. To convert 2938 to base 16   … 65,536 4096 256 16 1

$$\begin{array}{ccc} 11 = B & 7 & 10 = A \\ 256\overline{\smash)2938} & 16\overline{\smash)122} & 1\overline{\smash)10} \\ \underline{2816} & \underline{112} & \underline{10} \\ 122 & 10 & 0 \end{array}$$

$$2938 = B7A_{16}$$

16.
$$\begin{array}{r} 1101_2 \\ +\underline{1011_2} \\ 11000_2 \end{array}$$

17.
$$\begin{array}{r} 45_6 \\ \times\underline{23_6} \\ 223 \\ \underline{134} \\ 2003_6 \end{array}$$

18. $3_5 \times 1_5 = 3_5$
   $3_5 \times 2_5 = 11_5$
   $3_5 \times 3_5 = 14_5$
   $3_5 \times 4_5 = 22_5$

$$\begin{array}{r} 220_5 \\ 3_5\overline{\smash)1210_5} \\ \underline{11} \\ 11 \\ \underline{11} \\ 00 \\ \underline{00} \\ 0 \end{array}$$

19. 35 - 28
   17 - 56
   ~~8 112~~
   4 224
   ~~2 448~~
   1 - 896
   980

20.

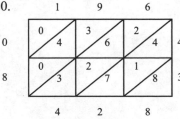

$$43 \times 196 = 8428$$

CHAPTER FIVE

NUMBER THEORY AND THE REAL NUMBER SYSTEM

Exercise Set 5.1

1. Theory

3. Zero

5. Composite

7. Multiple

9. Conjecture

11. The prime numbers between 1 and 100 are: 2, 3, 5, 7, 11, 13, 17, 19, 23, 29, 31, 37, 41, 43, 47, 53, 59, 61, 67, 71, 73, 79, 83, 89, 97.

13. True; since $35 \div 7 = 5$.

15. False; 81 is a multiple of 9.

17. True; $54 \div 6 = 9$.

19. False; 4 is a divisor of 32.

21. True; $42 \div 7 = 6$.

23. True; if a number is divisible by 10, then it is also divisible by 5.

25. True; since $2 \cdot 3 = 6$.

27. Divisible by 3 and 5.

29. Divisible by 2, 3, 4, 6, 8 and 9.

31. Divisible by 2, 3, 4, 5, 6, 8, and 10.

33. 60 (other answers are possible)

35.
$$\begin{array}{r|r} 2 & 20 \\ 2 & 10 \\ & 5 \end{array}$$

$20 = 2^2 \cdot 5$

37.
$$\begin{array}{r|r} 2 & 140 \\ 2 & 70 \\ 5 & 35 \\ & 7 \end{array}$$

$140 = 2^2 \cdot 5 \cdot 7$

39.
$$\begin{array}{r|r} 2 & 332 \\ 2 & 166 \\ & 83 \end{array}$$

$332 = 2^2 \cdot 83$

41.
$$\begin{array}{r|r} 3 & 513 \\ 3 & 171 \\ 3 & 57 \\ & 19 \end{array}$$

$513 = 3^3 \cdot 19$

43.
$$\begin{array}{r|r} 2 & 1336 \\ 2 & 668 \\ 2 & 334 \\ & 167 \end{array}$$

$1336 = 2^3 \cdot 167$

45. The prime factors of 10 and 14 are: $10 = 2 \cdot 5$, $14 = 2 \cdot 7$

a) The common factor is 2, thus, the GCD = 2.

b) The factors with the greatest exponent that appear in either are 2, 5, 7. Thus, the LCM = $2 \cdot 5 \cdot 7 = 70$.

47. The prime factors of 15 and 25 are: $15 = 3 \cdot 5$, $25 = 5^2$

a) The common factor is 5; thus the GCD = 5.

b) The factors with the greatest exponent that appear in either are: 3 and 5^2 ; thus, the LCM $= 3 \cdot 5^2 = 75$

49. The prime factors of 40 and 900 are: $40 = 2^3 \cdot 5$, $900 = 2^3 \cdot 3^2 \cdot 5^2$

a) The common factors are: 2^2, 5; thus, the GCD $= 2^2 \cdot 5 = 20$.

b) The factors with the greatest exponent that appear in either are: $2^3, 3^2, 5^2$; thus, the LCM $= 2^2 \cdot 3^2 \cdot 5^2 = 1800$

51. The prime factors of 96 and 212 are: $96 = 2^5 \cdot 3$, $212 = 2^2 \cdot 53$

a) The common factors are: 2^2; thus, the GCD $= 2^2 = 4$.

b) The factors with the greatest exponent that appear in either are: $2^5, 3, 53$; thus, the LCM $= 2^5 \cdot 3 \cdot 53 = 5088$

53. The prime factors of 24, 48, and 128 are: $24 = 2^3 \cdot 3$, $48 = 2^4 \cdot 3$, $128 = 2^7$

a) The common factors are: 2^3; thus, the GCD $= 2^3 = 8$.

b) The factors with the greatest exponent that appear in any are: $2^7, 3$; thus, LCM $= 2^7 \cdot 3 = 384$

55. The lcm of 6 and 16 is 48 days.

57. $2 \times 60, 3 \times 40, 4 \times 30, 5 \times 24, 6 \times 20, 8 \times 15, 10 \times 12,$ $12 \times 10, 15 \times 8, 20 \times 6, 24 \times 5, 30 \times 4, 40 \times 3, 60 \times 2$

59. The gcd of 70 and 175 is 35 cars.

61 The gcd of 150 and 180 is 30 trees.

63. The lcm of 5 and 6 is 30 days.

65. a) The prime factors of 8 and 9 are: $8 = 2^3$ and $9 = 3^2$
       Yes, they are relatively prime.

    b) The prime factors of 21 and 30 are: $21 = 3 \cdot 7$ and $30 = 2 \cdot 3 \cdot 5$
        No, they are not relatively prime.

    c) The prime factors of 39 and 52 are $39 = 3 \cdot 13$ and $52 = 2^2 \cdot 13$
        No, they are not relatively prime.

    d) The prime factors of 177 and 178 are $177 = 3 \cdot 59$ and $178 = 2 \cdot 89$
        Yes, they are relatively prime.

67. Use the list of primes generated in exercise 13. The next two sets of twin primes are: 17, 19, and 29, 31.

69. Fermat number $= 2^{2^n} + 1$, where n is a natural number. $2^{2^1} + 1 = 5$, $2^{2^2} + 1 = 2^4 + 1 = 17$, $2^{2^3} + 1 = 2^8 + 1 = 257$. These numbers are prime.

71. A number is divisible by 14 if both 2 and 7 divide the number.

73. $45 \div 12 = 3$ with rem. $= 9$
    $12 \div 9 = 1$ with rem. $= 3$
    $9 \div 3 = 3$ with rem. $= 0$
    Thus, gcd of 12 and 45 is 3.

75. $105 \div 35 = 3$ with rem. $= 0$.
    Thus, gcd of 105 and 35 is 35.

77. $180 \div 150 = 1$ with rem. $= 30$.
    $150 \div 30 = 5$ with rem. $= 0$.
    Thus, the gcd of 150 and 180 is 30.

79. The proper factors of 28 are: 1, 2, 4, 7, and 14.
    $1 + 2 + 4 + 7 + 14 = 28$
    Thus, 28 is a perfect number.

81. The proper factors of 72
    are:1, 2, 3, 4, 6, 8, 9, 12, 18,
    24, and 36.
    $1 + 2 + 3 + 4 + 6 + 8 + 9$
    $+12 + 18 + 24 + 36 = 123$
    Thus, 72 is not a perfect #.

83. a) $60 = 2^2 \cdot 3^1 \cdot 5^1$ Adding 1 to each exponent and then multiplying these numbers, we get
    $(2+1)(1+1)(1+1) = 3 \cdot 2 \cdot 2 = 12$ factors of 60   b) They are 1, 2, 3, 4, 5, 6, 10, 12, 15, 20, 30, and 60.

85. For any three consecutive natural numbers, one of the numbers is divisible by 2 and another number is divisible by 3. Therefore, the product of the three numbers would be divisible by 6.

87. $54036 = (54,000 + 36)$; $54,000 \div 18 = 3,000$ and $36 \div 18 = 2$
    Thus, since $18 \mid 54000$ and $18 \mid 36$, $18 \mid 54036$.

89. $8 = 2+3+3$, $9 = 3+3+3$, $10 = 2+3+5$, $11 = 2+2+7$, $12 = 2+5+5$, $13 = 3+3+7$, $14 = 2+5+7$, $15 = 3+5+7$,
    $16 = 2+7+7$, $17 = 5+5+7$, $18 = 2+5+11$, $19 = 3+5+11$, $20 = 2+7+11$.

91. a) $5 = 6 - 1$ $7 = 6 + 1$ $11 = 12 - 1$ $13 = 12 + 1$ $17 = 18 - 1$ $19 = 18 + 1$ $23 = 24 - 1$
    $29 = 30 - 1$

    b) Conjecture: Every prime number greater than 3 differs by 1 from a multiple of 6.

    c) The conjecture should appear to be correct.

Exercise Set 5.2
1. Whole
3. a) Positive
   b) Negative
5. a) $-3 < 2$ b) $-3 < -2$ c) $-3 < 0$ d) $-3 > -4$

7. $-9, -6, -3, 0, 3, 6$

9. $-6, -5, -4, -3, -2, -1$

11. $-3 + 5 = 2$

13. $-10 + 3 = -7$

15. $[6 + (-11)] + 0$
    $= -5 + 0 = -5$

17. $[(-3) + (-4)] + 9 =$
    $-7 + 9 = 2$

19. $3 - 5 = -2$

21. $-6 - 2 = -8$

23. $-5 - (-3) = -5 + 3 = -2$

25. $14 - 20 = 14 + (-20) = -6$

27. $-7 \cdot 8 = -56$

29. $(-9)(-9) = 81$

31. $[(-8)(-2)] \cdot 6 = 16 \cdot 6 = 96$

33. $(5 \cdot 6)(-2) = (30)(-2) = -60$

35. $-16 \div (-2) = 8$

37. $15 \div (-15) = -1$

39. $\dfrac{56}{-8} = -7$

41. $\dfrac{-210}{14} = -15$

43. a) $3^2 = 9$
    b) $2^3 = 8$

45. a) $(-5)^2 = 25$
    b) $-5^2 = -25$

47. a) $-2^4 = -16$
    b) $(-2^4) = 16$

49. a) $-4^3 = -64$
    b) $(-4^3) = -64$

51. $15 - 8 \cdot 2 = -1$
    $15 - 16 = -1$

53. $(24 - 8 \div 4) - (48 \div 4 \cdot 2)$
    $(24 - 2) - (12 \cdot 2)$
    $22 - 24 = -2$

55. $(-17+21)^2 \bullet 3 \div (13-19)$
$4^2 \bullet 3 \div (-6)$
$16 \bullet 3 \div (-6)$
$48 \div (-6) = -8$

57. $-3[(-5^2+5^3) \div (-2^2+2^3)]$
$-3[(-25+125) \div (-4+8)]$
$-3[100 \div 4]$
$-3[25] = -75$

59.
$$\frac{-[4-(6-12)^2]}{[9 \div 3+4]^2 - 34/2} = \frac{-4[-(-6)^2]}{[3+4]^2 - 17}$$
$$= \frac{-[4-36]}{[7]^2 - 17} = \frac{-[-32]}{49-17} = \frac{32}{32} = 1$$

61. $-600+200-400-300 = -1100$
1100 feet under water

63. $14,495 - (-282) =$
$14,495 + 282 = 14,777$ feet

65. a) $+1-(-8) = +1+8 = 9$.
There is a 9 hr. time diff.
b) $-5-(-7) = -5+7 = 2$.
There is a 2 hr. time diff.

67. True

69. False; the difference of two negative integers may be positive, negative, or zero.

71. True; the product of two integers with like signs is a positive integer.

73. True

75. False; the sum of a positive integer and a negative integer could be pos., neg., or zero.

77. Division by zero is undefined because $\frac{a}{0} = x$ always leads to a false statement, $a \neq 0$.

79.
$$\frac{-1+2-3+4-5+\dots 99+100}{1-2+3-4+5\dots +99-100} =$$
$$\frac{50}{-50} = -1$$

81. $0+1-2+3+4-5+6-7-8+9 = 1$ (other answers are possible)

Exercise Set 5.3

1. Integers
3. Denominator
5. Improper
7. Repeating
9. Ten-thousandths

11. GCD of 6 and 9 is 3
$$\frac{6}{9} = \frac{6 \div 3}{9 \div 3} = \frac{2}{3}$$

13. GCD of 28 and 63 is 7.
$$\frac{28}{63} = \frac{28 \div 7}{63 \div 7} = \frac{4}{9}$$

15. GCD of 95 and 125 is 5.
$$\frac{95}{125} = \frac{95 \div 5}{125 \div 5} = \frac{19}{25}$$

17.
$$1\frac{5}{8} = \frac{(8)(1)+5}{8} = \frac{8+5}{8} = \frac{13}{8}$$

19. $-5\frac{3}{4} = -\frac{((4)(5)+3)}{4}$
$$= -\frac{20+3}{4} = -\frac{23}{4}$$

21. $-4\frac{15}{16} = -\frac{(4)(16)+15}{16}$
$$= -\frac{64+15}{16} = -\frac{79}{16}$$

23. $1\frac{1}{2} = \frac{(1)(2)+1}{2} = \frac{2+1}{2} = \frac{3}{2}$

25. $1\frac{7}{8} = \frac{(1)(8)+7}{8} = \frac{8+7}{8} = \frac{15}{8}$

27. $\frac{13}{4} = \frac{12+1}{4} = \frac{3 \bullet 4+1}{4} = 3\frac{1}{4}$

29. $-\frac{73}{6} = \frac{-(72+1)}{6}$
$$= \frac{-(12 \bullet 6+1)}{6} = -12\frac{1}{6}$$

31. $-\dfrac{878}{15} = -\dfrac{870+8}{15}$

$= -\dfrac{(58)(15)+8}{15} = -58\dfrac{8}{15}$

33. $\dfrac{4}{5} = 0.8$

35. $\dfrac{2}{3} = 0.\overline{6}$

37. $\dfrac{3}{8} = 0.375$

39. $\dfrac{13}{6} = 2.1\overline{6}$

41. $0.6 = \dfrac{6}{10} = \dfrac{3}{5}$

43. $0.175 = \dfrac{175}{1000} = \dfrac{7}{40}$

45. $0.295 = \dfrac{295}{1000} = \dfrac{59}{200}$

47. $0.0131 = \dfrac{131}{10,000}$

49. Let $n = 0.\overline{1}$, $10n = 1.\overline{1}$

$100n = 23.\overline{23}$

$\underline{-n = 0.\overline{23}}$

$99n = 23$

$\dfrac{9n}{9} = \dfrac{1}{9} = n$

51. Let $n = 0.\overline{9}$, $10n = 9.\overline{9}$

$10n = 9.\overline{9}$

$\underline{-n = 0.\overline{9}}$

$9n = 9$

$\dfrac{9n}{9} = \dfrac{9}{9} = 1 = n$

53. Let

$n = 1.\overline{36}$, $100n = 136.\overline{36}$

$100n = 136.\overline{36}$

$\underline{-n = 1.\overline{36}}$

$99n = 135.0$

$\dfrac{99n}{99} = \dfrac{135}{99} = \dfrac{15}{11} = n$

55. Let $n = 2.0\overline{5}$,

$10n = 20.\overline{5}$, $100n = 205.\overline{5}$

$100n = 205.\overline{5}$

$\underline{-10n = \ \ 20.\overline{5}}$

$90n = 185.0$

$\dfrac{90n}{90} = \dfrac{185}{90} = \dfrac{37}{18} = n$

57.
$\dfrac{2}{5} \cdot \dfrac{10}{11} = \dfrac{2 \cdot 10}{5 \cdot 11} = \dfrac{20}{55} = \dfrac{20 \div 5}{55 \div 5} = \dfrac{4}{11}$

59. $\dfrac{-3}{8} \cdot \dfrac{-16}{15} = \dfrac{48}{120} = \dfrac{2}{5}$

61. $\dfrac{7}{8} \div \dfrac{8}{7} = \dfrac{7}{8} \cdot \dfrac{7}{8} = \dfrac{49}{64}$

63.
$\left(\dfrac{3}{5} \cdot \dfrac{4}{7}\right) \div \dfrac{1}{3} = \dfrac{12}{35} \div \dfrac{1}{3} = \dfrac{12}{35} \cdot \dfrac{3}{1} = \dfrac{36}{35}$

65.
$\left[\left(-\dfrac{2}{3}\right)\left(\dfrac{5}{8}\right)\right] \div \left(-\dfrac{7}{16}\right) = \left(-\dfrac{10}{24}\right) \cdot \left(-\dfrac{16}{7}\right) = \dfrac{160}{168} = \dfrac{20}{21}$

67. The lcm of 5 and 3 is 15.

$\dfrac{2}{5} + \dfrac{1}{3} = \left(\dfrac{2}{5} \cdot \dfrac{3}{3}\right) + \left(\dfrac{1}{3} \cdot \dfrac{5}{5}\right)$

$= \dfrac{6}{15} + \dfrac{5}{15} = \dfrac{11}{15}$

69. The lcm of 11 and 22 is 22.

$\dfrac{2}{11} + \dfrac{5}{22} = \left(\dfrac{2}{11} \cdot \dfrac{2}{2}\right)$

$= \dfrac{4}{22} + \dfrac{5}{29} = \dfrac{9}{22}$

71. The lcm of 14 and 21 is 42.

$$\frac{3}{14}+\frac{8}{21}=\left(\frac{3}{14}\cdot\frac{3}{3}\right)+\left(\frac{8}{21}\cdot\frac{2}{2}\right)=\frac{9}{42}+\frac{16}{42}=\frac{25}{42}$$

73. The lcm of 12, 48, and 72 is 144.

$$\frac{1}{12}+\frac{1}{48}+\frac{1}{72}=\left(\frac{1}{12}\cdot\frac{12}{12}\right)+\left(\frac{1}{48}\cdot\frac{3}{3}\right)+\left(\frac{1}{72}\cdot\frac{2}{2}\right)$$

$$=\frac{12}{144}+\frac{3}{144}+\frac{2}{144}=\frac{17}{144}$$

75. The lcm of 30, 40, and 50 is 600.

$$\frac{1}{30}-\frac{3}{40}-\frac{7}{50}=\left(\frac{1}{30}\cdot\frac{20}{20}\right)\left(\frac{3}{40}\cdot\frac{15}{15}\right)\left(\frac{7}{50}\cdot\frac{12}{12}\right)$$

$$=\frac{20}{600}-\frac{45}{600}-\frac{84}{600}=-\frac{109}{600}$$

77. $$\frac{5}{8}+\frac{1}{3}=\frac{5\cdot3+8\cdot1}{8\cdot3}=\frac{15+8}{24}=\frac{23}{24}$$

79. $$\frac{5}{6}-\frac{7}{8}=\frac{5\cdot8-6\cdot7}{6\cdot8}=\frac{40-42}{48}=-\frac{2}{48}=-\frac{1}{24}$$

81. $$\frac{1}{4}+\frac{1}{3}\cdot\frac{6}{7}=\frac{1}{4}+\frac{6}{21}=\left(\frac{1}{4}\cdot\frac{21}{21}\right)+\left(\frac{6}{21}\cdot\frac{4}{4}\right)=$$

$$=\frac{21}{84}+\frac{24}{84}=\frac{45}{84}=\frac{15}{28}$$

83. $$\frac{7}{8}-\frac{3}{4}\div\frac{5}{6}=\frac{7}{8}-\frac{3}{4}\cdot\frac{6}{5}=\frac{7}{8}-\frac{18}{20}$$

$$=\frac{7}{8}\cdot\frac{5}{5}-\frac{18}{20}\cdot\frac{2}{2}=\frac{7\cdot5-18\cdot2}{40}=\frac{35-36}{40}=-\frac{1}{40}$$

85. $$\frac{7}{8}\div\frac{3}{4}\cdot\frac{3}{16}=\frac{7}{8}\cdot\frac{4}{3}\cdot\frac{3}{16}=\frac{7\cdot4\cdot3}{8\cdot3\cdot16}$$

$$=\frac{7}{32}$$

87. $$\frac{1}{3}\cdot\frac{3}{7}+\frac{3}{5}\cdot\frac{10}{11}=\frac{1\cdot3}{3\cdot7}+\frac{3\cdot10}{5\cdot11}=\frac{3}{21}+\frac{30}{55}=\frac{1}{7}\cdot\frac{11}{11}+\frac{6}{11}\cdot\frac{7}{7}=\frac{11+42}{77}=\frac{53}{77}$$

89. $$\left(\frac{3}{4}+\frac{1}{6}\right)\div\left(2-\frac{7}{6}\right)=\left(\frac{3}{4}\cdot\frac{3}{3}+\frac{1}{6}\cdot\frac{2}{2}\right)\div\left(\frac{2}{1}\cdot\frac{6}{6}-\frac{7}{6}\right)=\left(\frac{9}{12}+\frac{2}{12}\right)\div\left(\frac{12}{6}-\frac{7}{6}\right)=\frac{11}{12}\div\frac{5}{6}=\frac{11}{12}\cdot\frac{6}{5}=\frac{11}{10}$$

91. $$71\frac{5}{8}\;\rightarrow\;70\frac{13}{8}$$

$$\underline{-69\frac{7}{8}}\;\rightarrow\;\underline{-69\frac{7}{8}}$$

$$1\frac{6}{8}\rightarrow1\frac{3}{4}\;\text{inches}$$

93. $$14\left(8\frac{5}{8}\right)=14\left(\frac{69}{8}\right)=\frac{966}{8}=\frac{966\div2}{8\div2}=\frac{483}{4}=120\frac{3}{4}"$$

95. $2\frac{1}{4}+3\frac{7}{8}+4\frac{1}{4}=2\frac{4}{16}+3\frac{14}{16}+4\frac{4}{16}$

$=9\frac{22}{16}=10\frac{6}{16}$

$20\frac{5}{16}-10\frac{6}{16}=19\frac{21}{16}-10\frac{6}{16}=9\frac{15}{16}$ "

97. $1-\left(\frac{1}{2}+\frac{2}{5}\right)=1-\left(\frac{5}{10}+\frac{4}{10}\right)=1-\frac{9}{10}=\frac{10}{10}-\frac{9}{10}=\frac{1}{10}$

Student tutors represent 0.1 of the budget.

99. $4\frac{1}{2}+30\frac{1}{4}+24\frac{1}{8}=4\frac{4}{8}+30\frac{2}{8}+24\frac{1}{8}=58\frac{7}{8}$ inches

101. a) $1\frac{49}{60},2\frac{48}{60},9\frac{6}{60},6\frac{3}{60},2\frac{9}{60},\frac{22}{60}$

b) $1\frac{49}{60}+2\frac{48}{60}+9\frac{6}{60}+6\frac{3}{60}+2\frac{9}{60}+\frac{22}{60}$

$=(1+2+9+6+2+0)+\dfrac{49+48+6+3+9+22}{60}$

$=20+\dfrac{137}{60}=22\frac{17}{60}$ or 22 hours, 17 minutes

103. $8\frac{3}{4}$ ft $=\left(\frac{35}{4}\cdot\frac{12}{1}\right)$ in. $=105$ in.

$\left[105-(3)\left(\frac{1}{8}\right)\right]\div 4=\left[\frac{840}{8}-\frac{3}{8}\right]\div 4=\frac{837}{8}\cdot\frac{1}{4}=\frac{837}{32}=26\frac{5}{32}$.   The length of each piece is $26\frac{5}{32}$ in.

105. a) $20+18\frac{3}{8}\div 2=20+9\frac{3}{16}=29\frac{3}{16}$ in.

b) $26\frac{1}{4}+6\frac{3}{4}=33$ in.

c) $26\frac{1}{4}+\left(6\frac{3}{4}-\frac{1}{4}\right)=26\frac{1}{4}+6\frac{2}{4}=32\frac{3}{4}$ in.

107. $\dfrac{0.21+0.22}{2}=\dfrac{0.43}{2}=0.215$

109. $\dfrac{-2.176+(-2.175)}{2}=\dfrac{-4.351}{2}=-2.1755$ 111. $\left(\frac{1}{4}+\frac{3}{4}\right)\div 2=\frac{4}{4}\cdot\frac{1}{2}=\frac{4}{8}=\frac{1}{2}$

113. $\left(\frac{1}{100}+\frac{1}{10}\right)\div 2=\frac{11}{100}\cdot\frac{1}{2}=\frac{11}{200}$

115. a) 1   b) $0.\overline{9}$

c) $\dfrac{1}{3}=0.\overline{3}$, $\dfrac{2}{3}=0.\overline{6}$, $\dfrac{1}{3}+\dfrac{2}{3}=\dfrac{3}{3}=1$

$0.\overline{3}+0.\overline{6}=0.\overline{9}$

d) $0.\overline{9}=1$

Exercise Set 5.4

1. Irrational

3. Radicand

5. Itself

7. Rationalized

9. $\sqrt{49}=7$ Rational

11. $\sqrt{10}$ Irrational

13. Irrational; non-terminating, non-repeating decimal

15. Rational; quotient of two integers

17. Rational; $\dfrac{\sqrt{75}}{\sqrt{3}}=5$

19. $\sqrt{0}=0$

21. $\sqrt{25}=5$

23. $-\sqrt{36}=-6$

25. $-\sqrt{100}=-10$

27. 0, Rational, integer

29. $\sqrt{13}$, Irrational

31. Rational

33. Rational

35. Irrational

37. $\sqrt{12}=\sqrt{4}\sqrt{3}=2\sqrt{3}$

39. $\sqrt{54}=\sqrt{9}\sqrt{6}=3\sqrt{6}$

41. $\sqrt{63}=\sqrt{9}\sqrt{7}=3\sqrt{7}$

43. $\sqrt{84}=\sqrt{4}\sqrt{21}=2\sqrt{21}$

45. $3\sqrt{7}+5\sqrt{7}=(3+5)\sqrt{7}=8\sqrt{7}$

47.
$\sqrt{18}+\sqrt{50}=\sqrt{9}\sqrt{2}$
$+\sqrt{25}\sqrt{2}=3\sqrt{2}+5\sqrt{2}=8\sqrt{2}$

49.
$4\sqrt{12}-7\sqrt{27}=4\sqrt{4}\sqrt{3}-7\sqrt{9}\sqrt{3}$
$=4\cdot2\sqrt{3}-7\cdot3\sqrt{3}=8\sqrt{3}-21\sqrt{3}$
$=-13\sqrt{3}$

51.
$5\sqrt{3}+7\sqrt{12}-3\sqrt{75}$
$=5\sqrt{3}+7\cdot2\sqrt{3}-3\cdot5\sqrt{3}$
$=5\sqrt{3}+14\sqrt{3}-15\sqrt{3}$
$=(5+14-15)\sqrt{3}=4\sqrt{3}$

53. $\sqrt{2}\sqrt{18}=\sqrt{2\cdot18}=\sqrt{36}=6$

55. $\sqrt{6}\sqrt{15}=\sqrt{6\cdot15}=\sqrt{90}$
$=\sqrt{9}\sqrt{10}=3\sqrt{10}$

57. $\dfrac{\sqrt{20}}{\sqrt{5}}=\sqrt{\dfrac{20}{5}}=\sqrt{4}=2$

59. $\dfrac{\sqrt{72}}{\sqrt{8}}=\sqrt{9}=3$

61. $\dfrac{1}{\sqrt{3}}=\dfrac{1}{\sqrt{3}}\cdot\dfrac{\sqrt{3}}{\sqrt{3}}=\dfrac{\sqrt{3}}{3}$

63. $\dfrac{\sqrt{3}}{\sqrt{11}}=\dfrac{\sqrt{3}}{\sqrt{11}}\cdot\dfrac{\sqrt{11}}{\sqrt{11}}=\dfrac{\sqrt{33}}{11}$

65. $\dfrac{\sqrt{20}}{\sqrt{3}}=\dfrac{\sqrt{20}}{\sqrt{3}}\dfrac{\sqrt{3}}{\sqrt{3}}=\dfrac{\sqrt{60}}{\sqrt{9}}$
$=\dfrac{\sqrt{4}\sqrt{15}}{3}=\dfrac{2\sqrt{15}}{3}$

67. $\dfrac{\sqrt{10}}{\sqrt{6}} \cdot \dfrac{\sqrt{6}}{\sqrt{6}} = \dfrac{\sqrt{60}}{6}$

    $= \dfrac{2\sqrt{15}}{6} = \dfrac{\sqrt{15}}{3}$

69. $\sqrt{37}$ is between 6 and 7 since $\sqrt{37}$ is between $\sqrt{36} = 6$ and $\sqrt{49} = 7$. $\sqrt{37}$ is between 6 and 6.5 since 37 is closer to 36 than to 49. Using a calculator $\sqrt{37} \approx 6.08$.

71. $\sqrt{97}$ is between 9 and 10 since $\sqrt{97}$ is between $\sqrt{81} = 9$ and $\sqrt{100} = 10$. $\sqrt{97}$ is between 9.5 and 10 since 97 is closer to 100 than to 81. Using a calculator $\sqrt{97} \approx 9.85$.

73. $\sqrt{170}$ is between 13 and 14 since $\sqrt{170}$ is between $\sqrt{169} = 13$ and $\sqrt{196} = 14$. $\sqrt{170}$ is between 13 and 13.5 since 170 is closer to 169 than to 196. Using a calculator $\sqrt{170} \approx 13.04$.

75. $H = 2.9\sqrt{30} + 20.1$
    $= 36.0$ inches

77. a) $s = \sqrt{\dfrac{4}{0.04}} = \sqrt{100} = 10$ mph

    b) $s = \sqrt{\dfrac{16}{0.04}} = \sqrt{400} = 20$ mph

    c) $s = \sqrt{\dfrac{64}{0.04}} = \sqrt{1600} = 40$ mph

    d) $s = \sqrt{\dfrac{256}{0.04}} = \sqrt{6400} = 80$ mph

79. False. $\sqrt{c}$ may be a rational number or an irrational number for a composite number c. For example, $\sqrt{25}$ is a rational number; $\sqrt{8}$ is an irrational number.

81. True

83. False. The product of a rational number and an irrational number may be a rational number or an irrational number.

85. $3\sqrt{2} + 5\sqrt{2} = 8\sqrt{2}$

87. $\sqrt{2} \cdot \sqrt{3} = \sqrt{6}$

89. No. $2 \neq 1.414$ since $\sqrt{2}$ is an irrational number and 1.414 is a rational number.

91. No. 3.14 and $\dfrac{22}{7}$ are rational numbers, π is an irrational number.

93. $\sqrt{4 \cdot 9} = \sqrt{4}\sqrt{9}$
    $\sqrt{36} = 2 \cdot 3$
    $6 = 6$

95. a) $\sqrt{0.04} = 0.2$ a terminating
    decimal and thus it is rational.

    b) $\sqrt{0.7} = \sqrt{\dfrac{7}{10}} = \dfrac{\sqrt{70}}{10}$; $\sqrt{70}$ is irrational since

    the only integers with rational square roots are
    the perfect squares and 70 is not a perfect square.

    Thus $\dfrac{\sqrt{70}}{10} = \sqrt{0.7}$ is irrational.

97. a) $\left(44 \div \sqrt{4}\right) \div \sqrt{4} = 11$

    b) $\left(44 \div 4\right) + \sqrt{4} = 13$

    c) $4 + 4 + 4 + \sqrt{4} = 14$

    d) $\sqrt{4}\left(4 + 4\right) + \sqrt{4} = 18$; Other answers are possible.

Exercise Set 5.5

1. Real

3. Closed

5. Commutative

7. Associative

9. Closed. The sum of two natural numbers is a natural number

11. Not closed. (e.g., $3 - 5 = -2$ is not a natural number).

13. Closed. The product of two integers is an integer.

15. Not closed. (e.g., $17 \div 2 = 8.5$ is not an integer).

17. Closed   19. Closed   21. Not closed   23. Not closed

25. Closed   27. Closed

29. Commutative property of addition. The order $5 + x$ is changed to $x + 5$. .

31. No. $4 - 3 = 1$, but $3 - 4 = -1$

33. $(-3) + (-4) = -7 = (-4) + (-3)$

35. $[(-2) + (-3)] + (-4) = (-5) + (-4) = -9$

    $(-2) + [(-3)] + (-4)] = (-2) + (-7) = -9$

37. No.

    $(16 \div 8) \div 2 = 2 \div 2 = 1$, but $16 \div (8 \div 2) = 16 \div 4 = 4$

39. No. $(81 \div 9) \div 3 = 9 \div 3 = 3$,

    but $81 \div (9 \div 3) = 81 \div 3 = 27$

41. $7(x + 5) = 7 \cdot x + 7 \cdot 5$

    Distributive property

43. $(7 \cdot 8) \cdot 9 = 7 \cdot (8 \cdot 9)$

    Associative property of multiplication

45. $(24 + 7) + 3 = 24 + (7 + 3)$

    Associative property of addition

47. $\sqrt{3} \cdot 7 = 7 \cdot \sqrt{3}$

    Commutative property of multiplication

49. $-1(x + 4) = (-1) \cdot x + (-1) \cdot 4$

    Distributive property

51. $(r + s) + t = t + (r + s)$

    Commutative property of addition

53. $5(a + 2) = 5a + 10$

55. $9(2c - 3) = 18c - 27$

57. $-3(2x-1) = -6x+3$

59. $32\left(\dfrac{1}{16}x - \dfrac{1}{32}\right) = \dfrac{32x}{16} - \dfrac{32}{32} = 2x-1$

61. $\sqrt{2}\left(\sqrt{8} - \sqrt{2}\right) = \sqrt{16} - \sqrt{4} = 4-2 = 2$

63. $5\left(\sqrt{2} + \sqrt{3}\right) = 5\sqrt{2} + 5\sqrt{3}$

65. Yes. You can either feed your cats first or give your cats water first.

67. No. The clothes must be washed first before being dried.

69. No. Pressing the keys will have no effect if there are no batteries in place.

71. Yes. The order does not matter.

73. Yes. The order does not matter

75. Baking pizzelles: mixing eggs into the batter, or mixing sugar into the batter.; Yard work: mowing the lawn, or trimming the bushes

77. No. $0 \div a = 0$ but $a \div 0$ is undefined.

79. a) No. (Man eating) tiger is a tiger that eats men, and man (eating tiger) is a man that is eating a tiger.
    b) No. (Horse riding) monkey is a monkey that rides a horse, and horse (riding monkey) is a horse that rides a monkey.
    c) Answers will vary.

Exercise Set 5.6

1. x^5

3. 1

5. x^6

7. a) $3^2 \cdot 3^3 = 3^{2+3} = 3^5 = 243$
   b) $(-3)^2 \cdot (-3)^3 = (-3)^{2+3} = (-3)^5 = -243$

9. a) $\dfrac{5^7}{5^5} = 5^{7-5} = 5^2 = 25$
   b) $\dfrac{(-5)^7}{(-5)^5} = (-5)^{7-5} = (-5)^2 = 25$

11. a) $6^0 = 1$
    b) $-6^0 = -1$

13. a) $(6x)^0 = 1$
    b) $6x^0 = 6$

15. a) $3^{-3} = \dfrac{1}{3^3} = \dfrac{1}{27}$
    b) $7^{-2} = \dfrac{1}{7^2} = \dfrac{1}{49}$

17. a) $-9^{-2} = -\dfrac{1}{9^2} = -\dfrac{1}{81}$
    b) $(-9)^{-2} = \dfrac{1}{(-9)^2} = \dfrac{1}{81}$

19. a) $\left(2^3\right)^2 = (8)^2 = 64$
    b) $(3^2)^3 = (9)^3 = 729$

21. a) $4^3 \cdot 4^{-2} = 4^{3-2} = 4^1 = 4$

   b) $2^{-2} \cdot 2^{-2} = 2^{-2-2} = 2^{-4} = \dfrac{1}{2^4} = \dfrac{1}{16}$

23. $201000 = 2.01 \times 10^5$

25. $0.00275 = 2.75 \times 10^{-3}$

27. $0.56 = 5.6 \times 10^{-1}$

29. $19000 = 1.9 \times 10^4$

31. $0.000186 = 1.86 \times 10^{-4}$

33. $1.6 \times 10^3 = 1600$

35. $1.32 \times 10^{-2} = 0.0132$

37. $8.62 \times 10^{-5} = 0.0000862$

39. $2.01 \times 10^0 = 2.01$

41. $\left(2.4 \times 10^3\right)\left(3 \times 10^2\right) = 720,000$

43. $\left(5.1 \times 10^1\right)\left(3.0 \times 10^{-4}\right)$
$= 15.3 \times 10^{-3} = 0.0153$

45. $\dfrac{7.5 \times 10^6}{3 \times 10^4} = 2.5 \times 10^2 = 250$

47. $\dfrac{8.4 \times 10^{-6}}{4.0 \times 10^{-3}} = 2.1 \times 10^{-3} = 0.0021$

49. $\left(2 \times 10^5\right)\left(3.6 \times 10^3\right) = 7.2 \times 10^8$

51. $\left(3.0 \times 10^{-3}\right)\left(1.5 \times 10^{-4}\right)$
$= 4.5 \times 10^{-7}$

53. $\dfrac{5.6 \times 10^6}{8 \times 10^4} = 0.7 \times 10^2 = 7 \times 10^1$

55. $\dfrac{4.0 \times 10^{-5}}{2.0 \times 10^2} = 2.0 \times 10^{-7}$

57. $3.6 \times 10^{-3}, 1.7, 9.8 \times 10^2,$
1.03×10^4

59. $8.3 \times 10^{-5}; 0.00079; 4.1 \times 10^3; 40,000$

   Note: $0.00079 = 7.9 \times 10^{-4}, 40,000$
$= 4 \times 10^4$

61. a) $\dfrac{1.308 \times 10^{13}}{3.095 \times 10^8} \approx .4226171 \times 10^5$
$= 42,261.71 \text{ or } \$42,261.71$

   b) $\$55,642.73 - \$42,261.71$
$= \$13,380.92$

63. $\dfrac{3.19 \times 10^8}{7.194 \times 10^9} \approx 0.44 \times 10^{-1}$
≈ 0.044

65. $\dfrac{1.68 \times 10^{13}}{3.19 \times 10^8} \approx 0.5266458 \times 10^5$
$= 52,664.58 \text{ or } \$52,664.58$

67. $t = \dfrac{d}{r} = \dfrac{239000 \text{ mi}}{20000 \text{ mph}} = 11.95$ 11.95 hrs

69. (100,000 cu ft/sec) (60 sec/min) (60 min/hr) (24 hr)
   $= 8,640,000,000 \text{ ft}^3$ or 8.64×10^9 cu ft

71. $(50)(5,800,000) = (5 \times 10^1)(5.8 \times 10^6) = 29 \times 10^7 = 2.9 \times 10^8$
   2.9×10^8 cells

73. $\dfrac{2 \times 10^{30}}{6 \times 10^{24}} = 0.\overline{3} \times 10^6 \approx 333,333$ times

75. 1,000 times, since 1 meter $= 10^3$ millimeters $= 1,000$ times

77. $(0.96)(326,000) = 312,960$ people

79. a) $1,000,000 = 1.0 \times 10^6$; $1,000,000,000 = 1.0 \times 10^9$; $1,000,000,000,000 = 1.0 \times 10^{12}$

   b) $\dfrac{1.0 \times 10^6}{1.0 \times 10^3} = 1.0 \times 10^3$ days or 1,000 days ≈ 2.74 years

   c) $\dfrac{1.0 \times 10^9}{1.0 \times 10^3} = 1.0 \times 10^6$ days or 1,000,000 days ≈ 2739.73 years

   d) $\dfrac{1.0 \times 10^{12}}{1.0 \times 10^3} = 1.0 \times 10^9$ days or 1,000,000,000 days $\approx 2,739,726.03$ years

e) $\dfrac{1 \text{ billion}}{1 \text{ million}} = \dfrac{1.0 \times 10^9}{1.0 \times 10^6} = 1.0 \times 10^3 = 1{,}000$ times greater

81. a) $E(0) = 2^{10} \times 2^0 = 2^{10} \times 1 = 1024$ bacteria   b) $E(1/2) = 2^{10} \times 2^{1/2} = 2^{10.5} \approx 1448$ bacteria

Exercise Set 5.7
1. Sequence
3. Arithmetic
5. Geometric
7. $a_1 = 7, d = 6$ 7, 13, 19, 25, 31
9. $a_1 = 25, d = -5$ 25, 20, 15, 10, 5
11. $a_1 = 5, d = -2$ 5, 3, 1, -1, -3
13. $a_9 = 4 + (9-1)(5) = 4 + (40) = 44$
15. $a_{11} = -20 + (11-1)(5) = -20 + 50 = 30$
17. $a_{20} = \dfrac{4}{5} + (19)(-1) = \dfrac{4}{5} - 19 = \dfrac{4}{5} - \dfrac{95}{5} = -\dfrac{91}{5}$
19. $a_n = n$ $(a_1 = 1, d = 1)$
21. $a_n = 2n - 1$ $(a_1 = 1, d = 2)$
23. $a_n = -6n + 26$ $(a_1 = 20, d = -6)$
25. $s_{50} = \dfrac{n(a_1 + a_n)}{2} = \dfrac{50(1 + 50)}{2} = \dfrac{50(51)}{2}$
    $= (25)(51) = 1275$
27. $s_{50} = \dfrac{50(1 + 99)}{2} = \dfrac{50(100)}{2} = (25)(100) = 2500$
29. $s_{17} = \dfrac{17(100 + (-60))}{2} = \dfrac{17 \cdot (40)}{2} = 340$
31. $a_1 = 3, r = 2$ 3, 6, 12, 24, 48
33. $a_1 = 5, r = 2$ 5, 10, 20, 40, 80
35. $a_1 = -3, r = -1$ -3, 3, -3, 3, -3
37. $a_5 = 7(2)^4 = (7)(16) = 112$
41. $a_7 = (-5) \cdot 3^6 = (-5)(729) = -3645$
39. $a_7 = 64\left(\dfrac{1}{2}\right)^6 = 64\left(\dfrac{1}{64}\right) = 1$
43. 5, 25, 125, 625 $a_n = 5^n$
45. $2, 1, \dfrac{1}{2}, \dfrac{1}{4}$ $a_n = 2\left(\dfrac{1}{2}\right)^{n-1}$
47. $-16, -8, -4, -2$ $a_n = a_1 r^{n-1} = -16\left(\dfrac{1}{2}\right)^{n-1}$
49. $s_6 = \dfrac{a_1(1 - r^6)}{1 - r} = \dfrac{3(1 - 2^6)}{1 - 2} = \dfrac{3(-63)}{-1} = 189$
51. $s_6 = \dfrac{a_1(1 - r^6)}{1 - r} = \dfrac{(-3)(1 - 4^6)}{1 - 4} = \dfrac{(-3)(-4095)}{-3}$
    $= -4095$
53. $s_{15} = \dfrac{a_1(1 - r^{15})}{1 - r} = \dfrac{2(1 - 3^{15})}{1 - 3} = \dfrac{2(-14{,}348{,}906)}{-2}$
    $= 14{,}348{,}906$
55. $s_{100} = \dfrac{(100)(1 + 100)}{2} = \dfrac{(100)(101)}{2}$
    $= 50(101) = 5050$
57. $s_{100} = \dfrac{(100)(2 + 200)}{2} = \dfrac{(100)(202)}{2}$
    $= 50(202) = 10{,}100$
59. $s_{12} = \dfrac{12(1 + 12)}{2} = \dfrac{12(13)}{2} = \dfrac{156}{2} = 78$ times
61. a) $a_{12} = 96 + (11)(-3) = 96 - 33 = 63$ in.
    b) $\dfrac{[12(96 + 63)]}{2} = \dfrac{(12)(159)}{2} = (6)(159) = 954$ in.
63. $a_{10} = (8000)(1.08)^9 = 15{,}992$ students
65. $a_{15} = 31{,}000(1.06)^{14} = \$70{,}088$

67. This is a geometric sequence where $a_1 = 1.4$ and $r = 1.005$. In sixteen years the population of China will be approximately:

$a_{16} = a_1 r^{16-1} = 1.4(1.005)^{16} = 1.5$ billion people

69. $\dfrac{82[1-(1/2)^6]}{1-(1/2)} = \dfrac{82[1-(1/64)]}{1/2}$

$= \dfrac{82}{1} \bullet \dfrac{63}{64} \bullet \dfrac{2}{1} = 161.4375$

71. 12, 18, 24, ... ,1608 is an arithmetic sequence with $a_1 = 12$ and $d = 6$. Using the expression for the nth term of an arithmetic sequence $a_n = a_1 + (n-1)d$ or $1608 = 12 + (n-1)6$ and dividing both sides by 6 gives $268 = 2 + n - 1$ or $n = 267$

73. The total distance is 30 plus twice the sum of the terms of the geometric sequence having $a_1 = (30)(0.8) = 24$

and $r = 0.8$. Thus $s_5 = \dfrac{24[1-(0.8)^5]}{(1-0.8)} = \dfrac{24[1-0.32768]}{0.2} = \dfrac{24(0.67232)}{0.2} = 80.6784$.

So the total distance is $30 + 2(80.6784) = 191.3568$ ft.

Exercise Set 5.8

1. Fibonacci

3. Divine

5. Ratio

7. Fibonacci type; $16 + 26 = 42$; $26 + 42 = 68$

9. Not Fibonacci type; it is not true that each term is the sum of the two preceding terms.

11. Fibonacci type: $1 + 2 = 3$, $2 + 3 = 5$ Each term is the sum of the two preceding terms.

13. Fibonacci type; $40 + 65 = 105$; $65 + 105 = 170$

15. 1, 1, 2, 3, 5, 8, 13, 21, 34, 55, 89, 144, 233, 377, 610

17. If the first ten are selected; $\dfrac{1+1+2+3+5+8+13+21+34+55}{11} = \dfrac{143}{11} = 13$

19. If 2, 3, 5, and 8 are selected the result is $5^2 - 3^2 = 2\bullet 8 \quad \rightarrow \quad 25 - 9 = 16 \quad \rightarrow \quad 16 = 16$

21. a) If 6 and 9 are selected the sequence is 6, 9, 15, 24, 39, 63, 102, ...

    b) $9/6 = 1.5$, $15/9 = 1.666$, $24/15 = 1.6$, $39/24 = 1.625$, $63/39 = 1.615$, $102/63 = 1.619$, ...

23. The sums of the numbers along the diagonals (parallel to the one shown) is a Fibonacci number.

25. a) $\dfrac{\sqrt{5}+1}{2} \approx 1.6180$

    b) $\dfrac{\sqrt{5}-1}{2} \approx 0.618$

    c) Differ by 1

27. Answers will vary.

29. 89, $\dfrac{1}{89} = .01\underline{12358}9551$, part of Fibonacci sequence

31. $55/34 = 1.6176$, $34/21 = 1.619$

33. $\dfrac{(a+b)}{a}=\dfrac{a}{b}$ Let $x=\dfrac{a}{b}$ $\dfrac{b}{a}=\dfrac{1}{x}$ $1+\dfrac{b}{a}=\dfrac{a}{b}$ $1+\dfrac{1}{x}=x$ multiply by x $x\left(1+\dfrac{1}{x}\right)=x(x)$

$x+1=x^2$ $x^2-x-1=0$

Solve for x using the quadratic formula, $x=\dfrac{-b\pm\sqrt{b^2-4ac}}{2a}=\dfrac{1\pm\sqrt{1-4(1)(-1)}}{2(1)}=\dfrac{1\pm\sqrt{5}}{2}$

35. a) – c) Answers will vary.

Review Exercises

1. Use the divisibility rules in section 5.1.
   94,380 is divisible by 2, 3, 4, 5, 6, and 10.

2. Use the divisibility rules in section 5.1.
   400,644 is divisible by 2, 3, 4, 6, and 9

3.
```
2 | 1260
2 | 630
3 | 315
3 | 105
5 | 35
    7
```
$1260 = 2^2 \cdot 3^2 \cdot 5 \cdot 7$

4.
```
2  | 1452
2  | 726
3  | 363
11 | 121
     11
```
$1452 = 2^2 \cdot 3 \cdot 11^2$

5. $42 = 2 \cdot 3 \cdot 7, \ 70 = 2 \cdot 5 \cdot 7$
   gcd $= 2 \cdot 7 = 14$; lcm $= 2 \cdot 3 \cdot 5 \cdot 7 = 210$

6. $63 = 3^2 \cdot 7; \ 108 = 2^2 \cdot 3^3$
   gcd $= 3^2 = 9$; lcm $= 2^2 \cdot 3^3 \cdot 7 = 756$

7. $15 = 3 \cdot 5, \ 9 = 3^2$; lcm $= 3^2 \cdot 5 = 45$. In 45 days the train stopped in both cities.

8. $1 + (-7) = -6$

9. $-2 + 5 = 3$

10. $(-2) + (-4) = -6$

11. $4 - 8 = 4 + (-8) = -4$

12. $-5 - 4 = -5 + (-4) = -9$

13. $-3 - (-6) = -3 + 6 = 3$

14. $8 \cdot (-8) = -64$

15. $(-2)(-12) = 24$

16. $-35/-7 = 5$

17. $12/-6 = -2$

18. $9 + 2 \cdot 3 = 9 + 6 = 15$

19. $5 \cdot 2^3 - 3^2 \cdot 6 = 5 \cdot 8 - 9 \cdot 6 = 40 - 54 = -14$

20. $(64 \div -4 \cdot 2) - (25 - 7 \cdot 4) = 32 - (-3) = 35$

21. $-2(5 \cdot 3^2 - 4^3 \div 8)^2 = -(5 \cdot 9 - 64 \div 8)^2$
    $= -(45 - 8)^2 = -2(37)^2 = -2(1369) = -2738$

22. $13/25 = 0.52$

23. $13/4 = 3.25$

24. $6/7 = 0.\overline{857142}$

25. $7/12 = 0.58\overline{3}$

26. $2.6 = \dfrac{26}{10} = \dfrac{13}{5}$

27. $0.6666...$ $10n = 6.6666....$

$\begin{array}{r} 10n = 6.\overline{6} \\ -n = 0.\overline{6} \\ \hline 9n = 6.0 \end{array}$ $\dfrac{9n}{9} = \dfrac{6}{9}$

$n = \dfrac{2}{3}$

28. $0.5151...$ $100n = 51.5151....$

$\begin{array}{r} 100n = 51.\overline{51} \\ -n = 0.\overline{51} \\ \hline 99n = 51 \end{array}$ $\dfrac{99n}{99} = \dfrac{51}{99}$

$n = \dfrac{51}{99}$

29. $0.083 = \dfrac{83}{1000}$

30. $4\dfrac{5}{8} = \dfrac{4 \cdot 8 + 5}{8} = \dfrac{37}{8}$

31. $-3\dfrac{1}{4} = \dfrac{((-3)(4)) - 1}{4} = -\dfrac{13}{4}$

32. $\dfrac{16}{7} = \dfrac{(2)(7) + 2}{7} = 2\dfrac{2}{7}$

33. $\dfrac{-136}{5} = \dfrac{(-27)(5) - 1}{5} = -27\dfrac{1}{5}$

34. $\dfrac{7}{15} - \dfrac{3}{5} = \dfrac{7}{15} - \dfrac{3}{5} \cdot \dfrac{3}{3} = \dfrac{7}{15} - \dfrac{9}{15} = -\dfrac{2}{15}$

35. $\dfrac{1}{6} + \dfrac{5}{4} = \dfrac{1}{6} \cdot \dfrac{2}{2} + \dfrac{5}{4} \cdot \dfrac{3}{3} = \dfrac{2}{12} + \dfrac{15}{12} = \dfrac{17}{12}$

36. $\dfrac{7}{16} \cdot \dfrac{12}{21} = \dfrac{84}{336} = \dfrac{84}{4 \cdot 84} = \dfrac{1}{4}$

37. $\dfrac{5}{9} \div \dfrac{6}{7} = \dfrac{5}{9} \cdot \dfrac{7}{6} = \dfrac{35}{54}$

38. $\dfrac{2}{3} + \dfrac{1}{5} \cdot \dfrac{10}{11} = \dfrac{2}{3} + \dfrac{2}{11} = \dfrac{2}{3} \cdot \dfrac{11}{11} + \dfrac{2}{11} \cdot \dfrac{3}{3}$

$= \dfrac{22 + 6}{33} = \dfrac{28}{33}$

39. $\dfrac{3}{4} - \dfrac{1}{8} \div \dfrac{5}{12} = \dfrac{3}{4} - \dfrac{1}{8} \cdot \dfrac{12}{5} = \dfrac{3}{4} - \dfrac{12}{40}$

$= \dfrac{3}{4} \cdot \dfrac{10}{10} - \dfrac{12}{40} = \dfrac{30 - 12}{40} = \dfrac{18}{40} = \dfrac{9}{20}$

40. $\dfrac{13}{16} \div \dfrac{7}{8} \cdot \dfrac{14}{15} = \dfrac{13}{16} \cdot \dfrac{8}{7} \cdot \dfrac{14}{15}$

$= \dfrac{13}{14} \cdot \dfrac{14}{15} = \dfrac{13}{15}$

41.
$$\left(\dfrac{1}{4} + \dfrac{5}{6}\right) \div \left(3 - \dfrac{1}{6}\right) = \left(\dfrac{1}{4} \cdot \dfrac{3}{3} + \dfrac{5}{6} \cdot \dfrac{2}{2}\right) \div \left(\dfrac{3}{1} \cdot \dfrac{3}{3} - \dfrac{1}{6}\right)$$
$$= \left(\dfrac{3}{12} + \dfrac{10}{12}\right) \div \left(\dfrac{9}{3} - \dfrac{1}{6}\right) = \dfrac{13}{12} \div \left(\dfrac{9}{3} \cdot \dfrac{2}{2} - \dfrac{1}{6}\right)$$
$$= \dfrac{13}{12} \div \left(\dfrac{18}{6} - \dfrac{1}{6}\right) = \dfrac{13}{12} \div \dfrac{17}{6} = \dfrac{13}{12} \cdot \dfrac{6}{17} = \dfrac{13}{34}$$

42. $\left(\dfrac{1}{8}\right)\left(17\dfrac{3}{4}\right) = \left(\dfrac{1}{8}\right)\left(\dfrac{71}{4}\right) = \dfrac{71}{32} = 2\dfrac{7}{32}$ teaspoons

43. $\sqrt{80} = \sqrt{16} \cdot \sqrt{5} = 4\sqrt{5}$

44. $\sqrt{2} - 4\sqrt{2} = (1 - 4)\sqrt{2} = -3\sqrt{2}$

45. $\sqrt{8} + 6\sqrt{2} = 2\sqrt{2} + 6\sqrt{2} = 8\sqrt{2}$

46. $\sqrt{3} - 7\sqrt{27} = \sqrt{3} - 21\sqrt{3} = -20\sqrt{3}$

47. $\sqrt{28} + \sqrt{63} = 2\sqrt{7} + 3\sqrt{7} = 5\sqrt{7}$

48. $\sqrt{3} \cdot \sqrt{6} = \sqrt{18} = \sqrt{9 \cdot 2} = \sqrt{9} \cdot \sqrt{2} = 3\sqrt{2}$

49. $\sqrt{8} \cdot \sqrt{6} = \sqrt{48} = \sqrt{16 \cdot 3} = \sqrt{16} \cdot \sqrt{3} = 4\sqrt{3}$

50. $\dfrac{\sqrt{300}}{\sqrt{3}} = \sqrt{\dfrac{300}{3}} = \sqrt{100} = 10$

51. $\dfrac{4}{\sqrt{3}} \cdot \dfrac{\sqrt{3}}{\sqrt{3}} = \dfrac{4\sqrt{3}}{3}$

52. $\dfrac{\sqrt{7}}{\sqrt{5}} \cdot \dfrac{\sqrt{5}}{\sqrt{5}} = \dfrac{\sqrt{35}}{5}$

53. $3(2 + \sqrt{7}) = 6 + 3\sqrt{7}$

54. $\sqrt{3}(4 + \sqrt{6}) = 4\sqrt{3} + \sqrt{18} = 4\sqrt{3} + 3\sqrt{2}$

55. Commutative property of addition

56. Commutative property of multiplication

57. Associative property of addition

58. Distributive property

59. Associative property of multiplication

60. Natural numbers – not closed for subtraction
$2 - 3 = -1$ and -1 is not a natural number.

61. Whole numbers – closed for multiplication

62. Not closed; $1 \div 2$ is not an integer

63. Closed

64. Not closed; $\sqrt{2} \cdot \sqrt{2} = 2$ is not irrational

65. $2^3 \cdot 2^4 = 2 \cdot 2 \cdot 2 \cdot 2 \cdot 2 \cdot 2 \cdot 2 = 128$

66. $\dfrac{7^5}{7^2} = 7^{5-2} = 7^3 = 343$

67. $9^0 = 1$

68. $5^{-3} = \dfrac{1}{5^3} = \dfrac{1}{5 \cdot 5 \cdot 5} = \dfrac{1}{125}$

69. $(2^3)^4 = 2^{3 \times 4} = 2^{12} = 4096$

70. $-8^0 = -1$

71. $-7^{-2} = -1/7^2 = -1/7 \cdot 7$
$-1/49$

72. $3 \cdot 2^{-3} = 3 \cdot 1/(2 \cdot 2 \cdot 2) = 1/8$

73. $90,100,000,000 = 9.01 \times 10^{10}$

74. $0.0000158 = 1.58 \times 10^{-5}$

75. $2.8 \times 10^5 = 280,000$

76. $1.39 \times 10^{-4} = 0.000139$

77. $(3 \times 10^4)(2 \times 10^{-9}) =$
$6.0 \times 10^{4-9} = 6.0 \times 10^{-5}$

78. $\dfrac{1.5 \times 10^{-3}}{5 \times 10^{-4}} = \dfrac{1.5}{5} \times \dfrac{10^{-3}}{10^{-4}} =$
$0.3 \times 10^1 = 3.0 \times 10^0$

79. $(550,000)(2,000,000) = (5.5 \times 10^5)(2 \times 10^6)$
$= (5.5)(2) \times 10^{5+6} = 11 \times 10^{11}$
$= 1,100,000,000,000$

80. $\dfrac{8,400,000}{70,000} = \dfrac{8.4 \times 10^6}{7 \times 10^4} = 1.2 \times 10^2 = 120$

81. $\dfrac{0.000002}{0.0000004} = \dfrac{2 \times 10^{-6}}{4 \times 10^{-7}} = 0.5 \times 10^1 = 5$

82. $\dfrac{1.49 \times 10^{11}}{3.84 \times 10^8} = .3880208333 \times 10^3 \approx 388.02$
388 times

83. $\dfrac{20,000,000}{3,600} = \dfrac{2.0 \times 10^7}{3.6 \times 10^3}$
$\approx 0.555556 \times 10^4 = \$5555.56$

84. Arithmetic   27, 33

85. Geometric   8, 16

86. $a_9 = -6 + 8(2) = -6 + 16 = 10$

87. $a_{10} = -20 + 9(5) = -20 + 45 = 25$

88. 3, 6, 12, 24, 48 $a_4 = 48$

89. $a_1 = -1, r = 3$ $a_{10} = -1(3)^9 = -19,683$

90. $s_{50} = \dfrac{50(3 + 150)}{2} = (25)(153) = 3825$

91. $s_{20} = \dfrac{20(0.5 + 5.25)}{2} = \dfrac{(20)(5.75)}{2} = 57.5$

92. $s_4 = \dfrac{3(1 - 2^4)}{1 - 2} = \dfrac{(3)(-15)}{-1} = 45$

93. $s_6 = \dfrac{1\left(1-(-2)^6\right)}{1-(-2)} = \dfrac{(1)(1-64)}{3} = \dfrac{(1)(-63)}{3} = -21$

94. Arithmetic: $a_n = 1 + (n-1)3 = 1 + 3n - 3$
$\qquad = 3n - 2$

95. Geometric: $a_n = 2(-1)^{n-1}$

96. 1, 1, 2, 3, 5, 8, 13, 21, 34, 55, 89, 144, 233, 377, 610

97. No

98. Yes; $-8, -13$

Chapter Test

1. 40,455 is divisible by: 3, 5, and 9

2.
$$
\begin{array}{r|r}
3 & 825 \\ \hline
5 & 275 \\ \hline
5 & 55 \\ \hline
 & 11
\end{array}
$$
$825 = 3 \bullet 5^2 \bullet 11$

3. $[(-3)+7]-(-4) = [4]+4 = 8$

4. $[(-70)(-5)] \div (8-10) = 350 \div [8+(-10)]$
$\qquad = 350 \div (-2) = -175$

5. $2\,\tfrac{7}{16} = \dfrac{(16)(2)+7}{16} = \dfrac{32+7}{16} = \dfrac{39}{16}$

6. $\dfrac{13}{40} = 0.325$

7. $6.45 = \dfrac{645}{100} = \dfrac{129}{20}$

8. $\dfrac{7}{20} - \dfrac{12}{25} \div \dfrac{9}{10} = \dfrac{7}{20} - \dfrac{12}{25} \bullet \dfrac{10}{9} = \dfrac{7}{20} - \dfrac{8}{15}$
$\qquad = \dfrac{21}{60} - \dfrac{32}{60} = -\dfrac{11}{60}$

9. $\sqrt{75}+\sqrt{48} = \sqrt{25}\sqrt{3}+\sqrt{16}\sqrt{3} = 5\sqrt{3}+4\sqrt{3} = 9\sqrt{3}$

10. $\dfrac{\sqrt{2}}{\sqrt{7}} = \dfrac{\sqrt{2}}{\sqrt{7}} \cdot \dfrac{\sqrt{7}}{\sqrt{7}} = \dfrac{\sqrt{14}}{\sqrt{49}} = \dfrac{\sqrt{14}}{7}$

11. Yes, the product of any two integers is an integer.

12. Distributive property of multiplication over addition

13. $\dfrac{4^5}{4^2} = 4^{5-2} = 4^3 = 64$

14. $4^3 \bullet 4^2 = 4^5 = 4 \cdot 4 \cdot 4 \cdot 4 \cdot 4 = 1024$

15. $3^{-4} = \dfrac{1}{3^4} = \dfrac{1}{81}$

16. $\dfrac{7.2 \times 10^6}{9.0 \times 10^{-6}} = 0.8 \times 10^{12} = 8.0 \times 10^{11}$

17. $a_n = -4n + 2$

18. $\dfrac{11\left[-2+(-32)\right]}{2} = \dfrac{11(-34)}{2} = -187$

19. $a_n = 3 \bullet (2)^{n-1}$

20. 1, 1, 2, 3, 5, 8, 13, 21, 34, 55

CHAPTER SIX

ALGEBRA, GRAPHS, AND FUNCTIONS

Exercise Set 6.1

1. Variable
3. Expression
5. Terms
7. Coefficient
9. Identity

11. $x = -3, \ x^2 = (-3)^2 = 9$

13. $x = -5, \ -x^2 = -(-5)^2 = -25$

15. $x = -7, -2x^3 = -2(-7)^3 = -2(-343) = 686$

17. $x = 4, x - 7 = 4 - 7 = -3$

19. $x = -2, \ -3x + 7 = -3(-2) + 7 = 6 + 7 = 13$

21. $x = 3, \ x^2 - 5x + 12 = (3)^2 - 5(3) + 12$
$$= 9 - 15 + 12 = 6$$

23. $x = \dfrac{2}{3}, \dfrac{1}{2}x^2 - 5x + 2 = \dfrac{1}{2}\left(\dfrac{2}{3}\right)^2 - 5\left(\dfrac{2}{3}\right) + 2$
$$= \dfrac{1}{2}\left(\dfrac{4}{9}\right) - \dfrac{10}{3} + 2$$
$$= \dfrac{4}{18} - \dfrac{60}{18} + \dfrac{36}{18}$$
$$= -\dfrac{20}{18} = -\dfrac{10}{9}$$

25. $x = 3, \ 4x + 5 = 17, 4(3) + 5 = 12 + 5 = 17$
   Yes, 3 is a solution.

27. $x = 2, \ -x^2 + 3x + 6 = 5, \ -(2)^2 + 3(2) + 6$
$$= -4 + 6 + 6 = 8$$
   No, 2 is not a solution.

29. $x = -4, \ 3x^2 + 2x = 40, 3(-4)^2 + 2(-4)$
$$= 3(16) - 8 = 48 - 8 = 40$$
   Yes, -4 is a solution.

31. $4x + 9x = 13x$

33. $-7x + 3x - 8 = -4x - 8$

35. $7x + 3y - 4x + 8y = 3x + 11y$

37. $-3x + 2 - 5x = -8x + 2$

39. $2 - 3x - 2x + 1 = -5x + 3$

41. $6.2x - 8.3 + 7.1x = 13.3x - 8.3$

43. $\dfrac{3}{5}x + \dfrac{3}{10}x - 8 = \dfrac{6}{10}x + \dfrac{3}{10}x - 8 = \dfrac{9}{10}x - 8$

45. $7x - 4y - 8y + 6x + 2 = 7x + 6x - 4y - 8y + 2$
$$= 13x - 12y + 2$$

47. $3(t + 3) + 5(t - 2) + 1$
   $3t + 9 + 5t - 10 + 1 = 8t$

101

49. $0.2(x+4)+1.2(x-3)=0.2x+0.8+1.2x-3.6$
$$=1.4x-2.8$$

51. $\dfrac{2}{3}(x+6)+\dfrac{1}{6}(x+6)=\dfrac{2}{3}x+4+\dfrac{1}{6}x+1$
$$=\dfrac{2}{3}\cdot\dfrac{2}{2}x+4+\dfrac{1}{6}x+1=\dfrac{5}{6}x+5$$

53. $\dfrac{1}{4}x+\dfrac{4}{5}-\dfrac{2}{3}x=\dfrac{3}{12}x-\dfrac{8}{12}x+\dfrac{4}{5}=-\dfrac{5}{12}x+\dfrac{4}{5}$

55.
$$y+7=9$$
$$y+7-7=9-7 \qquad \text{Subtract 7 from both sides of the equation.}$$
$$y=2$$

57.
$$16=-3t-2$$
$$16+2=-3t-2+2 \qquad \text{Add 2 to both sides of the equation.}$$
$$18=-3t$$
$$\dfrac{18}{-3}=\dfrac{-3t}{-3} \qquad \text{Divide both sides of the equation by } -3.$$
$$-6=t$$

59.
$$\dfrac{1}{2}x+\dfrac{1}{3}=\dfrac{2}{3}$$
$$6\left(\dfrac{1}{2}x+\dfrac{1}{3}\right)=6\left(\dfrac{2}{3}\right) \qquad \text{Multiply both sides of the equation by the LCD.}$$
$$3x+2=4 \qquad \text{Distributive Property}$$
$$3x+2-2=4-2 \qquad \text{Subtract 2 from both sides of the equation.}$$
$$3x=2$$
$$\dfrac{3x}{3}=\dfrac{2}{3} \qquad \text{Divide both sides of the equation by 3.}$$
$$x=\dfrac{2}{3}$$

61.
$$0.9x-1.2=2.4$$
$$0.9x+1.2-1.2=2.4+1.2 \qquad \text{Add 1.2 to both sides of the equation.}$$
$$0.9x=3.6$$
$$\dfrac{0.9x}{0.9}=\dfrac{3.6}{0.9} \qquad \text{Divide both sides of the equation by 0.9.}$$
$$x=4$$

63.
$$6t - 8 = 4t - 2$$
$$6t - 4t - 8 = 4t - 4t - 2$$ Subtract $4t$ from both sides of the equation.
$$2t - 8 = -2$$
$$2t - 8 + 8 = -2 + 8$$ Add 8 to both sides of the equation.
$$2t = 6$$
$$\frac{2t}{2} = \frac{6}{2}$$ Divide both sides of the equation by 2.
$$t = 3$$

65.
$$\frac{x}{4} + 2x = \frac{1}{3}$$
$$12\left(\frac{x}{4} + 2x\right) = 12\left(\frac{1}{3}\right)$$ Mulitply both sides of the equation by the LCD.
$$3x + 24x = 4$$ Distributive Property
$$27x = 4$$
$$\frac{27x}{27} = \frac{4}{27}$$ Divide both sides of the equation by 27.
$$x = \frac{4}{27}$$

67.
$$2(x + 3) - 4 = 2(x - 4)$$
$$2x + 6 - 4 = 2x - 8$$ Distributive Property
$$2x + 2 = 2x - 8$$
$$2x - 2x + 2 = 2x - 2x - 8$$ Subtract $2x$ from both sides of the equation.
$$2 = -8$$ False
$$\text{No solution}$$

69.
$$4(x - 4) + 12 = 4(x - 1)$$
$$4x - 16 + 12 = 4x - 4$$ Distributive Property
$$4x - 4 = 4x - 4$$
$$\text{All real numbers}$$

71.
$$\frac{1}{3}(x + 3) = \frac{2}{5}(x + 2)$$
$$15\left(\frac{1}{3}\right)(x + 3) = 15\left(\frac{2}{5}\right)(x + 2)$$ Multiply both sides of the equation by the LCD.
$$5(x + 3) = 6(x + 2)$$
$$5x + 15 = 6x + 12$$ Distributive Property
$$5x - 5x + 15 = 6x - 5x + 12$$ Subtract $5x$ from both sides of the equation.
$$15 = x + 12$$
$$15 - 12 = x + 12 - 12$$ Subtract 12 from both sides of the equation.
$$3 = x$$

73.
$$3x + 2 - 6x = -x - 15 + 8 - 5x$$
$$-3x + 2 = -6x - 7$$
$$-3x + 6x + 2 = -6x + 6x - 7 \qquad \text{Add } 6x \text{ to both sides of the equation.}$$
$$3x + 2 = -7$$
$$3x + 2 - 2 = -7 - 2 \qquad \text{Subtract 2 from both sides of the equation.}$$
$$3x = -9$$
$$\frac{3x}{3} = \frac{-9}{3} \qquad \text{Divide both sides of the equation by 3.}$$
$$x = -3$$

75.
$$4(3n + 1) = 5(4n - 6) + 9n$$
$$12n + 4 = 20n - 30 + 9n \qquad \text{Distributive Property}$$
$$12n + 4 = 29n - 30$$
$$12n - 12n + 4 = 29n - 12n - 30 \qquad \text{Subtract 12n from both sides of the equation.}$$
$$4 = 17n - 30$$
$$4 + 30 = 17n - 30 + 30 \qquad \text{Add 30 to both sides of the equation.}$$
$$34 = 17n \qquad \text{Divide both sides of the equation by 17.}$$
$$2 = n$$

77. $s = 0.08d$; Find s if $d = \$79$

$s = 0.08(79)$

$s = \$6.32$ in sales tax

79. $2(0.60)^2 + 80(0.60) + 40 = 0.72 + 48 + 40 = 88.72;$

88.72 min

81. $(-1)^n = 1$ for any even number, n, since there will be an even number of factors of (-1), and when these are multiplied, the product will always be 1.

83. $1^n = 1$ for all natural numbers since 1 multiplied by itself any number of times will always be 1.

85. a)
$$P = 14.70 + 0.43x$$
$$148 = 14.70 + 0.43x$$ Given $P = 148$, find x.
$$148 - 14.70 = 14.70 - 14.70 + 0.43x$$ Subtract 14.70 from both sides of the equation.
$$133.3 = 0.43x$$
$$\frac{133.3}{0.43} = \frac{0.43x}{0.43}$$ Divide both sides of the equation by 0.43.
$$x = 310 \text{ ft}$$

b)
$$P = 14.70 + 0.43x$$
$$128.65 = 14.70 + 0.43x$$ Given $P = 128.65$, find x.
$$128.65 - 14.70 = 14.70 - 14.70 + 0.43x$$ Subtract 14.70 from both sides of the equation.
$$113.95 = 0.43x$$
$$\frac{113.95}{0.43} = \frac{0.43x}{0.43}$$ Divide both sides of the equation by 0.43.
$$x = 265 \text{ ft below sea level}$$

Exercise Set 6.2

1. Subscripts

3. $P = 4s = 4(10) = 40$

5. $P = 2l + 2w$
$$P = 2(15) + 2(8) = 30 + 16 = 46$$

7.
$$K = \frac{1}{2}mv^2$$
$$4500 = \frac{1}{2}m(30)^2$$
$$4500 = 450m$$
$$\frac{4500}{450} = \frac{450m}{450}$$
$$10 = m$$

9.
$$S = \pi r(r + h)$$
$$S = \pi(8)(8 + 2)$$
$$S = \pi(8)(10)$$
$$S = \pi(80)$$
$$S = 251.33$$

11.
$$z = \frac{x - \mu}{\sigma}$$
$$\frac{1.62}{1} = \frac{36.7 - \mu}{3}$$
$$1.62(3) = 36.7 - \mu$$
$$4.86 = 36.7 - \mu$$
$$4.86 - 36.7 = 36.7 - 36.7 - \mu$$
$$-31.84 = -\mu$$
$$\frac{-31.84}{-1} = \frac{-\mu}{-1}$$
$$31.84 = \mu$$

13.
$$T = \frac{PV}{k}$$
$$\frac{80}{1} = \frac{P(20)}{0.5}$$
$$80(0.5) = 20P$$
$$40 = 20P$$
$$\frac{40}{20} = \frac{20P}{20}$$
$$2 = P$$

15.
$$V = -\frac{1}{2}at^2$$
$$2304 = -\frac{1}{2}a(12)^2$$
$$\frac{2304}{1} = -\frac{144a}{2}$$
$$2304(2) = -144a$$
$$4608 = -144a$$
$$\frac{4608}{144} = -\frac{144a}{144}$$
$$-32 = a$$

17.
$$V = \pi r^2 h$$
$$942 = 3.14(5)^2 h$$
$$942 = 3.14(25)h$$
$$942 = 78.5h$$
$$\frac{942}{78.5} = \frac{78.5h}{78.5}$$
$$11.99 = h$$

19.
$$F = \frac{9}{5}C + 32$$
$$F = \frac{9}{5}(7) + 32$$
$$F = \frac{63}{5} + 32 = 12.6 + 32 = 44.6$$

21.
$$m = \frac{y_2 - y_1}{x_2 - x_1}$$
$$m = \frac{8 - (-4)}{-3 - (-5)}$$
$$m = \frac{8 + 4}{-3 + 5} = \frac{12}{2} = 6$$

23.
$$E = a_1 p_1 + a_2 p_2 + a_3 p_3$$
$$E = 5(0.2) + 7(0.6) + 10(0.2)$$
$$E = 1 + 4.2 + 2 = 7.2$$

25.
$$s = -16t^2 + v_0 t + s_0$$
$$s = -16(4)^2 + 30(4) + 150$$
$$s = -16(16) + 120 + 150$$
$$s = -256 + 120 + 150 = 14$$

27.
$$R_T = \frac{R_1 R_2}{R_1 + R_2}$$
$$R_T = \frac{(100)(200)}{100 + 200}$$
$$R_T = \frac{20,000}{300}$$
$$R_T = 66.67$$

29.
$$s_n = \frac{a_1\left(1 - r^n\right)}{1 - r}$$
$$s_n = \frac{10\left(1 - \left(\frac{1}{2}\right)^4\right)}{1 - \frac{1}{2}}$$
$$s_n = \frac{10\left(1 - \frac{1}{16}\right)}{\frac{1}{2}}$$
$$s_n = \frac{10\left(\frac{15}{16}\right)}{\frac{1}{2}} = 20\left(\frac{15}{16}\right) = 18\frac{3}{4}$$

31.
$$6x + 3y = 9$$
$$6x - 6x + 3y = -6x + 9 \qquad \text{Subtract -6x from both sides of the equation.}$$
$$3y = -6x + 9$$
$$\frac{3y}{3} = \frac{-6x + 9}{3} \qquad \text{Divide both sides of the equation by } -3.$$
$$y = \frac{-6x + 9}{3} = -2x + 3$$

33.
$$8x + 7y = 21$$
$$-8x + 8x + 7y = -8x + 21 \qquad \text{Subtract 8x from both sides of the equation.}$$
$$7y = -8x + 21$$
$$\frac{7y}{7} = \frac{-8x + 21}{7} \qquad \text{Divide both sides of the equation by 7.}$$
$$y = \frac{-8x + 21}{7} = \frac{-8x}{7} + \frac{21}{7} = -\frac{8}{7}x + 3$$

35.
$$2x - 3y + 6 = 0$$
$$2x - 3y + 6 - 6 = 0 - 6 \qquad \text{Subtract 6 from both sides of the equation.}$$
$$2x - 3y = -6$$
$$-2x + 2x - 3y = -2x - 6 \qquad \text{Subtract 2x from both sides of the equation.}$$
$$-3y = -2x - 6$$
$$\frac{-3y}{-3} = \frac{-2x - 6}{-3} \qquad \text{Divide both sides of the equation by -3.}$$
$$y = \frac{-2x - 6}{-3} = \frac{-(-2x - 6)}{3} = \frac{2x + 6}{3} = \frac{2x}{3} + \frac{6}{3} = \frac{2}{3}x + 2$$

37. $d = rt$

$$\frac{d}{t} = \frac{rt}{t}$$ Divide both sides of the equation by t.

$$r = \frac{d}{t}$$

39. $p = a + b + c$

$p - b = a + b - b + c$ Subtract b from both sides of the equation.

$p - b = a + c$

$p - b - c = a + c - c$ Subtract c from both sides of the equation.

$a = p - b - c$

41. $A = \frac{1}{2}bh$

$2 \cdot A = 2 \cdot \frac{1}{2}bh$ Multiply both sides of the equation by 2.

$$\frac{2A}{h} = \frac{bh}{h}$$

$$\frac{2V}{h} = Ab$$ Divide both sides of the equation by h.

43. $V = \pi r^2 h$

$$\frac{V}{\pi} = \frac{\pi r^2 h}{\pi}$$ Divide both sides of the equation by π.

$$\frac{V}{\pi} = r^2 h$$

$$\frac{V}{\pi r^2} = \frac{r^2 h}{r^2}$$ Divide both sides of the equation by r^2.

$$h = \frac{V}{\pi r^2}$$

45. $y = mx + b$

$y - mx = mx - mx + b$ Subtract mx from both sides of the equation.

$b = y - mx$

47. $P = 2l + 2w$

$P - 2l = 2l - 2l + 2w$ Subtract $2l$ from both sides of the equation.

$P - 2l = 2w$

$$\frac{P - 2l}{2} = \frac{2w}{2}$$ Divide both sides of the equation by 2.

$$w = \frac{P - 2l}{2}$$

49.
$$F = \frac{9}{5}C + 32$$

$$F - 32 = \frac{9}{5}C + 32 - 32 \qquad \text{Subtract 32 from both sides of the equation.}$$

$$F - 32 = \frac{9}{5}C$$

$$\frac{5}{9}(F - 32) = \frac{5}{9}\left(\frac{9}{5}C\right) \qquad \text{Multiply both sides of the equation by } \frac{5}{9}.$$

$$C = \frac{5}{9}(F - 32)$$

51.
$$S = \pi r^2 + \pi rs$$

$$S - \pi r^2 = \pi r^2 - \pi r^2 + \pi rs \qquad \text{Subtract } \pi r^2 \text{ from both sides of the equation.}$$

$$S - \pi r^2 = \pi rs$$

$$\frac{S - \pi r^2}{\pi r} = \frac{\pi rs}{\pi r} \qquad \text{Divide both sides of the equation by } \pi r.$$

$$\frac{S - \pi r^2}{\pi r} = s$$

53. $d = rt$

$$403 = 62t$$

$$\frac{403}{62} = t$$

$$t = 6.5 \text{ hours}$$

55. $i = prt$

$$128 = 800(r)(2)$$

$$128 = 1600r$$

$$\frac{128}{1600} = \frac{1600r}{1600}$$

$$r = 0.08 = 8\%$$

57. a) $6 \text{ ft} = 6(12) = 72 \text{ in.}$

$$B = \frac{703w}{h^2}$$

$$B = \frac{703(200)}{(72)^2}$$

$$B = \frac{140,600}{5184} = 27.12191358 \approx 27.12$$

b)
$$B = \frac{703w}{h^2}$$

$$26 = \frac{703w}{(72)^2}$$

$$26 = \frac{703w}{5184}$$

$$134,784 = 703w$$

$$\frac{134,784}{703} = \frac{703w}{703}$$

$$w = 191.7268848 \text{ lb}$$

He would have to lose

$$200 - 191.7268848$$

$$= 8.2731152 \approx 8.27 \text{ lb}$$

Exercise Set 6.3

1. Ratio

3. $x + 5 : 5$ more than x

5. $2x - 3 :$ 2 times x, decreased by 3

7. $x + 3 : 3$ more than x

9. $6 - 4y :$ 6 decreased by 4 times y

11. $8w + 9$

13. $4x + 6$

15. $\dfrac{12 - s}{4}$

17. $3(x + 7)$

19.   Let $x =$ the number

     $x - 5 = 5$ less than the number

     $x - 5 = 20$

    $x - 5 + 5 = 20 + 5$

       $x = 25$

21.   Let $x =$ the number

     $x + 3 =$ the sum of the number and 3

     $x + 3 = 10$

    $x + 3 - 3 = 10 - 3$

       $x = 7$

23. Let $x =$ the number
$4x - 10 = 4$ times the number decreased by 10
$$4x - 10 = 42$$
$$4x - 10 + 10 = 42 + 10$$
$$4x = 52$$
$$\frac{4x}{4} = \frac{52}{4}$$
$$x = 13$$

25. Let $x =$ the number
$4x + 12 = 12$ more than 4 times the number
$$4x + 12 = 32$$
$$4x + 12 - 12 = 32 - 12$$
$$4x = 20$$
$$\frac{4x}{4} = \frac{20}{4}$$
$$x = 5$$

27. Let $x =$ the number
$x + 6 =$ the number increased by 6
$2x - 3 = 3$ less than twice the number
$$x + 6 = 2x - 3$$
$$x - x + 6 = 2x - x - 3$$
$$6 = x - 3$$
$$6 + 3 = x - 3 + 3$$
$$9 = x$$

29. Let $x =$ the number
$x + 10 =$ the number increased by 10
$2(x + 3) = 2$ times the sum of the number and 3
$$x + 10 = 2(x + 3)$$
$$x + 10 = 2x + 6$$
$$x - x + 10 = 2x - x + 6$$
$$10 = x + 6$$
$$10 - 6 = x + 6 - 6$$
$$4 = x$$

31. Let $x =$ the number of miles driven
$0.50x =$ reimbursement for mileage
$$150 + 0.50x = 255$$
$$0.50x = 105$$
$$\frac{0.50x}{0.50} = \frac{105}{0.50}$$
$$x = 210$$
210 miles

33. Let $x =$ Tito's dollar sales
$0.06x =$ the amount Tito made
on commission
$$400 + 0.06x = 790$$
$$400 - 400 + 0.06x = 790 - 400$$
$$0.06x = 390$$
$$\frac{0.06x}{0.06} = \frac{390}{0.06}$$
$$x = \$6500$$

35. Let $x =$ the number of copies Ronnie
must make
$0.08x =$ the amount spent on x copies
$$0.08x = 250$$
$$\frac{0.08x}{0.08} = \frac{250}{0.08}$$
$$x = 3125 \text{ copies}$$

37. Let $x =$ the amt. of $ invested in mutual funds
$3x =$ the amt. of $ invested in stocks
$$x + 3x = 20,000$$
$$4x = 20,000$$
$$\frac{4x}{4} = \frac{20,000}{4}$$
$$x = \$5000 \text{ invested in mutual funds}$$
$$3x = 3(5000) = \$15,000 \text{ invested in stocks}$$

39.   Let $w =$ the width
$$w + 2 = \text{the length}$$
$$2w + 2(w+2) = P$$
$$2w + 2(w+2) = 52$$
$$2w + 2w + 4 = 52$$
$$4w + 4 = 52$$
$$4w + 4 - 4 = 52 - 4$$
$$4w = 48$$
$$\frac{4w}{4} = \frac{48}{4}$$
width $= 12$ ft, length $= w + 2 = 12 + 2 = 14$ ft

41.   Let $r =$ regular fare
$$\frac{r}{2} = \text{half off regular fare}$$
$$0.07r = \text{tax on regular fare}$$
$$\frac{r}{2} + 0.07r = 257$$
$$2\left(\frac{r}{2} + 0.07r\right) = 2(257)$$
$$r + 0.14r = 514$$
$$1.14r = 514$$
$$\frac{1.14r}{1.14} = \frac{514}{1.14}$$
$$r = \$450.877193$$
$$\approx \$450.88$$

43. a)   Let c $=$ cost of the water bill
$$\frac{\$1.33}{748 \text{ gallons}} = \frac{c}{30,000}$$
$$748c = 1.33(30,000)$$
$$748c = 39,900$$
$$\frac{748c}{748} = \frac{39,900}{748}$$
$$c \approx \$53.34$$

b)   Let g $=$ gallons of water used
$$\frac{\$1.33}{748} = \frac{\$150}{g}$$
$$1.33g = 150(748)$$
$$1.33g = 112,200$$
$$\frac{1.33g}{1.33} = \frac{112,200}{1.33}$$
$$g \approx 84,361 \text{ gallons}$$

45. $$\frac{1}{1,700,000} = \frac{10.7}{x}$$
$$x = 1,700,000(10.7)$$
$$x \approx 18,190,000 \text{ households}$$

47. a) $$\frac{40}{12} = \frac{x}{480}$$
$$40(480) = 12x$$
$$\frac{40(480)}{12} = \frac{12x}{12}$$
$$x = 1600 \text{ lb}$$

b) $$\frac{1}{12} = \frac{x}{480}$$
$$12x = 480$$
$$x = \frac{480}{12} = 40 \text{ bags}$$

49. $$\frac{40}{0.6} = \frac{250}{x}$$
$$40x = 0.6(250)$$
$$40x = 150$$
$$\frac{40x}{40} = \frac{150}{40}$$
$$x = 3.75 \text{ m}\ell$$

51.
$$\frac{40}{1} = \frac{35}{x}$$
$$40x = 35$$
$$\frac{40x}{40} = \frac{35}{40}$$
$$x = 0.875 \text{ cc}$$

53. a)   Let $x =$ the number of months for the amount
         saved to equal the price of the course

$$0.10(100) = \$10 \text{ saved per year}$$
$$10x = 45$$
$$x = 4.5 \ = 4\frac{1}{2} \text{ months}$$

   b) $25 - 18 = 7$ years

$$7(12)(10) = \$840 \text{ saved before paying for course}$$
$$\$840 - \$45 = \$795 \text{ total savings}$$

Exercise Set 6.4

1. Direct
3. Joint
5. Direct
7. Inverse
9. Direct
11. Inverse
13. Direct
15. Inverse
17. Direct
19. Direct
21. Answers will vary.

23. a) $y = kx$

   b) $y = 9(12) = 108$

25. a) $m = \dfrac{k}{n^2}$

   b) $m = \dfrac{20}{(10)^2} = \dfrac{20}{100} = 0.20$

27. a) $A = \dfrac{kB}{C}$

   b) $A = \dfrac{(5)5}{10} = 2.5$

29. a) $F = kDE$
   b) $F = 7(3)(10) = 210$

31. a) $H = kL$

   b) $12 = k(40)$
$$\frac{12}{40} = \frac{40k}{40}$$
$$k = 0.3$$
$$H = 0.3L$$
$$H = 0.3(10) = 3$$

33. a) $y = \dfrac{k\sqrt{t}}{s}$

   b) $12 = \dfrac{k\sqrt{36}}{2}$

   $12 = \dfrac{6k}{2}$

   $24 = 6k$

   $\dfrac{24}{6} = \dfrac{6k}{6}$

   $k = 4$

   $y = \dfrac{4\sqrt{t}}{s}$

   $y = \dfrac{4\sqrt{81}}{4} = \dfrac{4(9)}{4} = \dfrac{36}{4} = 9$

35. a) $Z = kWY$

   b) $12 = k(9)(4)$

   $12 = 36k$

   $\dfrac{12}{36} = \dfrac{36k}{36}$

   $k = \dfrac{1}{3}$

   $Z = \dfrac{1}{3}WY$

   $Z = \dfrac{1}{3}(50)(6) = \dfrac{300}{3} = 100$

37. a) $F = \dfrac{kM_1M_2}{d^2}$

   b) $20 = \dfrac{k(5)(10)}{(0.2)^2}$

   $20 = \dfrac{50k}{0.04}$

   $50k = 0.8$

   $k = \dfrac{0.8}{50} = 0.016$

   $F = \dfrac{0.016M_1M_2}{d^2}$

   $F = \dfrac{0.016(10)(20)}{(0.4)^2} = \dfrac{3.2}{0.16} = 20$

39. a) $p = kn$

   b) $450 = k(8)$

   $\dfrac{450}{8} = \dfrac{8k}{8}$

   $k = 56.25$

   $p = 56.25n$

   $p = 56.25(18) = \$1012.50$

41. a) $t = \dfrac{k}{T}$

   b) $2 = \dfrac{k}{75}$

   $k = 75(2) = 150$

   $t = \dfrac{150}{T}$

   $t = \dfrac{150}{80} = 1.875$ minutes

43. a) $d = kt^2$

   b) $64 = k(2)^2$

   $64 = 4k$

   $k = \dfrac{64}{4} = 16$

   $d = 16t^2$

   $d = 16(3)^2 = 16(9)$

   $= 144$ feet

45.   a) $R = \dfrac{kL}{A}$

   b) $0.2 = \dfrac{k(200)}{0.05}$

   $200k = 0.01$

   $k = \dfrac{0.01}{200} = 0.00005$

   $R = \dfrac{0.00005L}{A}$

   $R = \dfrac{0.00005(5000)}{0.01} = \dfrac{0.25}{0.01} = 25$ ohms

47.   a) $W = kI^2R$

   b) $6 = k(0.2)^2(150)$

   $6k = 6$

   $k = 1$

   $W = 1(0.3)^2(100)$

   $W = (.09)(100) = 9$ watts

49.   a) $y = kx$

   $y = 2x$

   $\dfrac{y}{2} = \dfrac{2x}{2}$

   $x = \dfrac{y}{2} = 0.5y$

   Varies directly

   b) $k = 0.5$

51. a) y will double

   b) y will be half as large

53. $I = \dfrac{k}{d^2}$

   $\dfrac{1}{16} = \dfrac{k}{(4)^2}$

   $\dfrac{1}{16} = \dfrac{k}{16}$

   $k = 1$

   $I = \dfrac{1}{d^2}$

   $I = \dfrac{1}{(3)^2} = \dfrac{1}{9}$

Exercise Set 6.5

1. Inequality

3. Compound

5. All

7. Closed

9. $x \geq 4$

11. $x + 9 \geq 4$

$x + 9 - 9 \geq 4 - 9$

$x \geq -5$

13. $-6x \leq 36$

$\dfrac{-6x}{-6} \geq \dfrac{36}{-6}$

$x \geq -6$

15 . $\dfrac{x}{6} < 3$

$6\left(\dfrac{x}{6}\right) < 6(3)$

$x < 18$

17. $\dfrac{-x}{3} \geq 3$

$-3\left(\dfrac{-x}{3}\right) \leq -3(3)$

$x \leq -9$

19. $2x + 6 \geq 14$

$2x + 6 - 6 \geq 14 - 6$

$2x \geq 8$

$\dfrac{2x}{2} \geq \dfrac{8}{2}$

$x \geq 4$

21. $4(2x - 1) < 2(4x - 3)$

$8x - 4 < 8x - 6$

$-4 < -6$

False, no solution

23. $-1 \leq x \leq 3$

25. $2 < x - 4 \leq 6$

$2 + 4 < x - 4 + 4 \leq 6 + 4$

$6 < x \leq 10$

27 $x > 1$

29. $-4x \leq 36$

$\dfrac{-4x}{-4} \geq \dfrac{36}{-4}$

$x \geq -9$

31. $x - 6 < -4$

$x - 6 + 6 < -4 + 6$

$x < 2$

33. $\dfrac{x}{3} \leq -2$

$3\left(\dfrac{x}{3}\right) \leq 3(-2)$

$x \leq -6$

35. $-\dfrac{x}{4} \geq 2$

$(-4)\left(-\dfrac{x}{4}\right) \leq (-4)(2)$

$x \leq -8$

37. $-15 < -4x - 3$

$-12 < -4x$

$\dfrac{-12}{-4} > \dfrac{-4x}{-4}$

$3 > x$

39. a) $3(x + 4) \geq 4x + 13$

$3x + 12 \geq 4x + 13$

$3x - 4x + 12 \geq 4x - 4x + 13$

$-x + 12 \geq 13$

$-x + 12 - 12 \geq 13 - 12$

$-x \geq 1$

$\dfrac{-x}{-1} \leq \dfrac{1}{-1}$

$x \leq -1$

41. $5(x + 4) - 6 \leq 2x + 8$

$5x + 20 - 6 \leq 2x + 8$

$5x + 14 \leq 2x + 8$

$5x - 2x + 14 \leq 2x - 2x + 8$

$3x + 14 \leq 8$

$3x + 14 - 14 \leq 8 - 14$

$3x \leq -6$

$\dfrac{3x}{3} \leq \dfrac{-6}{3}$

$x \leq -2$

43. $1 > -x > -6$

$\dfrac{1}{-1} < \dfrac{-x}{-1} < \dfrac{-6}{-1}$

$-1 < x < 6$

45.
$$0.3 \le \frac{x+2}{10} \le 0.5$$

$$10(0.3) \le 10\left(\frac{x+2}{10}\right) \le 10(0.5)$$

$$3 \le x+2 \le 5$$

$$3-2 \le x+2-2 \le 5-2$$

$$1 \le x \le 3$$

47. a) 2006, 2007, 2012, 2013
b) 2008, 2009, 2010, 2011
c) 2006, 2007, 2013
d) 2009

49. Let $x =$ the number of months
Option A: 150+20x

Option B: $35x$

$$35x > 150+20x$$

$$35x-20x > 150+20x-20x$$

$$15x > 150$$

$$\frac{15x}{15} > \frac{150}{15}$$

$$x > 10$$

The number of months it will take for Option A to cost less than Option B is 11.

51. Let $x =$ the number hours she works as cashier

$$7.25x + 662.5 \le 3200$$

$$7.25x \le 2537.5$$

$$\frac{7.25x}{7.25} \le \frac{2537.5}{7.25}$$

$$x \le 350$$

The maximum number of hours that Samantha can work as a cashier is 350 hours.

53. Let $x =$ the length of time Tom can park.

$$1.75 + 0.50(x-1) \le 4.25$$

$$1.75 + 0.50x - 0.50 \le 4.25$$

$$0.50x \le 3$$

$$\frac{0.50x}{0.50} \le \frac{3}{0.50}$$

$$x \le 6$$

He can park for at most 6 hours.

55.
$$36 < 84 - 32t < 68$$

$$36 - 84 < 84 - 84 - 32t < 68 - 84$$

$$-48 < -32t < -16$$

$$\frac{-48}{-32} > \frac{-32t}{-32} > \frac{-16}{-32}$$

$$1.5 > t > 0.5$$

$$0.5 < t < 1.5$$

The velocity will be between $36\frac{\text{ft}}{\text{sec}}$ and $68\frac{\text{ft}}{\text{sec}}$ when t is between $0.5\,\text{sec}$ and $1.5\,\text{sec}$.

57. Let $x =$ Devon's grade on the fifth test

$$80 \le \frac{78+64+88+76+x}{5} < 90$$

$$80 \le \frac{306+x}{5} < 90$$

$$5(80) \le 5\left(\frac{306+x}{5}\right) < 5(90)$$

$$400 \le 306+x < 450$$

$$400-306 \le 306-306+x < 450-306$$

$$94 \le x < 144$$

Devon must have a score of $94 \le x \le 100$, assuming 100 is the highest grade possible.

59. Let $x =$ the number of gallons

$$250x = 2750 \text{ and } 400x = 2750$$

$$x = \frac{2750}{250}, \ x = \frac{2750}{400}$$

$$x = 11, \ x = 6.875$$

$$6.875 \le x \le 11$$

61. Student's answer: $-\dfrac{1}{3}x \le 4$

$$-3\left(-\dfrac{1}{3}x\right) \le -3(4)$$

$$x \le -12$$

Correct answer: $-\dfrac{1}{3}x \le 4$

$$-3\left(-\dfrac{1}{3}x\right) \ge -3(4)$$

$$x \ge -12$$

Yes, -12 is in both solution sets.

Exercise Set 6.6

1. Graph
3. x
5. Plotting points, using intercepts, and using the slope and y-intercept
7. m

9. − 16.

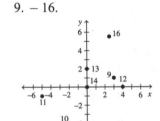

17. − 24.

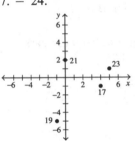

25. $(0, 2)$ 27. $(-2, 0)$ 29. $(-3, -4)$ 31. $(2, -2)$ 33. $(2, 2)$

35.   Substituting (5, 2) into $x + 3y = 11$,

$$5 + 3(2) = 11$$
$$5 + 6 = 11$$
$$11 = 11$$

Therefore, (5, 2) satisfies $x + 3y = 11$.

Substituting (8, 1) into $x + 3y = 11$,

$$8 + 3(1) = 11$$
$$8 + 3 = 11$$
$$11 = 11$$

Therefore, (8, 1) satisfies $x + 3y = 11$.

Substituting (0, 5) into $x + 3y = 11$,

$$0 + 3(5) = 11$$
$$15 \neq 11$$

Therefore, (0, 5) does not satisfy $x + 3y = 11$.

37.   Substituting (8, 7) into $3x - 2y = 10$,

$$3(8) - 2(7) = 10$$
$$24 - 14 = 10$$
$$10 = 10$$

Therefore, (8, 7) satisfies $3x - 2y = 10$.

Substituting (−1, 4) into $3x - 2y = 10$,

$$3(-1) - 2(4) = 10$$
$$-3 - 8 = 10$$
$$-11 \neq 10$$

Therefore, (−1, 4) does not satisfy $3x - 2y = 10$.

Substituting $\left(\dfrac{10}{3}, 0\right)$ into $3x - 2y = 10$,

$$3\left(\frac{10}{3}\right) - 2(0) = 10$$
$$10 - 0 = 10$$
$$10 = 10$$

Therefore, $\left(\dfrac{10}{3}, 0\right)$ satisfies $3x - 2y = 10$.

39. Substituting $(1, -1)$ into $6y = 3x + 6$,

$$6(-1) = 3(1) + 6$$
$$-6 = 3 + 6$$
$$-6 \neq 9$$

Therefore, $(1, -1)$ does not satisfy $6y = 3x + 6$.

Substituting $(4, 3)$ into $6y = 3x + 6$,

$$6(3) = 3(4) + 6$$
$$18 = 12 + 6$$
$$18 = 18$$

Therefore, $(4, 3)$ satisfies $6y = 3x + 6$.

Substituting $(2, 5)$ into $6y = 3x + 6$,

$$6(5) = 3(2) + 6$$
$$30 = 6 + 6$$
$$35 \neq 12$$

Therefore, $(2, 5)$ does not satisfy $6y = 3x + 6$.

41.

Substituting $(8, 0)$ into $\dfrac{x}{4} + \dfrac{2y}{3} = 2$,

$$\frac{8}{4} + \frac{2(0)}{3} = 2$$
$$2 + 0 = 2$$
$$2 = 2$$

Therefore, $(8, 0)$ satisfies

$\dfrac{x}{4} + \dfrac{2y}{3} = 2$.

Substituting $\left(1, \dfrac{1}{2}\right)$ into $\dfrac{x}{4} + \dfrac{2y}{3} = 2$,

$$\frac{1}{4} + \frac{2}{3}\left(\frac{1}{2}\right) = 2$$
$$\frac{7}{12} \neq 2$$

Therefore, $\left(1, \dfrac{1}{2}\right)$ does not satisfy

$\dfrac{x}{4} + \dfrac{2y}{3} = 2$.

$$\frac{1}{4} + \frac{2}{3}\left(\frac{1}{2}\right) = 2$$
$$\frac{7}{12} \neq 2$$

Therefore, $\left(1, \dfrac{1}{2}\right)$ does not satisfy

$\dfrac{x}{4} + \dfrac{2y}{3} = 2$.

Substituting $(0, 3)$ into $\dfrac{x}{4} + \dfrac{2y}{3} = 2$,

$$\frac{0}{4} + \frac{3(2)}{3} = 2$$
$$2 = 2$$

Therefore, $(0, 3)$ satisfies $\dfrac{x}{4} + \dfrac{2y}{3} = 2$.

43. Since the line is vertical, its slope is undefined.

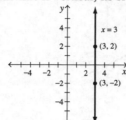

45. Since the line is horizontal, its slope is 0.

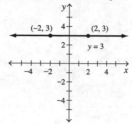

47.

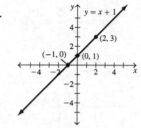

49.

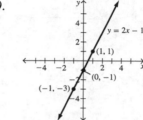

51.

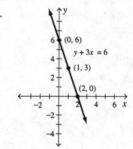

53.

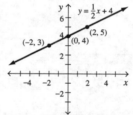

55.

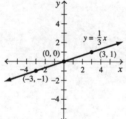

57.

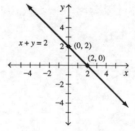

59.

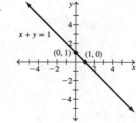

61.

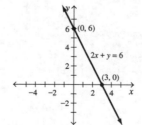

63.

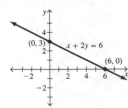

65.

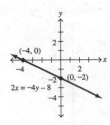

67. $(3,5),(5,9)$ $\quad m = \dfrac{9-5}{5-3} = \dfrac{4}{2} = 2$

69. $(5,1),(8,8)$ $\quad m = \dfrac{8-1}{8-5} = \dfrac{7}{3}$

70. $(-2,6),(-4,9)$ $\quad m = \dfrac{9-6}{-4-(-2)} = \dfrac{3}{-2} = -\dfrac{3}{2}$

71. $(1,5),(4,5)$ $\quad m = \dfrac{5-5}{4-1} = \dfrac{0}{3} = 0$

73. $(8,-3),(8,3)$ $\quad m = \dfrac{3-(-3)}{8-8} = \dfrac{6}{0}$ Undefined

75. $(-3,-1),(8,-9)$

$m = \dfrac{-9-(-1)}{8-(-3)} = \dfrac{-8}{11} = -\dfrac{8}{11}$

77.

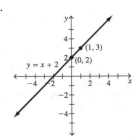

79.

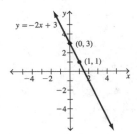

81.

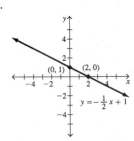

83.

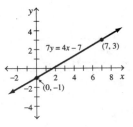

85.

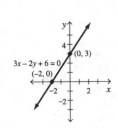

87. The y-intercept is 3; thus $b = 3$. The slope is negative since the graph falls from left to right. The change in y is 3, while the change in x is 4. Thus m, the slope, is $-\dfrac{3}{4}$. The equation is $y = -\dfrac{3}{4}x + 3$.

89. The y-intercept is 2; thus $b = 2$. The slope is positive since the graph rises from left to right. The change in y is 3, while the change in x is 1. Thus m, the slope, is $\dfrac{3}{1} = 3$. The equation is $y = 3x + 2$.

91. a)

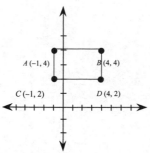

93.

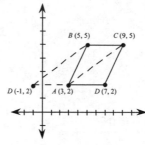

b) $A = lw = 5(2) = 10$ square units

95. For the line joining points P and Q to be parallel to the x-axis, both ordered pairs must have the same y-value. Thus, $b = -3$.

97. For the line joining points P and Q to be parallel to the y-axis, both ordered pairs must have the same x-value.
$$3b - 1 = 8$$
$$3b = 9$$
$$b = 3$$

99. a)

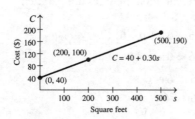

b) $130

c)
$$70 = 40 + 0.3s$$
$$70 - 40 = 40 - 40 + 0.3s$$
$$30 = 0.3s$$
$$s = 100 \text{ square feet}$$

101. a)

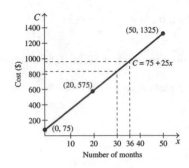

b) $825

c) 36 months

103. a) $m = \dfrac{19-9}{5-0} = \dfrac{10}{5} = 2$

b) $y = 2x + 9$

c) $y = 2(3) + 9 = 6 + 9 = 15$ defects

d) $17 = 2x + 9$

$17 - 9 = 2x + 9 - 9$

$8 = 2x$

$x = 4$ workers

105. a) slope $\approx \dfrac{4.3 - 2.7}{8 - 0} = \dfrac{1.6}{9} = 0.2$

b) $y = 0.2x + 2.7$

c) $y = 0.2x + 2.7$

$y = 0.2(5) + 2.7 = 3.7$

$3.7 trillion

d) $3.9 = 0.2x + 2.7$

$1.2 = 0.2x$

$6 = x$

6 years after 2012, or in 2018

107. a) Solve the equations for y to put them in slope-intercept form. Then compare the slopes and y-intercepts. If the slopes are equal but the y-intercepts are different, then the lines are parallel.

b) $2x - 3y = 6$

$2x - 2x - 3y = -2x + 6$

$-3y = -2x + 6$

$\dfrac{-3y}{-3} = \dfrac{-2x}{-3} + \dfrac{6}{-3}$

$y = \dfrac{2}{3}x - 2$

$4x = 6y + 6$

$4x - 6 = 6y + 6 - 6$

$4x - 6 = 6y$

$\dfrac{4x}{6} - \dfrac{6}{6} = \dfrac{6y}{6}$

$\dfrac{2}{3}x - 1 = y$

Since the two equations have the same slope, $m = \dfrac{2}{3}$, and different y-intercepts, the graphs of the equations are parallel lines.

Exercise Set 6.7

1. System

3. Inconsistent

5. Dependent

7. One

9. Infinite

11. $(2, 6)$

$$y = 4x - 2 \qquad y = -x + 8$$
$$6 = 4(2) - 2 \qquad 6 = -(2) + 8$$
$$6 = 8 - 2 \qquad\quad 6 = 6$$
$$6 = 6$$

Therefore, $(2, 6)$ is a solution.

$(3, -10)$

$$y = 4x - 2 \qquad y = -x + 8$$
$$-10 = 4(3) - 2 \qquad -10 = -(3) + 8$$
$$-10 = 12 - 2 \qquad\quad -10 \neq 5$$
$$-10 \neq 10$$

Therefore, $(3, -10)$ is not a solution.

$(1, 7)$

$$y = 4x - 2 \qquad y = -x + 8$$
$$7 = 4(1) - 2 \qquad 7 = -(1) + 8$$
$$7 = 4 - 2 \qquad\quad 7 = 7$$
$$7 \neq 2$$

Therefore $(1, 7)$ is not a solution.

13.

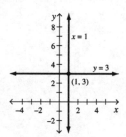

15.

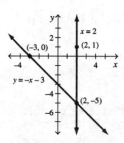

17.

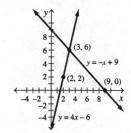

19.

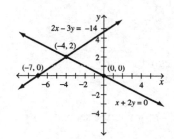

21.

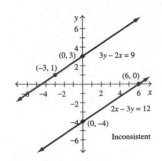

23. $y = x + 9$

   $y = -x + 11$

   Substitute $(x + 9)$ in place of y in the second equation.

   $x + 9 = -x + 11$ (solve for x)

   $x + 9 + x = 11$

   $2x + 9 = 11$

   $2x + 9 - 9 = 11 - 9$

   $2x = 2, \quad x = 1$

   Now substitute 1 for x in the 1st equation

   $y = 1 + 9 \qquad y = 10$

   The solution is $(1, 10)$.   Consistent

25. $6x + 3y = 3$

   $x - 3y = 4 \;\rightarrow\; x = 3y + 4$

   Substitute $(3y + 4)$ in place of x in the first equation.

   $6(3y + 4) + 3y = 3$ (solve for y)

   $18y + 24 + 3y = 3$

   $21y = -21 \qquad y = -1$

   Now substitute -1 for y in the 1st equation

   $6x + 3(-1) = 3$

   $6x = 6 \qquad x = 1$

   The solution is $(1, -1)$.   Consistent

27. $y - x = 4$

   $x - y = 3$

   Solve the first equation for y.

   $y - x + x = x + 4$

   $y = x + 4$

   Substitute $(x + 4)$ for y in the second equation.

   $x - (x + 4) = 3$ (combine like terms)

   $-4 = 3 \qquad$ False

   Since -4 does not equal 3, there is no solution to this system. The system is inconsistent.

29. $3y + 2x = 4$
    $3y = 6 - x$
    Solve the second equation for x.
    $3y = 6 - x$
    $3y - 6 = 6 - 6 - x$
    $3y - 6 = -x$
    $-3y + 6 = x$
    Now substitute $(-3y + 6)$ for x in the 1st
    eq'n.
    $3y + 2(6 - 3y) = 4$ (solve for y)
    $3y + 12 - 6y = 4$
    $-3y = -8$ (div. by -3) $y = 8/3$
    Substitute 8/3 for y in the 2nd eq'n.
    $3(8/3) = 6 - x$
    $8 = 6 - x$ $x = -2$
    The solution is $(-2, 8/3)$.   Consistent

31. $x + 2y = 6$
    $y = 2x + 3$
    Substitute $(2x + 3)$ for y in the first
    equation.
    $x + 2(2x + 3) = 6$
    $x + 4x + 6 = 6$
    $5x + 6 - 6 = 6 - 6$
    $5x = 0$
    $\dfrac{5x}{5} = \dfrac{0}{5}$ $x = 0$

    Now substitute 0 for x in the second
    equation.
    $y = 2(0) + 3 = 0 + 3 = 3$
    The solution is $(0,3)$.   Consistent

33. $3x + y = 15$
    $x = (-1/3)y + 5$
    Substitute $(-1/3)y + 5$ for x in the 1st
    equation.
    $3(-1/3y+5) + y = 15$
    $-y + 15 + y = 15$
    $15 = 15$
    This statement is true for all values of x.
    The system is dependent.

35. $x + y = 4$

$3x - y = 8$

Add the equations to eliminate y.

$4x = 12 \quad x = 3$

Substitute 3 for x in either eq'n.

$3(3) + y = 8 \quad$ (solve for y)

$9 - y = 8 \quad y = 1$

The solution is (3, 1)   Consistent

37. $2x - y = -4$

$-3x - y = 6$

Multiply the second equation by -1,

$2x - y = -4$

$3x + y = -6 \quad$ add the equations to eliminate y

$5x = -10 \quad x = -2$

Substitute -2 in place of x in the first equation.

$2(-2) - y = -4$

$-4 - y = -4$

$-y = 0 \quad y = 0$

The solution is $(-2, 0)$.   Consistent

39. $4x + 3y = -1$

$2x - y = -13$

Multiply the second equation by -2,

$4x + 3y = -1$

$-4x + 2y = 26 \quad$ add the equations to eliminate x

$5y = 25 \quad y = 5$

Substitute 5 for y in the 2nd equation.

$2x - 5 = -13$

$2x = -8 \quad x = -4$

The solution is $(-4, 5)$.   Consistent

41. $2x + y = 11$

$x + 3y = 18$

Multiply the second equation by -2,

$2x + y = 11$

$-2x - 6y = -36 \quad$ add the equations to elim. x

$-5y = -25 \quad y = 5$

Substitute 5 for y in the 2nd equation.

$x + 3(5) = 18$

$x + 15 = 18 \quad x = 3$

The solution is (3, 5).   Consistent.

43. $3x - 5y = -6$

   $2x + 4y = 18$

   Multiply the first equation by 2, and the second equation by -3,

   $6x - 10y = -12$

   $-6x - 12y = -54$ add the equations to elim. x

   $-22y = -66$ $y = 3$

   Substitute 3 for y in the 1st equation.

   $3x - 5(3) = -6$

   $3x - 15 = -6$

   $3x = 9$ $x = 3$

   The solution is $(3, 3)$.   Consistent

45. $4x - 2y = 6$

   $4y = 8x - 12$ or $8x - 4y = 12$

   Multiply the first equation by (-2),

   $-8x + 4y = -12$

   $8x - 4y = -12$ add the equations to elim. y

   $0 = 0$ True

   This statement is true for all values of x.

   This system is dependent.

47. $2x - 4y = 4$

   $3x - 2y = 4$

   Multiply the second equation by -2.

   $2x - 4y = 4$

   $-6x + 4y = -8$ add the equations to elim. y

   $-4x = -4$ $x = 1$

   Substitute 1 for x in the first equation.

   $2(1) - 4y = 4$

   $2 - 4y = 4$

   $-4y = 2$ $y = -1/2$

   The solution is $(1, -1/2)$.   Consistent

49. a)
Let x equal the number of miles driven, C equal cost

Cost (U-Haul rentals): $C_U = 0.79x + 30$

Cost (Discount rentals): $C_D = 0.85x + 24$

b)

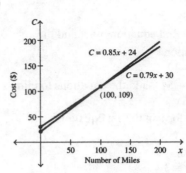

c) 100 miles

51. a) Let C = cost , R = revenue

$C(x) = 15x + 400$

$R(x) = 25x$

b)

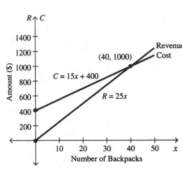

53. a) Let $C(x)$ = cost , $R(x)$ = revenue

$C(x) = 15x + 4050$

$R(x) = 40x$

b)

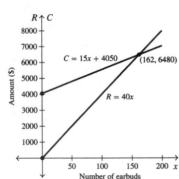

51. c) The cost and revenue graphs intersect when $x = 40$ so 40 is the number of backpacks Benjamin's must sell to break even.

d) $P = R(x) - C(x) = 25x - (15x + 400)$

$P = 10x - 400$

e) $P = 10(30) - 400 = 300 - 400 = -\$100$ (loss)

f) $1000 = 10x - 400$ → $10x = 1400$

$x = 140$ backpacks

53. c) The cost and revenue graphs intersect when $x = 162$ so 162 is the number of units the manufacturer must sell to break even.

d) $P = R(x) - C(x) = 40x - (15x + 4050)$

$P = 25x - 4050$

e) $P = 25(155) - 4050 = 3875 - 4050$

$= -\$175$ (loss)

f) $575 = 25x - 4050$ → $25x = 4625$

$x = 185$ units

55. Let x = # of shares of Under Armour stock
Let y = # of shares of Hershey Company stock.
$$x + y = 50$$
$$73x + 85y = 4070$$

$-73x - 73y = -3650$ Multiply top eqn by -73
$\underline{73x + 85y = 4070}$
$\qquad 12y = 420$

$\dfrac{12y}{12} = \dfrac{420}{12}$ so $y = 35$ shares of Hershey
company stock

$$x + y = 50$$
$$x + 35 = 50$$
$$x = 15 \text{ Shares of Under Armour}$$

59. a) Let s = weekly salary
x = weekly sales in dollars
Harbor Sales: $s = 249 + 0.15x$
Industrial Sales.: $s = 300 + 0.12x$ set eq'ns.
equal
$$249 + 0.15x = 300 + 0.12x$$
$$0.03x = 51 \qquad x = \$1700$$

Nathan's weekly sales volume would need to be \$1700
for the total income from both companies to be equal.

b) Harbor Sales: $s = 249 + 0.15(2500)$
$$s = 249 + 375$$
$$s = \$624$$

Industrial Sales: $s = 300 + 0.12(2500)$
$$s = 300 + 300$$
$$s = \$600$$
Harbor Sales would give the greater salary.

57. Let x = # of liters at 10%
$15 - x$ = # of liters at 40%

$$0.10x + 0.40(15 - x) = 0.25(15)$$
$$0.10x + 6 - 0.40x = 3.75$$
$$-0.30x = -2.25 \qquad x = 7.5$$
Substitute 7.5 for x in 2nd statement
$$15 - x = 15 - (7.5) = 7.5$$

7.5 liters of 10% solution and
7.5 liters of 40% solution

61. HD cost $= 2.65x + 468.75$
    HG cost $= 3.10x + 412.50$
    where x is the number of square feet installed

    a) $y = 2.65x + 468.75$ and $y = 3.10x + 412.50$

    $$2.65x + 468.75 = 3.10x + 412.50$$
    $$\underline{-2.65x - 412.50 \quad -2.65x - 412.50}$$
    $$56.25 \quad = \quad 0.45x$$

    $$\frac{6.25}{0.45} = \frac{0.45x}{0.45}$$
    $$125 = x$$

    The costs are equal when Roberto has
    125 square feet of flooring installed.

    b) 195 square feet is 70 square feet more than
    125 square feet. The costs are equal for
    125 square feet, and since Home Depot has
    the smaller cost per square foot, the
    additional 70 square feet will c0st less with
    Home Depot.

63. SG cost $= 200 + 50x$
    GG cost $= 375 + 25x$
    where x is the number of hours of labor

    $$200 + 50x = 375 + 25x$$
    $$25x = 175$$
    $$x = 7 \text{ hours}$$

65. a) Two lines with different slopes are not
    parallel, nor are they on the same line, and
    therefore have exactly one point of
    intersection giving one solution.

    b) Two lines with the same slope and
    different y-intercepts are distinct parallel
    lines so the system will have no solution.

    c) Two lines with the same slopes and
    y-intercepts have infinitely many solutions,
    each point on the line.

69. a) $(2) + (1) + (4) = 7$ $(2) - (1) + 2(4) = 9$
         $7 = 7$ $9 = 9$
        $-(2) + 2(1) + (4) = 4$
         $4 = 4$ $(2,1,4)$ is a solution.

    b) Add eq'ns. 1 and 2 to yield eq'n. 4
    Multiply eq'n. 2 by 2, then add eq'ns. 2 and
    3 to yield eq'n. 5. Combine eq'ns. 4 and 5 to
    find one variable. Substitute back into
    various equations to find the other 2
    variables.

67. $(1/u) + (2/v) = 8$
    $(3/u) - (1/v) = 3$

    Substitute x for $\dfrac{1}{u}$ and y for $\dfrac{1}{v}$.

    (1) $x + 2y = 8$
    (2) $3x - y = 3$
    Multiply eq'n. (2) by 2,
    $x + 2y = 8$
    $6x - 2y = 6$ add to eliminate y
    $7x = 14$ $x = 2$, thus $u = \frac{1}{2}$
    Substitute 2 for x in eq. (1).
    $(2) + 2y = 8$
    $2y = 6$ $y = 3$, thus $v = 1/3$

    Answer: $(1/2, 1/3)$

71. $y = 3x + 3$

    $(1/3)y = x + 1$

    If we multiply the 2nd eq'n. by 3, we get the eq'n.

    $y = 3x + 1$, the same as eq'n. # 1.

    2 lines that lie on top of one another have an infinite number of solutions.

73. a) – d) Answers will vary.

Exercise Set 6.8
1. Half-plane
3. Solid
5. Ordered
7. Feasible
9. Objective

11. a) $4x + 2y < 10$

        $4(2) + 2(1) < 10$

            $8 + 2 < 10$

                $10 < 10$ False

        No

   b) $4x + 2y < 10$

        $4(2) + 2(1) < 10$

            $8 + 2 < 10$

                $10 < 10$ False

        Yes

   c) $4x + 2y \geq 10$

        $4(2) + 2(1) \geq 10$

            $8 + 2 \geq 10$

                $10 \geq 10$ True

        Yes

   d) $4x + 2y > 10$

      $4(2) + 2(1) > 10$

         $8 + 2 > 10$

            $10 > 10$ False

        No

13. Graph $y = x + 1$. Since the original statement is greater than, a dashed line is drawn. Since the point (0, 0) does not satisfy the inequality $y > x + 1$, all points in the half-plane above the line $y = x + 1$ are in the solution set.

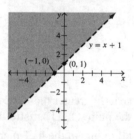

15. Graph $y = 2x - 6$. Since the original statement is greater than or equal to, a solid line is drawn. Since the point (0, 0) satisfies the inequality $y \geq 2x - 6$, all points on the line and in the half-plane above the line $y = 2x - 6$ are in the solution set.

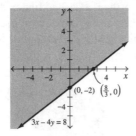

17. Graph $2x - 3y = 6$. Since the original statement is strictly greater than, a dashed line is drawn. Since the point (0, 0) does not satisfy the inequality $2x - 3y > 6$, all points in the half-plane below the line $2x - 3y = 6$ are in the solution set.

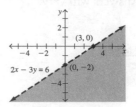

19. Graph $3x - 4y = 8$. Since the original statement is less than or equal to, a solid line is drawn. Since the point (0, 0) satisfies the inequality $3x - 4y \leq 8$, all points in the half-plane above the line $3x - 4y = 8$ are in the solution set.

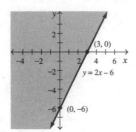

21. Graph $3x + 2y = 6$. Since the original statement is strictly less than, a dashed line is drawn. Since the point (0, 0) satisfies the inequality $3x + 2y < 6$, all points in the half-plane to the left of the line $3x + 2y = 6$ are in the solution set.

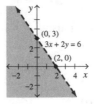

23. Graph $x + y = 0$. Since the original statement is strictly greater than, a dashed line is drawn. Since the point (1, 1) satisfies the inequality $x + y > 0$, all points in the half-plane above the line $x + y = 0$ are in the solution set.

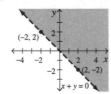

25. Graph $5x + 2y = 10$. Since the original statement is greater than or equal to, a solid line is drawn. Since the point (0, 0) does not satisfy the inequality $5x + 2y \geq 10$, all points on the line and in the half-plane above the line $5x + 2y = 10$ are in the solution set.

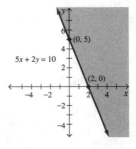

27. Graph $3x - 2y = 12$. Since the original
statement is strictly less than, a
dashed line is drawn. Since the
point (0, 0) satisfies the inequality
$3x - 2y < 12$, all points in the half-plane
above the line $3x - 2y = 12$ are
in the solution set.

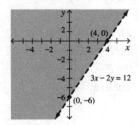

29.

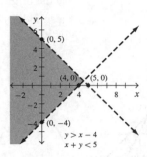

$y > x - 4$
$x + y < 5$

31.

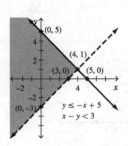

$y \le -x + 5$
$x - y < 3$

33.

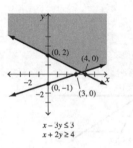

$x - 3y \le 3$
$x + 2y \ge 4$

35.

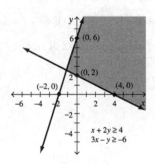

$x + 2y \ge 4$
$3x - y \ge -6$

37.

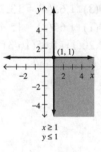

$x \ge 1$
$y \le 1$

39. At (0, 0), $K = 6(0) + 4(0) = 0$
At (0, 4), $K = 6(0) + 4(4) = 16$
At (2, 3), $K = 6(2) + 4(3) = 24$
At (5, 0), $K = 6(5) + 4(0) = 30$
The maximum value is 30 at (5, 0); the minimum
value is 0 at (0, 0).

41. At (10, 20), $K = 2(10) + 3(20) = 80$
At (10, 40), $K = 2(10) + 3(40) = 140$
At (50, 30), $K = 2(50) + 3(30) = 190$
At (50, 10), $K = 2(50) + 3(10) = 130$
At (20, 10), $K = 2(20) + 3(10) = 70$
The maximum value is 190
at (50, 30); the minimum value is 70
at (20, 10).

43. a)

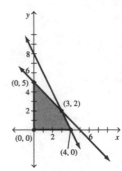

b) $x + y \le 5$ $2x + y \le 8$ $x \ge 0$ $y \ge 0$

$P = 4x + 3y$

At (0,0), $P = 4(0) + 3(0) = 0$ min. at (0, 0)

At (0,5), $P = 4(0) + 3(5) = 15$

At (3, 2), $P = 4(3) + 3(2) = 18$ max. at (3, 2)

At (4,0), $P = 4(4) + 3(0) = 16$

45. a)

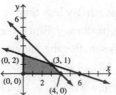

b) $P = 7x + 6y$

At (0,0), $P = 7(0) + 6(0)$

$= 0$ min. at (0, 0)

At (0,2), $P = 7(0) + 6(2) = 12$

At (3,1), $P = 7(3) + 6(1) = 27$

At (4,0), $P = 7(4) + 6(0)$

$= 28$ max. at (4, 0)

47. a)

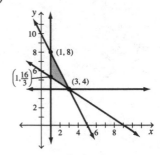

b) $P = 2.20x + 1.65y$

At (3, 4), $P = 2.20(3) + 1.65(4) = 13.2$

At (1, 8), $P = 2.20(1) + 1.65(8) = 15.4$

At (1, 16/3), $P = 2.20(1) + 1.65(16/3) = 11$

Maximum of 15.4 at (1, 8) and

Minimum of 11 at (1, 16/3)

49. a) $x + y < 500$, $x \ge 150$, $y \ge 150$

b)

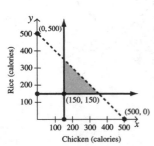

c) One example is (220, 220).

$$\frac{220}{180}(3) \approx 3.7, \quad \frac{220}{200}(8) = 8.8$$

3.7 oz of chicken, 8.8 oz of rice

51. a) Let x = number of skateboards
      y = number of in-line skates
      $$x + y \leq 20 \quad x \geq 3 \quad x \leq 6 \quad y \geq 2$$

 b) $P = 25x + 20y$

 c)

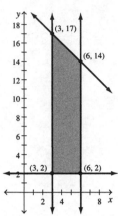

 d) (3,2) (3,17) (6,14) (6,2)

 e) At (3,2), $P = 25(3) + 20(2) = 115$
    At (3,17), $P = 25(3) + 20(17) = 415$
    At (6,14), $P = 25(6) + 20(14) = 430$
    At (6,2), $P = 25(6) + 20(2) = 190$
    Six skateboards and 14 pairs of in-line skates.

 f) Max. cost = $430

53. Let x = hours machine I operates
    y = hours machine II operates

 (a) $3x + 4y \geq 60 \quad x \geq 0$
     $10x + 5y \geq 100 \quad y \geq 0$

 (b) $C = 28x + 33y$

 (c)

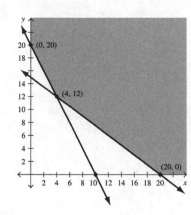

 d) (0, 20), (20, 0), (4, 12)

 e) At (0,20), $C = 28(0) + 33(20) = 660$
    At (20,0), $C = 28(20) + 33(0) = 560$
    At (4,12), $C = 28(4) + 33(12) = 508$
    4 hours on Mach. 1 and 12 hours on Mach. 2

 f) Min. profit = $ 508.00

55. Let x = pounds of all-beef hot dogs
    y = pounds of regular hot dogs
    $$x + (1/2)y \leq 200$$
    $$(1/2)y \leq 150 \quad x \geq 0 \quad y \geq 0$$

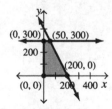

 $P = 0.30y + 0.40x$
 Maximum profit occurs at (50,300).
 Thus the manufacturer should make 50 lb.
 of the all-beef hot dogs and 300 lb. of the
 regular hot dogs for a profit of $110.

57. a) No, if the lines are parallel there may not be
       a solution to the system.

  b) Example: $y \geq x \quad y \leq x - 2$
     This system has no solution.

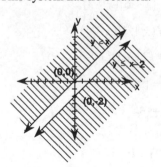

59. a)
$$2x - y < 8$$
$$2x - 2x - y < -2x + 8$$
$$-y < -2x + 8$$
$$\frac{-y}{-1} > \frac{-2x}{-1} + \frac{2}{-1}$$
$$y > 2x - 8$$

b)
$$-2x + y > -8$$
$$-2x + 2x + y > 2x - 8$$
$$y > 2x - 8$$

c)
$$2x - 4y < 16$$
$$2x - 2x - 4y < -2x + 16$$
$$-4y < -2x + 16$$
$$\frac{-4y}{-4} > \frac{-2x}{-4} + \frac{16}{-4}$$
$$y > \frac{1}{2}x - 4$$

d) $y > 2x - 8$

a, b, and d are equivalent to the same inequality.

Exercise Set 6.9

1. Binomial

3. FOIL

5. Quadratic

7. $x^2 + 5x + 6 = (x + 2)(x + 3)$

9. $x^2 - x - 6 = (x - 3)(x + 2)$

11. $x^2 + 3x - 10 = (x + 5)(x - 2)$

13. $x^2 - 2x - 3 = (x + 1)(x - 3)$

15. $x^2 - 9x + 18 = (x - 6)(x - 3)$

17. $x^2 - 16 = (x - 4)(x + 4)$

19. $x^2 + 3x - 28 = (x + 7)(x - 4)$

21. $x^2 - 2x - 63 = (x - 9)(x + 7)$

23. $2x^2 + 7x + 3 = (2x + 1)(x + 3)$

25. $3x^2 + x - 2 = (3x - 2)(x + 1)$

27. $5x^2 + 12x + 4 = (5x + 2)(x + 2)$

29. $4x^2 + 11x + 6 = (4x + 3)(x + 2)$

31. $4x^2 - 11x + 6 = (4x - 3)(x - 2)$

33. $8x^2 - 2x - 3 = (4x - 3)(2x + 1)$

35.
$$(x - 4)(x + 3) = 0$$
$$x - 4 = 0 \quad \text{or} \quad x + 3 = 0$$
$$x = 4 \qquad\qquad x = -3$$

37.
$$(3x + 4)(2x - 1) = 0$$
$$3x + 4 = 0 \quad \text{or} \quad 2x - 1 = 0$$
$$3x = -4 \qquad\qquad 2x = 1$$
$$x = -\frac{4}{3} \qquad\qquad x = \frac{1}{2}$$

39. $x^2 + 8x + 15 = 0$

    $(x+5)(x+3) = 0$

    $x + 5 = 0$ or $x + 3 = 0$

    $x = -5$ $x = -3$

41. $x^2 - 8x + 7 = 0$

    $(x-7)(x-1) = 0$

    $x - 7 = 0$ or $x - 1 = 0$

    $x = 7$ $x = 1$

43. $x^2 - 15 = 2x$

    $x^2 - 2x - 15 = 0$

    $(x-5)(x+3) = 0$

    $x - 5 = 0$ or $x + 3 = 0$

    $x = 5$ $x = -3$

45. $x^2 = 4x - 3$

    $x^2 - 4x + 3 = 0$

    $(x-3)(x-1) = 0$

    $x - 3 = 0$ or $x - 1 = 0$

    $x = 3$ $x = 1$

47. $x^2 - 81 = 0$

    $(x-9)(x+9) = 0$

    $x - 9 = 0$ or $x + 9 = 0$

    $x = 9$ $x = -9$

49. $x^2 + 5x - 36 = 0$

    $(x+9)(x-4) = 0$

    $x + 9 = 0$ or $x - 4 = 0$

    $x = -9$ $x = 4$

51. $3x^2 + 10x = 8$

    $3x^2 + 10x - 8 = 0$

    $(3x-2)(x+4) = 0$

    $3x - 2 = 0$ or $x + 4 = 0$

    $3x = 2$ $x = -4$

    $x = \dfrac{2}{3}$

53. $5x^2 + 11x = -2$

    $5x^2 + 11x + 2 = 0$

    $(5x+1)(x+2) = 0$

    $5x + 1 = 0$ or $x + 2 = 0$

    $5x = -1$ $x = -2$

    $x = -\dfrac{1}{5}$

55. $3x^2 = -5x - 2$

    $3x^2 + 5x + 2 = 0$

    $(3x+2)(x+1) = 0$

    $3x + 2 = 0$ or $x + 1 = 0$

    $3x = -2$ $x = -1$

    $x = -\dfrac{2}{3}$

57. $6x^2 - 19x + 3 = 0$

    $(6x-1)(x-3) = 0$

    $6x - 1 = 0$ or $x - 3 = 0$

    $6x = 1$ $x = 3$

    $x = \dfrac{1}{6}$

59. $x^2 + 2x - 3 = 0$

    $a = 1, \ b = 2, \ c = -3$

    $$x = \frac{-2 \pm \sqrt{(2)^2 - 4(1)(-3)}}{2(1)}$$

    $$x = \frac{-2 \pm \sqrt{4 + 12}}{2} = \frac{-2 \pm \sqrt{16}}{2} = \frac{-2 \pm 4}{2}$$

    $$x = \frac{2}{2} = 1 \text{ or } x = \frac{-6}{2} = -3$$

61. $x^2 - 3x - 18 = 0$

    $a = 1, \ b = -3, \ c = -18$

    $$x = \frac{-(-3) \pm \sqrt{(-3)^2 - 4(1)(-18)}}{2(1)}$$

    $$x = \frac{3 \pm \sqrt{9 + 72}}{2} = \frac{3 \pm \sqrt{81}}{2} = \frac{3 \pm 9}{2}$$

    $$x = \frac{12}{2} = 6 \text{ or } x = \frac{-6}{2} = -3$$

63. $x^2 - 4x + 2 = 0$
$a = 1, b = -4, c = 2$

$$x = \frac{-(-4) \pm \sqrt{(-4)^2 - 4(1)(2)}}{2(1)}$$

$$x = \frac{4 \pm \sqrt{16 - 8}}{2} = \frac{4 \pm \sqrt{8}}{2} = \frac{4 \pm 2\sqrt{2}}{2}$$

$$x = 2 \pm \sqrt{2}$$

65. $x^2 - 2x + 3 = 0$
$a = 1, b = -2, c = 3$

$$x = \frac{-(-2) \pm \sqrt{(-2)^2 - 4(1)(3)}}{2(1)}$$

$$x = \frac{2 \pm \sqrt{4 - 12}}{2} = \frac{2 \pm \sqrt{-8}}{2}$$

No real solution

67. $3x^2 + 9x + 5 = 0$
$a = 3, b = 9, c = 5$

$$x = \frac{-9 \pm \sqrt{(9)^2 - 4(3)(5)}}{2(3)}$$

$$x = \frac{-9 \pm \sqrt{81 - 60}}{6} = \frac{-9 \pm \sqrt{21}}{6}$$

69. $3x^2 - 8x + 1 = 0$
$a = 3, \ b = -8, \ c = 1$

$$x = \frac{-(-8) \pm \sqrt{(-8)^2 - 4(3)(1)}}{2(3)}$$

$$x = \frac{8 \pm \sqrt{64 - 12}}{6} = \frac{8 \pm \sqrt{52}}{6} = \frac{8 \pm 2\sqrt{13}}{6}$$

$$x = \frac{4 \pm \sqrt{13}}{3}$$

71. $4x^2 - 5x - 3 = 0$
$a = 4, b = -5, c = -3$

$$x = \frac{-(-5) \pm \sqrt{(-5)^2 - 4(4)(-3)}}{2(4)}$$

$$x = \frac{5 \pm \sqrt{25 + 48}}{8} = \frac{5 \pm \sqrt{73}}{8}$$

73. $2x^2 + 7x + 5 = 0$
$a = 2, b = 7, c = 5$

$$x = \frac{-7 \pm \sqrt{(7)^2 - 4(2)(5)}}{2(2)}$$

$$x = \frac{-7 \pm \sqrt{49 - 40}}{4} = \frac{-7 \pm \sqrt{9}}{4} = \frac{-7 \pm 3}{4}$$

$$x = \frac{-4}{4} = -1 \ \text{ or } \ x = \frac{-10}{4} = -\frac{5}{2}$$

75. $3x^2 - 10x + 7 = 0$
$a = 3, b = -10, c = 7$

$$x = \frac{-(-10) \pm \sqrt{(-10)^2 - 4(3)(7)}}{2(3)}$$

$$x = \frac{10 \pm \sqrt{100 - 84}}{6} = \frac{10 \pm \sqrt{16}}{6} = \frac{10 \pm 4}{6}$$

$$x = \frac{14}{6} = \frac{7}{3} \ \text{ or } \ x = \frac{6}{6} = 1$$

77. $4x^2 + 6x + 5 = 0$
$a = 4, b = 6, c = 5$

$$x = \frac{-6 \pm \sqrt{6^2 - 4(4)(5)}}{2(4)}$$

$$x = \frac{-6 \pm \sqrt{36 - 80}}{8} = \frac{6 \pm \sqrt{-44}}{8}$$

No real solution

79. Area of backyard $= lw = 30(20) = 600$ m^2

Let $x =$ width of grass around all sides of the flower garden

Width of flower garden $= 20 - 2x$

Length of flower garden $= 30 - 2x$

Area of flower garden $= lw = (30 - 2x)(20 - 2x)$

Area of grass $= 600 - (30 - 2x)(20 - 2x)$

$600 - (30 - 2x)(20 - 2x) = 336$

$600 - (600 - 100x + 4x^2) = 336$

$600 - 600 + 100x - 4x^2 = 336$

$-4x^2 + 100x = 336$

$4x^2 - 100x + 336 = 0$

$x^2 - 25x + 84 = 0$

$(x - 21)(x - 4) = 0$

$x - 21 = 0$ or $x - 4 = 0$

$x = 21$ $x = 4$

$x \neq 21$ since the width of the backyard is 20 m.

Width of grass $= 4$ m

Width of flower garden

$= 20 - 2x = 20 - 2(4) = 20 - 8 = 12$ m

Length of flower garden

$= 30 - 2x = 30 - 2(4) = 30 - 8 = 22$ m

81. $-16t^2 + 128t = 256$

$-16t^2 + 128t - 256 = 0$

$16(t - 4)(t - 4) = 0$

$x - 4 = 0$

$x = 4$

At 4 seconds, the ball will be 256 ft above ground.

83. a) Since the equation is equal to 6 and not 0, the zero-factor property cannot be used.

b) $(x - 4)(x - 7) = 6$

$x^2 - 11x + 28 = 6$

$x^2 - 11x + 22 = 0$

$a = 1, b = -11, c = 22$

$x = \dfrac{-(-11) \pm \sqrt{(-11)^2 - 4(1)(22)}}{2(1)}$

$x = \dfrac{11 \pm \sqrt{121 - 88}}{2} = \dfrac{11 \pm \sqrt{33}}{2}$

85. $(x + 1)(x - 3) = 0$

$x^2 - 2x - 3 = 0$, Other answers are possible.

Exercise Set 6.10

1. Relation

3. Domain

5. Upward

7. $x = -\dfrac{b}{2a}$

9. Function since each value of x is paired with a unique value of y.

   D: $x = -2, -1, 1, 2, 3$ R: $y = -1, 1, 2, 3$

11. Function since each vertical line intersects the graph at only one point.

    D: all real numbers   R: all real numbers

13. Not a function since $x = -1$ is not paired with a unique value of y.

15. Function since each vertical line intersects the graph at only one point.

    D: all real numbers   R: $y \geq -4$

17. Not a function since it is possible to draw a vertical line that intersects the graph at more than one point.

19. Function since each vertical line intersects the graph at only one point.

    D: all real numbers   R: all real numbers

21. Function since each vertical line intersects the graph at only one point.

    D: $0 \leq x \leq 10$ R: $-1 \leq y \leq 3$

23. Function since each value of x is paired with a unique value of y.

25. Not a function since $x = 2$ is paired with four different values of y.

27. Function since each value of x is paired with a unique value of y.

29. $f(x) = x + 3, \quad x = 5$

    $f(5) = 5 + 3 = 8$

31. $f(x) = -3x + 3, \ x = -1$

    $f(-1) = -3(-1) + 3 = 3 + 3 = 6$

33. $f(x) = x^2 - 3x + 1, \quad x = 4$

    $f(4) = (4)^2 - 3(4) + 1 = 16 - 12 + 1 = 5$

35. $f(x) = -x^2 - 2x + 1, \ x = 4$

    $f(4) = -(4)^2 - 2(4) + 1 = -16 - 8 + 1 = -23$

37. $f(x) = 4x^2 - 6x - 9, \ x = -3$

    $f(-3) = 4(-3)^2 - 6(-3) - 9 = 36 + 18 - 9 = 45$

39. $f(x) = -5x^2 + 3x - 9, \quad x = -1$

    $f(-1) = -5(-1)^2 + 3(-1) - 9 = -5 - 3 - 9 = -17$

41.

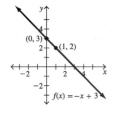

43.

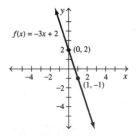

45.

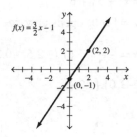

47. $y = x^2 - 1$

a) $a = 1 > 0$, opens upward

b) $x = 0$ c) $(0, -1)$ d) $(0, -1)$

e) $(1, 0), (-1, 0)$

f)

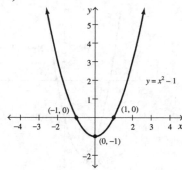

g) D: all real numbers   R: $y \geq -1$

49. $y = -x^2 + 4$

a) $a = -1 < 0$, opens downward

b) $x = 0$ c) $(0, 4)$ d) $(0, 4)$

e) $(-2, 0), (2, 0)$

f)

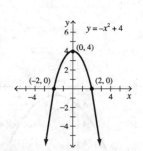

g) D: all real numbers   R: $y \leq 4$

51. $y = -2x^2 - 8$

a) $a = -2 < 0$, opens downward

b) $x = 0$ c) $(0, -8)$ d) $(0, -8)$

e) no x-intercepts

f)

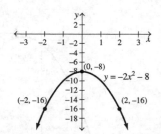

g) D: all real numbers   R: $y \leq -8$

53. $f(x) = x^2 + 4x - 4$

   a) $a = 1 > 0,$ opens upward

   b) $x = -2$ c) $(-2, -8)$ d) $(0, -4)$

   e) $(0.83, 0), (-4.83, 0)$

   f)

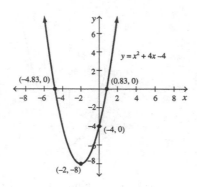

   g) D: all real numbers   R: $y \geq -8$

55. $y = x^2 + 5x + 6$

   a) $a = 1 > 0,$ opens upward

   b) $x = -\dfrac{5}{2}$ c) $(-2.5, -0.25)$ d) $(0, 6)$

   e) $(-3, 0), (-2, 0)$

   f)

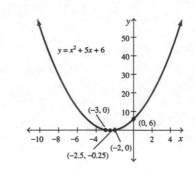

   g) D: all real numbers   R: $y \geq -0.25$

57. $y = -x^2 + 4x - 6$

   a) $a = -1 < 0,$ opens downward

   b) $x = 2$ c) $(2, -2)$ d) $(0, -6)$

   e) no x-intercepts

   f)

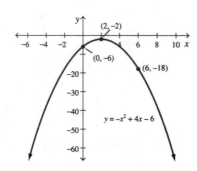

   g) D: all real numbers   R: $y \leq -2$

59. $y = -2x^2 + 3x - 2$

   a) $a = -2 < 0,$ opens downward

   b) $x = \dfrac{3}{4}$ c) $\left(\dfrac{3}{4}, -\dfrac{7}{8}\right)$ d) $(0, -2)$

   e) None

   f)

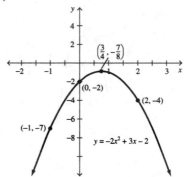

   g) D: all real numbers   R: $y \leq -\dfrac{7}{8}$

61.

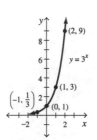

   D: all real numbers   R: $y > 0$

63.

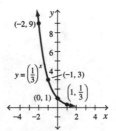

   D: all real numbers   R: $y > 0$

65.

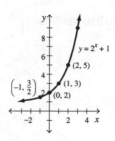

D: all real numbers   R: $y > 1$

67.

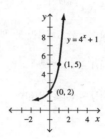

D: all real numbers   R: $y > 1$

69.

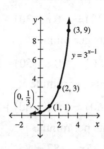

D: all real numbers   R: $y > 0$

71. $p(x) = 3.5x - 15,000$

$$p(20,000) = 3.5(20,000) - 15,000$$
$$= 70,000 - 15,000 = \$55,000$$

73. a) $2014 \to x = 94$

$$f(94) = 0.004(94)^2 - 0.34(94) + 13.79$$
$$\approx 17.17 \approx 17\%$$

b) 1970

c) $x = -\dfrac{b}{2a} = -\dfrac{-0.34}{2(0.004)} \approx 43$

$$f(43) = 0.004(43)^2 - 0.34(43) + 13.79$$
$$\approx 6.57 \approx 7\%$$

75. $y = 2000(3)^x$

$x = 5, \ y = 2000(3)^5$

$= 2000(243) = 486,000$ bacteria

77. $P(x) = 4000(1.3)^{0.1x}$

a) $x = 10, \ P(10) = 4000(1.3)^{0.1(10)}$

$= 4000(1.3) = 5200$ people

b) $x = 50, \ P(50) = 4000(1.3)^{0.1(50)}$

$= 4000(3.71293)$

$= 14,851.72 \approx 14,852$ people

79. $P(t) = 1.26e^{0.0075t}$

$P(5) = 1.26e^{0.0075(5)}$

$P(5) = 1.26e^{0.0375}$

$P(5) \approx 1.31$ billion people

81. $R(t) = R_0 e^{-0.000428t}$

   $R(1000) = 10e^{-0.000428(1000)}$

   $R(1000) = 10e^{-0.428}$

   $R(1000) \approx 6.52$ grams

83. a) Yes

   b) $\approx \$1010$ million

85. $d = (21.9)(2)^{(20-x)/12}$

   a) $x = 19, d = (21.9)(2)^{(20-19)/12}$

   $= (21.9)(1.059463094) \approx 23.2$ cm

   b) $x = 4, d = (21.9)(2)^{(20-4)/12}$

   $= (21.9)(2.519842099) \approx 55.2$ cm

   c) $x = 0, d = (21.9)(2)^{(20-0)/12}$

   $= (21.9)(3.174802104) \approx 69.5$ cm

87. $f(x) = -0.85x + 187$

   a) $f(20) = -0.85(20) + 187 = 170$
   beats per minute

   b) $f(30) = -0.85(30) + 187 = 161.5 \approx 162$
   beats per minute

   c) $f(50) = -0.85(50) + 187 = 144.5 \approx 145$
   beats per minute

   d) $f(60) = -0.85(60) + 187 = 136$
   beats per minute

   e) $-0.85x + 187 = 85$

   $-0.85x = -102$

   $x = 120$ years of age

Review Exercises

1. $x = 3,\ x^2 + 10x = (3)^2 + 10(3) = 9 + 30 = 39$

2. $x = -2,$
   $-x^2 - 5 = -(-2)^2 - 5 = -4 - 5 = -9$

3. $x = 2,\ 3x^2 - 2x + 7 = 3(2)^2 - 2(2) + 7$
   $= 12 - 4 + 7 = 15$

4. $x = -1,\ 4x^3 - 7x^2 + 3x + 1$
   $= 4(-1)^3 - 7(-1)^2 + 3(-1) + 1$
   $= -4 - 7 - 3 + 1 = -13$

5. $3x - 9 + x + 6 = 4x - 3$

6. $2x + 3(x - 2) + 7x$
   $2x + 3x - 6 + 7x$
   $12x - 6$

7. $4t + 8 = -12$
   $4t + 8 - 8 = -12 - 8$
   $4t = -20$
   $\dfrac{4t}{4} = \dfrac{-20}{4}$
   $t = -5$

8. $4(x - 2) = 3 + 5(x + 4)$
   $4x - 8 = 3 + 5x + 20$
   $4x - 8 = 5x + 23$
   $4x - 4x - 8 = 5x - 4x + 23$
   $-8 = x + 23$
   $-8 - 23 = x + 23 - 23$
   $-31 = x$

9.
$$\frac{x+5}{2} = \frac{x-3}{4}$$
$$4(x+5) = 2(x-3)$$
$$4x+20 = 2x-6$$
$$4x-2x+20 = 2x-2x-6$$
$$2x+20 = -6$$
$$2x+20-20 = -6-20$$
$$2x = -26$$
$$\frac{2x}{2} = \frac{-26}{2}$$
$$x = -13$$

10.
$$\frac{x+2}{5} = \frac{x+1}{4}$$
$$4(x+2) = 5(x+1)$$
$$4x+8 = 5x+5$$
$$4x-4x+8 = 5x-4x+5$$
$$8 = x+5$$
$$8 = x+5-5$$
$$x = 3$$

11. $F = ma$
$$F = (40)(5) = 200$$

12. $V = lwh$
$$V = (10)(7)(3) = 210$$

13.
$$z = \frac{\overline{x} - \mu}{\frac{\sigma}{\sqrt{n}}}$$
$$2 = \frac{\overline{x} - 100}{\frac{3}{\sqrt{16}}}$$
$$\frac{2}{1} = \frac{\overline{x} - 100}{\frac{3}{4}}$$
$$2\left(\frac{3}{4}\right) = 1(\overline{x} - 100)$$
$$\frac{3}{2} = \overline{x} - 100$$
$$\frac{3}{2} + 100 = \overline{x} - 100 + 100$$
$$\frac{3}{2} + \frac{200}{2} = \overline{x}$$
$$\frac{203}{2} = \overline{x}$$
$$101.5 = \overline{x}$$

14. $E = mc^2$
$$400 = m(4)^2$$
$$400 = 16m$$
$$\frac{400}{16} = \frac{16m}{16}$$
$$25 = m$$

15. $8x - 4y = 16$

$8x - 8x - 4y = -8x + 16$

$-4y = -8x + 16$

$\dfrac{-4y}{-4} = \dfrac{-8x}{-4} + \dfrac{16}{-4}$

$y = 2x - 4$

16. $2x + 9y = 17$

$2x - 2x + 9y = -2x + 17$

$9y = -2x + 17$

$\dfrac{9y}{9} = -\dfrac{2x}{9} + \dfrac{17}{9}$

17. $A = lw$

$\dfrac{A}{l} = \dfrac{lw}{l}$

$\dfrac{A}{l} = w$

18. $L = 2(wh + lh)$

$L = 2wh + 2lh$

$L - 2wh = 2wh - 2wh + 2lh$

$L - 2wh = 2lh$

$\dfrac{L - 2wh}{2h} = \dfrac{2lh}{2h}$

$\dfrac{L - 2wh}{2h} = l$ or $l = \dfrac{L}{2h} - \dfrac{2wh}{2h} = \dfrac{L}{2h} - w$

19. $5 + 3x$

20. $\dfrac{9}{q} - 15$

21.   Let $x =$ the number

$3 + 7x = 17$

$3 - 3 + 7x = 17 - 3$

$7x = 14$

$\dfrac{7x}{7} = \dfrac{14}{7}$

$x = 2$

22. Let $x =$ the number

$3x =$ the product of 3 and the number

$3x + 8 =$ the product of 3 and the number
          increased by 8

$x - 6 = 6$ less than the number

$3x + 8 = x - 6$

$3x - x + 8 = x - x - 6$

$2x + 8 = -6$

$2x + 8 - 8 = -6 - 8$

$2x = -14$

$\dfrac{2x}{2} = \dfrac{-14}{2}$

$x = -7$

23. Let $x =$ the number

$x - 4 =$ the difference of the number and 4

$5(x - 4) = 5$ times the difference of the number

and 4

$5(x - 4) = 45$

$5x - 20 = 45$

$5x - 20 + 20 = 45 + 20$

$5x = 65$

$\dfrac{5x}{5} = \dfrac{65}{5}$

$x = 13$

24. Let $x =$ the number

$10x = 10$ times the number

$10x + 14 = 14$ more than 10 times the number

$x + 12 =$ the sum of the number and 12

$8(x + 12) = 8$ times the sum of the number and 12

$10x + 14 = 8(x + 12)$

$10x + 14 = 8x + 96$

$10x - 8x + 14 = 8x - 8x + 96$

$2x + 14 = 96$

$2x + 14 - 14 = 96 - 14$

$2x = 82$

$\dfrac{2x}{2} = \dfrac{82}{2}$

$x = 41$

25. Let $x =$ the amount invested in bonds

$2x =$ the amount invested in mutual funds

$x + 2x = 15,000$

$3x = 15,000$

$\dfrac{3x}{3} = \dfrac{15,000}{3}$

$x = \$5000$ in bonds

$2x = 2(5000) = \$10,000$ in mutual funds

26. Let $x =$ profit at restaurant B

$x + 15,000 =$ profit at restaurant A

$x + (x + 15,000) = 75,000$

$2x + 15,000 = 75,000$

$2x + 15,000 - 15,000 = 75,000 - 15,000$

$2x = 60,000$

$\dfrac{2x}{2} = \dfrac{60,000}{2}$

$x = \$30,000$ for restaurant B

$x + 15,000 = 30,000 + 15,000 = \$45,000$ for restaurant A

27. $\dfrac{2}{\frac{1}{3}} = \dfrac{3}{x}$

$2x = 3\left(\dfrac{1}{3}\right)$

$2x = 1$

$\dfrac{2x}{2} = \dfrac{1}{2}$

$x = \dfrac{1}{2}$ cup

28. 1 hr 40 min $= 60$ min $+ 40$ min $= 100$ min

$\dfrac{120}{100} = \dfrac{300}{x}$

$120x = 100(300)$

$120x = 30{,}000$

$\dfrac{120x}{120} = \dfrac{30{,}000}{120}$

$x = 250$ min, or 4 hr 10 min

29. $s = kt$

$60 = k(10)$

$k = \dfrac{60}{10} = 6$

$s = 6t$

$s = 6(12)$

$s = 72$

30. $J = \dfrac{k}{A^2}$

$25 = \dfrac{k}{2^2}$

$k = 2^2(25)$

$k = 100$

$J = \dfrac{100}{A^2}$

$J = \dfrac{100}{5^2}$

$J = \dfrac{100}{25} = 4$

31. $W = \dfrac{kL}{A}$

$80 = \dfrac{k(100)}{20}$

$100k = 1600$

$\dfrac{100k}{100} = \dfrac{1600}{100}$

$k = 16$

$W = \dfrac{16L}{A}$

$W = \dfrac{16(50)}{40} = \dfrac{800}{40} = 20$

32. $z = \dfrac{kxy}{r^2}$

$12 = \dfrac{k(20)(8)}{(8)^2}$

$160k = 768$

$\dfrac{160k}{160} = \dfrac{768}{160}$

$k = 4.8$

$z = \dfrac{4.8xy}{r^2}$

$z = \dfrac{4.8(10)(80)}{(3)^2} = \dfrac{3840}{9} = 426.\overline{6} \approx 426.7$

33. $t = kv$

$\qquad 2325 = k\,(155{,}000)$

$\qquad k = \dfrac{2325}{155{,}000}$

$\qquad k = 0.015$

$\qquad t = 0.015v$

$\qquad t = 0.015\,(210{,}000)$

$\qquad t = \$3150$

34. $\dfrac{1\ kWh}{\$0.162} = \dfrac{740\ kWh}{x}$

$\qquad x = \$119.88$

35. $\qquad 8 + 9x \le 6x - 4$

$\qquad 8 - 8 + 9x \le 6x - 4 - 8$

$\qquad 9x \le 6x - 12$

$\qquad 9x - 6x \le 6x - 6x - 12$

$\qquad 3x \le -12$

$\qquad \dfrac{3x}{3} \le \dfrac{-12}{3}$

$\qquad x \le -4$

36. $\qquad 3(x + 9) \le 4x + 11$

$\qquad 3x + 27 \le 4x + 11$

$\qquad 3x - 4x + 27 \le 4x - 4x + 11$

$\qquad -x + 27 \le 11$

$\qquad -x + 27 - 27 \le 11 - 27$

$\qquad -x \le -16$

$\qquad \dfrac{-x}{-1} \ge \dfrac{-16}{-1}$

$\qquad x \ge 16$

37. $\qquad 5x + 13 \ge -22$

$\qquad 5x + 13 - 13 \ge -22 - 13$

$\qquad 5x \ge -35$

$\qquad \dfrac{5x}{5} \ge \dfrac{-35}{5}$

$\qquad x \ge -7$

38. $\qquad -8 \le x + 2 \le 7$

$\qquad -8 - 2 \le x + 2 - 2 \le 7 - 2$

$\qquad -10 \le x \le 5$

39. – 42.

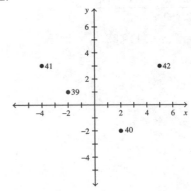

43.

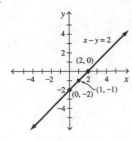

44.
$$3x + 8 \geq 7x + 12$$
$$3x - 7x + 8 \geq 7x - 7x + 12$$
$$-4x + 8 \geq 12$$
$$-4x + 8 - 8 \geq 12 - 8$$
$$-4x \geq 4$$
$$\frac{-4x}{-4} \leq \frac{4}{-4}$$
$$x \leq -1$$

45.

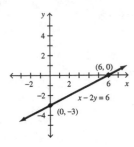

46.

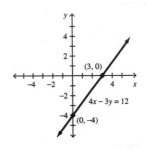

47. $m = \dfrac{7-3}{4-1} = \dfrac{4}{3}$

48. $m = \dfrac{-4-(-1)}{5-3} = \dfrac{-4+1}{5-3} = -\dfrac{3}{2}$

49. $m = \dfrac{3-(-4)}{2-(-1)} = \dfrac{3+4}{2+1} = \dfrac{7}{3}$

50. $m = \dfrac{-2-2}{6-6} = \dfrac{-4}{0}$ Undefined

51.

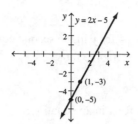

52.

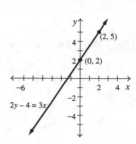

53. The y-intercept is 4, thus $b = 4$. Since the graph rises from left to right, the slope is positive. The change in y is 4 units while the change in x is 2. Thus, m, the slope is $\frac{4}{2}$ or 2. The equation is $y = 2x + 4$.

54. The y-intercept is 1, thus $b = 1$. Since the graph falls from left to right, the slope is negative. The change in y is 3 units while the change in x is 3. Thus, m, the slope is $\frac{-3}{3}$ or -1. The equation is $y = -x + 1$.

55. a)

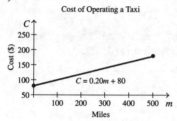

b) $C = 0.20(150) + 80$
$C = 30 + 80 = \$110$

c) $104 = 0.20m + 80$
$104 - 80 = 0.20m$
$24 = 0.20m$
$\dfrac{24}{0.20} = \dfrac{0.20m}{0.20}$
$120 = m;\ 120$ miles

56.

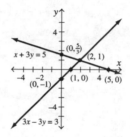

57.

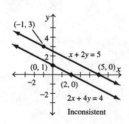

58. (1) $-x + y = -2$
    (2) $\underline{x + 2y = 5}$ (add)
        $3y = 3 \qquad\qquad y = 1$
    Substitute 1 in place of y in the first equation.
    $-x + 1 = -2$
    $-x = -3 \qquad x = 3$
    The solution is (3, 1).

59. $x - 2y = 9$
   $y = 2x - 3$
   Substitute $(2x - 3)$ in place of y in the 1st
   equation.
   $x - 2(2x - 3) = -9$ (solve for x)
   $x - 4x + 6 = -9$
   $-3x + 6 = 9$
   $-3x = 3$ $x = -1$
   Substitute (-1) in place of x in the 2nd equation.
   $y = 2(-1) - 3 = -2 - 3 = -5$
   The solution is $(-1, -5)$.

60. $2x - y = 4$ $y = 2x - 4$
   $3x - y = 2$
   Substitute $2x - 4$ for y in the second equation.
   $3x - (2x - 4) = 2$ (solve for x)
   $3x - 2x + 4 = 2$
   $x + 4 = 2$ $x = -2$
   Substitute -2 for x in an equation.
   $2(-2) - y = 4$
   $-4 - y = 4$ $y = -8$
   The solution is $(-2, -8)$.

61. $3x + y = 1$ $y = -3x + 1$
   $3y = -9x - 4$
   Substitute $-3x + 1$ for y in the second equation.
   $3(-3x + 1) = -9x - 4$ (solve for x)
   $-9x + 3 = -9x - 4$
   $3 \neq 4$ False   There is no solution to this
   system.
   The system is inconsistent.

62. (1) $x + y = 2$
   (2) $x + 3y = -2$
   Multiply the first equation by -1.
   $-x - y = -2$
   $\underline{x + 3y = -2}$ (add)
   $2y = -4$ $y = -2$
   Substitute (-2) for y in equation (2).
   $x + 3(-2) = -2$
   $x - 6 = -2$ $x = 4$
   The solution is $(4, -2)$.

63. (1) $4x - 8y = 16$
   (2) $x - 2y = 4$ $x = 2y + 4$
   Substitute $2y + 4$ for x in the first equation.
   $4(2y + 4) - 8y = 16$
   $8y + 16 - 8y = 16$
   $16 = 16$ True
   There are an infinite number of solutions.
   The system is dependent.

64. (1) $3x - 4y = 10$
   (2) $5x + 3y = 7$
   Multiply the first equation by 3, and the 2nd
   equation by 4.
   $9x - 12y = 30$
   $\underline{20x + 12y = 28}$ (add)
   $29x = 58$ $x = 2$
   Substitute 2 for x in the second equation.
   $5(2) + 3y = 7$
   $3y = -3$ $y = -1$
   The solution is $(2, -1)$.

65. (1) $3x + 4y = 6$
(2) $2x - 3y = 4$
Multiply the first equation by 2, and the second equation by -3.
$6x + 8y = 12$
$\underline{-6x + 9y = -12}$ (add)
$\qquad 17y = 0 \qquad y = 0$
Substitute 0 for y in the first equation.
$3x + 4(0) = 6$
$3x = 6 \qquad\qquad x = 2$
The solution is (2,0).

66. Let x = amount borrowed at 3%
$\quad y$ = amount borrowed at 6%
$\quad 0.03x + 0.06y = 16500 \qquad x + y = 400000$

$$\begin{bmatrix} .03 & .06 & | & 16500 \\ 1 & 1 & | & 400000 \end{bmatrix} \overset{(r_1 \cdot 100/3)}{=} \begin{bmatrix} 1 & 2 & | & 550000 \\ 1 & 1 & | & 400000 \end{bmatrix}$$

$$\overset{=}{(r_2 - r_1)} \begin{bmatrix} 1 & 2 & | & 550000 \\ 0 & -1 & | & -150000 \end{bmatrix}$$

$$\overset{=}{(-1 \cdot r_2)} \begin{bmatrix} 1 & 2 & | & 550000 \\ 0 & 1 & | & 150000 \end{bmatrix} \overset{(r_1 - 2r_2)}{=} \begin{bmatrix} 1 & 0 & | & 250000 \\ 0 & 1 & | & 150000 \end{bmatrix}$$

$250,000 borrowed at 3% and $150,000 borrowed at 6%

67. Let s = liters of 80% acid solution
$\quad w$ = liters of 50% acid solution
$s + w = 100$
$0.80s + 0.50w = 100(0.75)$
$0.80s + 0.50w = 75$
$s = 100 - w$
$0.80(100 - w) + 0.50w = 75$
$80 - 0.80w + 0.50w = 75$
$-0.30w = -5$
$w = -5/(-0.30) = 16\,\dfrac{2}{3}$ liters
$s = 100 - 16\,2/3 = 83\,\dfrac{1}{3}$ liters

Mix $83\,\dfrac{1}{3}$ liters of 80% solution with $16\,\dfrac{2}{3}$ liters of 50% solution.

68. Let c = total cost
$\quad x$ = no. of months to operate
a) model 1600A: $c_A = 950 + 32x$
$\quad$ model 6070B: $c_B = 1275 + 22x$
$\quad 950 + 32x = 1275 + 22x$
$\quad 10x = 325 \qquad x = 32.5$ months
After 32.5 months of operation the total cost of the units will be equal.
b) After 32.5 months or 2.7 years, the most cost effective unit is the unit with the lower per month cost to operate. Thus, model 6070B is the better deal in the long run.

69. a) Let C = total cost for parking
$\quad x$ = number of additional hours
All-Day: $C = 5 + 0.50x$
Sav-A-Lot: $C = 4.25 + 0.75x$
$5 + 0.50x = 4.25 + 0.75x$
$0.75 = 0.25x \qquad 3 = x$
The total cost will be the same after 3 additional hours or 4 hours total.
b) After 5 hours or $x = 4$ additional hours:
All-Day: $C = 5 + 0.50(4) = \$7.00$
Sav-A-Lot: $C = 4.25 + 0.75(4) = \$7.25$
All-Day would be less expensive.

70. Graph $2x + 3y = 12$. Since the original inequality is less than or equal to, a solid line is drawn. Since the point $(0, 0)$ satisfies the inequality $2x + 3y \le 12$, all points on the line and in the half-plane below the line $2x + 3y = 12$ are in the solution set.

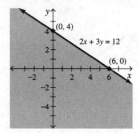

71. Graph $4x + 2y = 12$. Since the original inequality is greater than or equal to, a solid line is drawn. Since the point $(0, 0)$ does not satisfy the inequality $4x + 2y = 12$, all points on the line and in the half plane above the line $4x + 2y = 12$ are in the solution set.

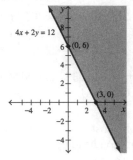

72.

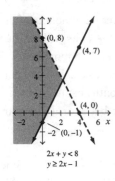

$2x + y < 8$
$y \geq 2x - 1$

73.

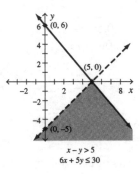

$x - y > 5$
$6x + 5y \leq 30$

74. a)

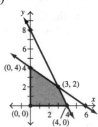

b) $P = 5x + 3y$
   At $(0, 0)$ $P = 5(0) + 3(0) = 0$
   At $(0, 4)$, $P = 5(0) + 3(4) = 12$
   At $(3, 2)$, $P = 5(3) + 3(2) = 21$
   At $(4, 0)$, $P = 5(4) + 3(0) = 20$
   The maximum is 21 at $(3, 2)$ and the minimum is 0 at $(0, 0)$.

75. $x^2 + 6x + 9 = (x + 3)(x + 3)$

76. $x^2 + 2x - 15 = (x + 5)(x - 3)$

77. $x^2 - 10x + 24 = (x-6)(x-4)$

78. $6x^2 + 7x - 3 = (3x-1)(2x+3)$

79. $x^2 + 9x + 20 = 0$

$(x+4)(x+5) = 0$

$x+4 = 0$ or $x+5 = 0$

$x = -4$ $x = -5$

80. $x^2 + 3x = 10$

$x^2 + 3x - 10 = 0$

$(x+5)(x-2) = 0$

$x+5 = 0$ or $x-2 = 0$

$x = -5$ $x = 2$

81. $3x^2 - 17x + 10 = 0$

$(3x-2)(x-5) = 0$

$3x - 2 = 0$ or $x-5 = 0$

$3x = 2$ $x = 5$

$x = \dfrac{2}{3}$

82. $3x^2 = -7x - 2$

$3x^2 + 7x + 2 = 0$

$(x+2)(3x+1) = 0$

$x+2 = 0$ or $3x+1 = 0$

$x = -2$ $3x = -1$

$x = -\dfrac{1}{3}$

83. $x^2 - 6x - 16 = 0$

$a = 1, b = -6, c = -6$

$x = \dfrac{-(-6) \pm \sqrt{(-6)^2 - 4(1)(-16)}}{2(1)}$

$x = \dfrac{6 \pm \sqrt{36+64}}{2} = \dfrac{6 \pm \sqrt{100}}{2} = \dfrac{6 \pm 10}{2}$

$x = \dfrac{16}{2} = 8$ or $x = \dfrac{-4}{2} = -2$

84. $2x^2 - x - 3 = 0$

$a = 2, b = -1, c = -3$

$x = \dfrac{-(-1) \pm \sqrt{(-1)^2 - 4(2)(-3)}}{2(2)}$

$x = \dfrac{1 \pm \sqrt{1+24}}{4} = \dfrac{1 \pm \sqrt{25}}{4} = \dfrac{1 \pm 5}{4}$

$x = \dfrac{6}{4} = \dfrac{3}{2}$ or $x = \dfrac{-4}{4} = -1$

85. $2x^2 - 3x + 4 = 0$

$a = 2, b = -3, c = 4$

$x = \dfrac{-(-3) \pm \sqrt{(-3)^2 - 4(2)(4)}}{2(2)}$

$x = \dfrac{3 \pm \sqrt{9-32}}{4} = \dfrac{3 \pm \sqrt{-23}}{4}$

No real solution

86. $x^2 - 3x - 2 = 0$

$a = 1, b = -3, c = -2$

$x = \dfrac{-(-3) \pm \sqrt{(-3)^2 - 4(1)(-2)}}{2(1)}$

$x = \dfrac{3 \pm \sqrt{9+8}}{2} = \dfrac{3 \pm \sqrt{17}}{2}$

87.   Function since each value of x is paired with a unique value of y.

D: $x = -2, -1, 2, 3$ R: $y = -1, 0, 2$

88.   Not a function since it is possible to draw a vertical line that intersects the graph at more than one point.

89.   Not a function since it is possible to draw a vertical line that intersects the graph at more than one point.

90.   Function since each vertical line intersects the graph at only one point.

D: all real numbers   R: all real numbers

91. $f(x) = 4x + 3, \quad x = 4$

$f(4) = 4(4) + 3 = 16 + 3 = 19$

92. $f(x) = -2x + 5, \quad x = -3$

$f(-3) = -2(-3) + 5 = 6 + 5 = 11$

93   . $f(x) = 3x^2 - 2x + 1, \ x = 5$

$f(5) = 3(5)^2 - 2(5) + 1 = 75 - 10 + 1 = 66$

94. $f(x) = -4x^2 + 7x + 9, \ x = -1$

95. $y = -x^2 - 4x + 21$

a) $a = -1 < 0$, opens downward

b) $x = -2$ c) $(-2, 25)$ d) $(0, 21)$

e) $(-7, 0), (3, 0)$

f)

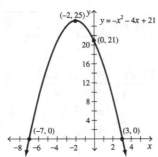

g)   D: all real numbers   R: $y \leq 25$

96. $f(x) = 2x^2 + 8x + 6$

a) $a = 2 > 0$, opens upward

b) $x = -\dfrac{8}{2(2)} = -2$ c) $(-2, -2)$

d) $(0, 6)$

e) $(-3, 0), (-1, 0)$

f)

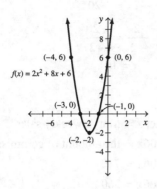

g) D: all real numbers   R: $y \geq -2$

97.

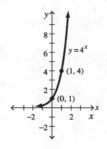

D: all real numbers   R: $y > 0$

98.

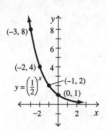

D: all real numbers   R: $y > 0$

99. $m = 30 - 0.002n^2$, $n = 60$

$m = 30 - 0.002(60)^2 = 30 - 0.002(3600)$

$- 30 - 7.2 - 22.8$ mpg

100. $P = 100(0.92)^x$, $x = 4.5$

$P = 100(0.92)^{4.5}$

$= 100(0.6871399881) = 68.71399881$

$\approx 68.7\%$

Chapter Test

1. $3x^2 + 6x - 1,\quad x = -2$

$3(-2)^2 + 6(-2) - 1 = 12 - 12 - 1 = -1$

2. $\quad 4x + 6 = 2(3x - 7)$

$\quad 4x + 6 = 6x - 14$

$\quad 4x - 6x + 6 = 6x - 6x - 14$

$\quad -2x + 6 = -14$

$\quad -2x + 6 - 6 = -14 - 6$

$\quad -2x = -20$

$\quad \dfrac{-2x}{-2} = \dfrac{-20}{-2}$

$\quad x = 10$

3. $-2(x - 3) + 6x = 2x + 3(x - 4)$

$\quad -2x + 6 + 6x = 2x + 3x - 12$

$\quad 4x + 6 = 5x - 12$

$\quad 4x - 5x + 6 = 5x - 5x - 12$

$\quad -x + 6 = -12$

$\quad -x + 6 - 6 = -12 - 6$

$\quad -x = -18$

$\quad \dfrac{-x}{-1} = \dfrac{-18}{-1}$

$\quad x = 18$

4. $\quad$ Let $x =$ Mary's weekly sales

$\quad 0.06x =$ the amount of commission

$\quad 350 + 0.06x = 710$

$\quad 0.06x = 360$

$\quad \dfrac{0.06x}{0.06} = \dfrac{360}{0.06}$

$\quad x = \$6000$

5. $L = ah + bh + ch;\ a = 2,\ b = 5,\ c = 4,\ h = 7$

$\quad L = 2(7) + 5(7) + 4(7)$

$\quad = 14 + 35 + 28 = 77$

6. $\quad 3x + 5y = 11$

$\quad 3x - 3x + 5y = -3x + 11$

$\quad 5y = -3x + 11$

$\quad \dfrac{5y}{5} = \dfrac{-3x + 11}{5}$

$\quad y = \dfrac{-3x + 11}{5} = -\dfrac{3}{5}x + \dfrac{11}{5}$

7. $l = \dfrac{k}{w}$

$15 = \dfrac{k}{9}$

$k = 15(9) = 135$

$l = \dfrac{135}{w}$

$l = \dfrac{135}{20} = 6.75 \text{ ft}$

8. $\quad -5x + 14 \le 2x + 35$

$\quad -5x - 2x + 14 \le 2x - 2x + 35$

$\quad -7x + 14 \le 35$

$\quad -7x + 14 - 14 \le 35 - 14$

$\quad -7x \le 21$

$\quad \dfrac{-7x}{-7} \ge \dfrac{21}{-7}$

$\quad x \ge -3$

9. $m = \dfrac{14-8}{1-(-2)} = \dfrac{6}{3} = 2$

10.

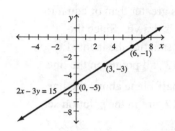

11.

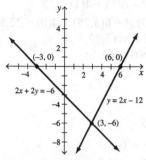

The solution is (3, –6).

12. $x + y = -1$ $x = -y - 1$
    $2x + 3y = -5$
    Substitute $(-y - 1)$ for x in the second equation.
    $2(-y - 1) + 3y = -5$ (solve for y)
    $-2y - 2 + 3y = -5$
    $y = -3$
    Substitute (-3) for y in the equation $x = -y - 1$.
    $x = -(-3) - 1 = 2$ The solution is (2, –3).

13. $4x + 3y = 5$
     $2x + 4y = 10$
     Multiply the second equation by (-2).
     $4x + 3y = 5$
     $\underline{-4x - 8y = -20}$ (add)
     $-5y = -15$ $y = 3$
     Substitute 3 for y in the first equation.
     $4x + 3(3) = 5$
     $4x + 9 = 5$
     $4x = -4$ $x = -1$
     The solution is (–1,3).

14. Let x = daily fee
        y = mileage charge
     $3x + 150y = 132$
     $2x + 400y = 142$ $x = 71 - 200y$
     Substitute $(71 - 200y)$ for x in the 1st equation.
     $3(71 - 200y) + 150y = 132$
     $213 - 600y + 150y = 132$
     $-450y = -81$
     $y = 0.18$
     Substitute 0.18 for y in the first equation.
     $3x + 150(0.18) = 132$
     $3x + 27 = 132$
     $3x = 105$ so $x = 35$
     The daily fee is \$35 and the mileage charge
     is 18 cents per mile.

15. Graph $3y = 5x - 12$. Since the original statement is greater than or equal to, a solid line is drawn. Since the point $(0, 0)$ satisfies the inequality $3y \geq 5x - 12$, all points on the line and in the half-plane above the line $3y = 5x - 12$ are in the solution set.

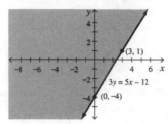

16. a)

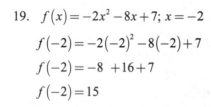

b) $P = 6x + 4y$
At $(0, 0)\ P = 6(0) + 4(0) = 0$
At $(0, 2)\ P = 6(0) + 4(2) = 8$
At $(3, 1)\ P = 6(3) + 4(1) = 22$
At $(3.75, 0)\ P = 6(3.75) + 4(0) = 22.5$
Max. is 22.5 at $(3.75, 0)$
and min. is 0 at $(0, 0)$

17. $x^2 + 7x = -6$
$x^2 + 7x + 6 = 0$
$(x + 6)(x + 1) = 0$
$x + 6 = 0$ and $x + 1 = 0$
$x = -6$ and $x = -1$

18. $3x^2 + 2x = 8$
$3x^2 + 2x - 8 = 0$
$a = 3, b = 2, c = -8$
$$x = \frac{-2 \pm \sqrt{(2)^2 - 4(3)(-8)}}{2(3)}$$
$$x = \frac{-2 \pm \sqrt{4 + 96}}{6} = \frac{-2 \pm \sqrt{100}}{6} = \frac{-2 \pm 10}{6}$$
$$x = \frac{8}{6} = \frac{4}{3} \text{ or } x = \frac{-12}{6} = -2$$

19. $f(x) = -2x^2 - 8x + 7;\ x = -2$
$f(-2) = -2(-2)^2 - 8(-2) + 7$
$f(-2) = -8 + 16 + 7$
$f(-2) = 15$

20. $y = x^2 - 2x + 4$
a) $a = 1 > 0$, opens upward
b) $x = 1$ c) $(1, 3)$ d) $(0, 4)$
e) no x-intercepts
f)

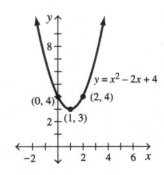

g) D: all real numbers   R: $y \geq 3$

CHAPTER SEVEN

THE METRIC SYSTEM

Exercise Set 7.1

1. Metric

3. a) Meter   b) Kilogram   c) Liter   d) Celsius

5. a) Deka

   b) Deci

7. a) Kilo   b) Milli

9. a) $0°$ C   b) $100°$ C   c) $37°$ C

11. milli b

13. hecto c

15. deci f

17. a) 100 grams   b) 0.001 gram   c) 1000 grams   d) 0.01 gram   e) 10 grams   f) 0.1 gram

19. cg   0.01 g

21. dg   0.1 g

23. kg   1000 g

25. Max. mass 3000 kg = (3000 x 1000) g
    $\qquad$ = 3 000 000 g

27. 5 m = (5 x 1000) mm = 5000 mm

29. 0.063 ℓ = (0.063)(0.001) kℓ = 0.000063 kℓ

31. 186.2 cm = (186.2)(0.01) m = 1.862hm

33. 6.5 km = (6.5)(1,000,000) mm = 6 500 000 mm

35. 24 hm = (24)(0.1) km = 2.4 km

37. 40,302 mℓ = (40,302)(0.0001) daℓ
    = 4.0302 daℓ

39. 2 m = (2)(100) cm = 200 cm

41. 37 m = (37) (100) cm = 3700 cm

43. 200 kg = (200)(1000) g
    $\qquad$ 200 000 g

45. 1.5 m/sec = (1.5)(10)dm = 15 dm/sec

47. 620 cm, 4.4 dam, 0.52 km

49. 1.4 kg, 1600 g, 16,300 dg

51. 105,000 mm, 2.6 km, 52.6 hm

53. 1 hectometer in 10 min. 1 hm > 1 dam

55. The pump that removes 1 daℓ of
    water per min. 1 dekaliter > 1deciliter

57. a) 4302m – 3776 m = 526 m
    b) 526 m = (526)(100) cm = 52 600 cm
    c) 526 m = (526)(0.01) hm = 5.26 hm

59. a) (2)(250)(7) = 3,500 mg / week
    b) 3,500 mg / week =
       $(3500)(0.001)$ g = 3.5 g / week

61. a) 6(360) m ℓ = 2,160 m ℓ
    b) 2160(0.001) ℓ = 2.16 ℓ
    c) 2.45 / 2.16 = \$1.13 per liter

63. a) $(4)(27\text{ m}) = 108\text{ m}$

   b) b) $108\text{ m} = (106)(0.001)\text{ km} = 0.108\text{ km}$

   c) $108\text{ m} = (108)(1000)\text{ mm} = 108\ 000\text{ mm}$

65. 1 gigameter = 1000 megameters

67. 1 teraliter = 1×10^{24} picoliters

69. 9000 cm = 9 dam

71. 0.00006 hg = 6 mg

73. 0.02 kℓ = 2 daℓ

Exercise Set 7.2

1. Length
3. Area
5. Volume
7. Volume
9. Volume
11. Centimeters
13. Meters
15. Centimeters
17. Kilometers
19. c   120 m
21. c   490 km
23. b   168 m
25. cm Answers will vary.
27. m Answers will vary.
29. cm Answers will vary.
31. sq m
33. sq. cm or sq. m
35. sq. mm or sq. cm
37. a   50 cm^2
39. b   1/8 ha
41. a   200 cm^2
43. Answers will vary.
45. Answers will vary.
47. Answers will vary.
49. Milliliters
51. Kiloliters
53. Milliliters
55. Cubic meters
57. c   7780 cm3
59. b   45 m^3
61. b   120 mℓ
63. b   355 mℓ
65. a) Answers will vary.

   b) $(46)(45)(5) = 10\ 580\text{ cm}^3$

67. a) Answers will vary.

   b) $V \approx (3.14)(0.25)^2\,(1) \approx 0.20\text{ m}^3$

69. 1 cubic decimeter

71. A cubic centimeter

73. Longer side = 4cm; Shorter side = 2.2 cm

   $A = lw = (4)(2.2) = 8.8\text{ cm}^2$

75. $A = \pi r^2 \approx 3.14\left(16^2\right) \approx 804.25\text{ m}^2$

77. a) $(3.2)(1.9) = 6.08\text{ km}^2$

   b) $(6.08)(100)\text{ha} = 608\text{ ha}$

79. $(37)(28) = 1036\text{ cm}^2$

   $2540 - 1036 = 1504\text{ cm}^2$

81. a) $V = lwh = (70)(40)(20) = 56\ 000\text{ cm}^3$ b) $56{,}000\text{ cm}^3 = 56\ 000\text{ m}\ell$ c) $56\ 000\text{ m}\ell = \left(\dfrac{56000}{1000}\right)\ell = 56\ \ell$

83. $10^2 = 100$ times larger

85. $10^3 = 1000$ times larger

87. 100 mm^2

89. 0.0001 m2

91. 1 000 000 cm3

93. 501 cm3 = 501 mℓ

95. 263 kl = 263 m^3

97. 60 m^3 = 60 kℓ

99. 6.7 kl = 6.7 m^3 = (6.7×10^3) dm^3 = 6 700 dm^3

101. a) $1 \text{ sq mi} = (1 \text{ mi}^2)(5280)^2 \dfrac{\text{ft}^2}{\text{mi}^2} = 27{,}878{,}400 \text{ ft}^2$

$27{,}878{,}400 \text{ ft}^2 \times (12)^2 \dfrac{\text{in}^2}{\text{ft}^2} = 4{,}014{,}489{,}600 \text{ in}^2$

b) Answers will vary. It is easier to convert in the metric system because it is a base 10 system.

Exercise Set 7.3

1. Kilogram   3. Kilogram   5. Celsius
7. $0\,^\circ$ C   9. Kilograms   11. Grams or kilograms
13. Grams   15. Kilograms   17. Kilograms
    or metric tonnes
19. b   18 kg   21. b   1.4 kg   23. b   2300 kg
25. Answers will vary.   27. Answers will vary.   29. c $0\,^\circ$ C
31. b   Dress warmly   33. c $1200\,^\circ$ C   35. b $-5\,^\circ$ C
    and walk.

37. $F = \dfrac{9}{5}(15) + 32 = 27 + 32 = 59^\circ \text{F}$ 39. $C = \dfrac{5}{9}(-20 - 32) = \dfrac{5}{9}(-52) \approx -28.9^\circ \text{ C}$

41. $C = \dfrac{5}{9}(0 - 32) = \dfrac{5}{9}(-32) \approx -17.8^\circ \text{ C}$ 43. $F = \dfrac{9}{5}(37) + 32 = 66.6 + 32 = 98.6^\circ \text{ F}$

45. $C = \dfrac{5}{9}(10 - 32) = \dfrac{5}{9}(-22) \approx -12.2^\circ \text{ C}$ 47. $F = \dfrac{9}{5}(0) + 32 = 0 + 32 = 32^\circ \text{ F}$

49. $C = \dfrac{5}{9}(-10 - 32) = \dfrac{5}{9}(-42) \approx -23.3^\circ \text{ C}$ 51. $F - \dfrac{9}{5}(22) + 32 - 39.6 + 32 = 71.6^\circ \text{ F}$

53. $F = \dfrac{9}{5}(15.6) + 32 = 28.08 + 32 = 60.08^\circ \text{ F}$ 55. low: $F = \dfrac{9}{5}(17.8) + 32 = 32.04 + 32 = 64.04^\circ \text{ F}$

high: $F = \dfrac{9}{5}(23.5) + 32 = 42.3 + 32 = 74.3^\circ \text{ F}$

57. total mass = 43 g + 30 g + 120 mℓ =
43 g + 30 g + 120 g = 193 g

59. a) cost = (1 kg)(1000 g)(7 euros/100g)

= 70 euros

b) cost = (500 g)(7 euros/100g) = 35 euros

c) 300 g

61. a) $V = lwh$, $l = 16$ m, $w = 12$ m, $h = 12$ m

$V = (16)(12)(12) = 2304$ m^3

b) 2304 m^3 = 2304 kℓ

c) 2304 kℓ = 2304 t

63. 6.2 kg = (6.2 kg)$\left(\dfrac{1 \text{ t}}{1000 \text{ kg}}\right) = 0.0062$ t

65. 1,460,000 mg = 1.46 kg = (1.46 kg)$\left(\dfrac{1 \text{ t}}{1000 \text{ kg}}\right)$

= 0.00146 t

67. a) Yes; mass is a measure of the amount of matter in an object.

b) No; weight is a measure of gravitational force.

69. 1.2 ℓ = 1200 mℓ a) 1200 g   b) 1200 cm

71. $\underline{3 \text{ kg}} \qquad \qquad x$

$$(3\,\text{kg})(2\,\text{m}) = (400\ \text{cm})(?\,\text{g})$$
$$(3000\ \text{g})(200\ \text{cm}) = (400\ \text{cm})(?\,\text{g})$$
$$? = 1500\ \text{g}$$

Exercise Set 7.4

1. Dimensional

3. $\dfrac{60 \text{ seconds}}{1 \text{ minute}}$ or $\dfrac{1 \text{ minute}}{60 \text{ seconds}}$ because $60 \text{ seconds} = 1 \text{ minute}$

5. $\dfrac{100 \text{ cm}}{1 \text{ m}}, \dfrac{1 \text{ m}}{100 \text{ cm}}$

7. a) $\dfrac{1 \text{ lb}}{0.45 \text{ kg}}$ Since we need to eliminate kilograms, kg must appear in the denominator. Since we

   need to convert to pounds, lb must appear in the numerator.

   b) $\dfrac{0.45 \text{ kg}}{1 \text{ lb}}$ Since we need to eliminate pounds; lb must appear in the denominator. Since we

   need to convert to kilograms, kg must appear in the numerator.

9. a) $\dfrac{3.8 \ \ell}{1 \text{ gal}}$ Since we need to eliminate gallons, gal must appear in the denominator. Since we

   need to convert to liters, ℓ must appear in the numerator.

   b) $\dfrac{1 \text{ gal}}{3.8 \ \ell}$ Since we need to eliminate liters, ℓ must appear in the denominator. Since we

   need to convert to gallons, gal must appear in the numerator.

11. $12 \text{ lb} = (12 \text{ lb})\left(\dfrac{0.45 \text{ kg}}{1 \text{ lb}}\right) = 5.4 \text{ kg}$

13. $1060 \text{ g} = (1060 \text{ g})\left(\dfrac{1 \text{ oz}}{28 \text{ g}}\right) \approx 37.86 \text{ oz}$

15. $162 \text{ kg} = (162 \text{ kg})\left(\dfrac{1 \text{ lb}}{0.45 \text{ kg}}\right) = 360 \text{ lb}$

17. $39 \text{ mi} = (39 \text{ mi})\left(\dfrac{1.6 \text{ km}}{1 \text{ mi}}\right) = 62.4 \text{ km}$

19. $675 \text{ ha} = (675 \text{ ha})\left(\dfrac{1 \text{ acre}}{0.4 \text{ ha}}\right) = 1687.5 \text{ acres}$

21. $20.3\ \ell = \left(20.3\ \ell\right)\left(\dfrac{1\ \text{pt}}{0.47\ \ell}\right) = 43.19148936 \approx 43.19$ pints

23. $3.8\ \text{km}^2 = \left(3.8\ \text{km}^2\right)\left(\dfrac{1\ \text{mi}^2}{2.6\ \text{km}^2}\right) = 1.4615\ldots \approx 1.46\ \text{mi}^2$

25. $120\ \text{lb} = \left(120\ \text{lb}\right)\left(\dfrac{0.45\ \text{kg}}{1\ \text{lb}}\right) = 54\ \text{kg}$

27. $414\ \text{m} = \left(414\ \text{m}\right)\left(\dfrac{1\ \text{yd}}{0.9\ \text{m}}\right) \approx 460\ \text{yd}$

29. $357\ \text{m} = \left(357\ \text{m}\right)\left(\dfrac{100\ \text{cm}}{1\ \text{m}}\right)\left(\dfrac{1\ \text{ft}}{30\ \text{cm}}\right) = 1190\ \text{ft}$

31. $70\ \text{kph} = \left(70\ \text{km}\right)\left(\dfrac{1\ \text{mi}}{1.6\ \text{km}}\right) = 43.75\ \text{mph}$

33. $\left(6\ \text{yd}\right)\left(9\ \text{yd}\right) = 54\ \text{yd}^2$

   $54\ \text{yd}^2 = \left(54\ \text{yd}^2\right)\left(\dfrac{0.8\ \text{m}^2}{1\ \text{yd}^2}\right) = 43.2\ \text{m}^2$

35. $6\ \text{g} = \left(6\ \text{g}\right)\left(\dfrac{1\ \text{oz}}{28\ \text{g}}\right) \approx 0.21\ \text{oz}$

37. $\left(91{,}696\ \text{acres}\right)\left(\dfrac{0.4\ \text{ha}}{1\ \text{acre}}\right) = 36\ 678.4\ \text{ha}$

39. $1\ \text{kg} = \left(1\ \text{kg}\right)\left(\dfrac{1\ \text{lb}}{0.45\ \text{kg}}\right) = 2.\overline{2}\ \text{lb}$

   $\dfrac{\$1.60}{2.\overline{2}} = \$0.72$ per pound

41. $34.5\ \text{k}\ell = \left(34.5\ \text{k}\ell\right)\left(\dfrac{1000\ \ell}{1\ \text{k}\ell}\right)\left(\dfrac{1\ \text{gal}}{3.8\ \ell}\right) \approx 9078.95\ \text{gal}$

43. a) $15\ \text{mi} + 1800\ \text{mi} + 1400\ \text{mi} + 760\ \text{mi} = 3975\ \text{mi}$

   b) $3975\ \text{mi} = \left(3975\ \text{mi}\right)\left(\dfrac{1.6\ \text{km}}{1\ \text{mi}}\right) = 6360\ \text{km}$

   c) $12.2\ \text{km} = \left(12.2\ \text{km}\right)\left(\dfrac{1\ \text{mi}}{1.6\ \text{km}}\right) = 7.625 \approx 7.63\ \text{mi}$

   d) $24\ \text{km} = \left(24\ \text{km}\right)\left(\dfrac{1\ \text{mi}}{1.6\ \text{km}}\right) = 15\ \text{mi}$

45. a) $15.3 \text{ acres} = (15.3 \text{ acres})\left(\dfrac{0.4 \text{ ha}}{1 \text{ acre}}\right) = 6.12 \text{ ha}$

    b) $6.12 \text{ ha} = (6.12 \text{ ha})\left(\dfrac{10{,}000 \text{ m}^2}{1 \text{ ha}}\right) = 61\ 200 \text{ m}^2$

47. $120 \text{ lb} = (120 \text{ lb})\left(\dfrac{0.45 \text{ kg}}{1 \text{ lb}}\right)\left(\dfrac{1.5 \text{ mg}}{1 \text{ kg}}\right) = 81 \text{ mg}$

49. $76 \text{ lb} = (76 \text{ lb})\left(\dfrac{0.45 \text{ kg}}{1 \text{ lb}}\right)\left(\dfrac{200 \text{ mg}}{1 \text{ kg}}\right) = 6840 \text{ mg};\ 6840 \text{ mg} = (6840 \text{ mg})\left(\dfrac{1 \text{ g}}{1000 \text{ mg}}\right) = 6.84 \text{ g}$

51. a) $2 \text{ tablespoons} = (2 \text{ tablespoons})\left(\dfrac{236 \text{ mg}}{1 \text{ tablespoon}}\right) = 472 \text{ mg}$

    b) $8 \text{ fl oz} = (8 \text{ fl oz})\left(\dfrac{30 \text{ m}\ell}{1 \text{ fl oz}}\right)\left(\dfrac{1 \text{ tablespoon}}{15 \text{ m}\ell}\right)\left(\dfrac{236 \text{ mg}}{1 \text{ tablespoon}}\right) = 3776 \text{ mg}$

53. a) $32 \text{ kph} = (32 \text{ kph})\left(\dfrac{1 \text{ mi}}{1.6 \text{ km}}\right) = 20 \text{ mph}$ for the roadrunner's speed

         Usain Bolt ran faster.

    b) $42 \text{ kph} = (42 \text{ kph})\left(\dfrac{1 \text{ mi}}{1.6 \text{ km}}\right) = 26.25 \text{ mph}$ for the roadrunner's speed

         The roadrunner ran faster.

55. a) $9730 \text{ m} = (9730 \text{ m})\left(\dfrac{1 \text{ km}}{1000 \text{ m}}\right)\left(\dfrac{1 \text{ mi}}{1.6 \text{ km}}\right) \approx 6.08 \text{ mi}$

    b) $709 \text{ kph} = (709 \text{ kph})\left(\dfrac{1 \text{ mi}}{1.6 \text{ km}}\right) \approx 443.13 \text{ mph}$

    c) $181 \text{ kph} = (181 \text{ kph})\left(\dfrac{1 \text{ mi}}{1.6 \text{ km}}\right) \approx 113.13 \text{ mph}$

    d) $-45^\circ \text{ C} = \dfrac{9}{5}(-45) + 32 = -49^\circ\text{F}$

57. a) $\dfrac{2.50 \text{ €}}{1 \text{ kg}} = \left(\dfrac{2.50 \text{ €}}{1 \text{ kg}}\right)\left(\dfrac{1 \text{ kg}}{2.\overline{2}\text{lb}}\right) \approx \dfrac{1.13 \text{ €}}{1 \text{ lb}}$ b) $\dfrac{1.13 \text{ €}}{1 \text{ lb}} = \left(\dfrac{1.13 \text{ €}}{1 \text{ lb}}\right)\left(\dfrac{\$1.34}{1 \text{ €}}\right) = \dfrac{\$1.51}{1 \text{ lb}}$

59. a) $\dfrac{745 \text{ kph}}{402 \text{ knots}} \approx \dfrac{1.85 \text{ kph}}{1 \text{ knot}}$

    b) $\dfrac{402 \text{ knots}}{463 \text{ mph}} \approx \dfrac{0.87 \text{ knots}}{1 \text{ mph}}$

    c) $\dfrac{463 \text{ mph}}{745 \text{ kph}} \approx \dfrac{0.62 \text{ mph}}{1 \text{ kph}}$

61. $(0.2 \text{ mg})\left(\dfrac{1 \text{ grain}}{60 \text{ mg}}\right)\left(\dfrac{1 \text{ m}\ell}{\dfrac{1}{300} \text{ grain}}\right) = 1.0 \text{ cc, or b)}$

63. a) $(7.0 \ \ell)\left(\dfrac{1000 \text{ m}\ell}{1 \text{ l}}\right)\left(\dfrac{1 \text{ cm}^3}{1 \text{ m}\ell}\right) = 7000 \text{ cc}$

    b) $\left(7000 \text{ cm}^3\right)\left(\dfrac{1 \text{ in.}^3}{(2.54)^3 \text{ cm}^3}\right) = \dfrac{7000}{16.387064} = 427.17... \approx 427.17 \text{ in.}^3$

Review Exercises

1. $1000\times$ basic unit

2. 1/100 of basic unit

3. 1/1000 of basic unit

4. $100\times$ basic unit

5. 10 times basic unit

6. 1/10 of basic unit

7. $12 \text{ g} \cdot \dfrac{1000 \text{ mg}}{1 \text{ g}} = 12\,000 \text{ mg}$

8. $4.5\ell \cdot \dfrac{0.001 \text{ k}\ell}{1\ell} = .0045 \text{ k}\ell$

9. $5700 \text{ cm} \cdot \dfrac{0.0001 \text{ hm}}{1 \text{ cm}} = 0.57 \text{ hm}$

10. $1\,000\,000 \text{ mg} \cdot \dfrac{0.000001 \text{ kg}}{1 \text{ mg}} = 1 \text{ kg}$

11. $4.6 \text{ k}\ell \cdot \dfrac{1000\ell}{\text{k}\ell} = 4620\ell$

12. $192.6 \text{ dag} \cdot \dfrac{100 \text{ dm}}{1 \text{ dag}} = 19\,260 \text{ dg}$

13. $2.67 \text{ k}\ell = 2\,670\,000 \text{ m}\ell$
    $14\,630 \text{ c}\ell = 146\,300 \text{ m}\ell$
    3000 mℓ, 14 630 cl, 2.67 kℓ

14. 0.047 km = 47 m
    47 000 cm = 470 m
    0.047 km, 47 000 cm,
    4700 m

15. Degrees Celsius

16. Millimeters

17. Square meters

18. Milliliters or liters

19. Kilograms or tonnes

20. Kilometers

21. a) and b) Answers will vary.

22. a) and b) Answers will vary.

23. b

24. a

25. c

26. a

27. a

28. b

29. $1500 \text{ kg} = (1500 \text{ kg})\left(\dfrac{1 \text{ t}}{1000 \text{ kg}}\right) = 1.5 \text{ t}$

30. $7.6 \text{ t} = (7.6 \text{ t})\left(\dfrac{1000 \text{ kg}}{1 \text{ t}}\right)\left(\dfrac{1000 \text{ g}}{1 \text{ kg}}\right)$
    $= 7\,600\,000 \text{ g}$

31. $24° \text{ C} = \dfrac{9}{5}(24) + 32 = 75.2° \text{ F}$

32. $68° \text{ F} = \dfrac{5}{9}(68 - 32) = 20° \text{ C}$

33. $-6° \text{ F} = \dfrac{5}{9}(-6 - 32) = -21.\overline{1} \approx -21.1° \text{ C}$

34. $39° \text{ C} = \dfrac{9}{5}(39) + 32 = 102.2° \text{ F}$

35. $l = 4 \text{ cm}, \ w = 1.6 \text{ cm}; A = lw = 4(1.6) = 6.4 \text{ cm}^2$

36. $r = 1.5 \text{ cm}; A = \pi r^2 \approx 3.14(1.5)^2 = 7.065 \approx 7.07 \text{ cm}^2$

37. a) $V = \pi r^2 h = \pi(1)(1) \approx 3.14 \text{ m}^3$

    b) $\left(3.14 \text{ m}^3\right)\left(\dfrac{1000 \text{ kg}}{1 \text{ m}^3}\right) \approx 3140 \text{ kg}$

38. a) $A = lw = (33.7)(26.7) = 899.79 \text{ cm}^2$

   b) $899.79 \text{ cm}^2 = \left(899.79 \text{ cm}^2\right)\left(\dfrac{1 \text{ m}^2}{10,000 \text{ cm}^2}\right) = 0.089979 \text{ m}^2$

39. a) $V = lwh = (90)(50)(40) = 180\ 000 \text{ cm}^3$

   b) $180\ 000 \text{ cm}^3 = \left(180\ 000 \text{ cm}^3\right)\dfrac{1 \text{ m}^3}{(100)^3 \text{ cm}^3} = 0.18 \text{ m}^3$

   c) $180\ 000 \text{ cm}^3 = \left(180\ 000 \text{ cm}^3\right)\left(\dfrac{1 \text{ m}\ell}{1 \text{ cm}^3}\right) = 180\ 000 \text{ m}\ell$

   d) $0.18 \text{ m}^3 = \left(0.18 \text{ m}^3\right)\left(\dfrac{1 \text{ k}\ell}{1 \text{ m}^3}\right) = 0.18 \text{ k}\ell$

40. Since $1 \text{ km} = 100 \times 1 \text{ dam}, \ 1 \text{ km}^2 = 100^2 \times 1 \text{ dam}^2 = 10\ 000 \text{ dam}^2$.

   Thus, 1 square kilometer is 10,000 times larger than a square dekameter.

41. $(15 \text{ kg})\left(\dfrac{1 \text{ lb}}{0.45 \text{ kg}}\right) \approx 33.33 \text{ lb}$

42. $(106 \text{ cm})\left(\dfrac{1 \text{ in.}}{2.54 \text{ cm}}\right) = 41.732... \approx 41.73 \text{ in.}$

43. $(37 \text{ yd})\left(\dfrac{0.9 \text{ m}}{1 \text{ yd}}\right) = 33.3 \text{ m}$

44. $(100 \text{ m})\left(\dfrac{1 \text{ yd}}{0.9 \text{ m}}\right) = 111.\overline{1} \approx 111.11 \text{ yd}$

45. $(45 \text{ mi})\left(\dfrac{1.6 \text{ km}}{1 \text{ mi}}\right) = 72 \text{ km}$

46. $(42 \ \ell)\left(\dfrac{1 \text{ qt}}{0.95 \ \ell}\right) = 44.210... \approx 44.21 \text{ qt}$

47. $(20 \text{ gal})\left(\dfrac{3.8 \ \ell}{1 \text{ gal}}\right) = 76 \ \ell$

48. $\left(60 \text{ m}^3\right)\left(\dfrac{1 \text{ yd}^3}{0.76 \text{ m}^3}\right) = 78.947... \approx 78.95 \text{ yd}^3$

49. $\left(96 \text{ cm}^2\right)\left(\dfrac{1 \text{ in.}^2}{6.5 \text{ cm}^2}\right) \approx 14.77 \text{ in.}^2$

50. $(4 \text{ qt})\left(\dfrac{0.95 \ \ell}{1 \text{ qt}}\right) = 3.8 \ \ell$

51. a) $(137 \text{ cm})\left(\dfrac{1 \text{ in.}}{2.54 \text{ cm}}\right) \approx 53.94 \text{ in.}$

   b) $(44.2 \text{ kg})\left(\dfrac{1 \text{ lb}}{0.45 \text{ kg}}\right) \approx 98.22 \text{ lb}$

52. $(241 \text{ mi})\left(\dfrac{1.6 \text{ km}}{1 \text{ mi}}\right) = 385.6 \text{ km}$

53. $A = lw = (24)(15) = 360 \text{ ft}^2; \ 360 \text{ ft}^2 = \left(360 \text{ ft}^2\right)\left(\dfrac{0.09 \text{ m}^2}{1 \text{ ft}^2}\right) = 32.4 \text{ m}^2$

54. a) $(50,000 \text{ gal})\left(\dfrac{3.8 \ \ell}{1 \text{ gal}}\right)\left(\dfrac{1 \text{ k}\ell}{1000 \ \ell}\right) = 190 \text{ k}\ell$

   b) $(190 \text{ k}\ell)\left(\dfrac{1000 \ \ell}{1 \text{ k}\ell}\right)\left(\dfrac{1 \text{ kg}}{1 \ \ell}\right) = 190\ 000 \text{ kg}$

55. a) $70 \text{ km/hr} = (70 \text{ km/hr})\left(\dfrac{1 \text{ m}}{1.6 \text{ km}}\right) = 43.75 \text{ mi/hr}$

   b) $70 \text{ km/hr} = (70 \text{ km/hr})\left(\dfrac{1000 \text{ m}}{1 \text{ km}}\right) = 70\ 000 \text{ m/hr}$

56. a) $V = lwh = (90)(70)(40) = 252\,000 \text{ cm}^3; 252\,000 \text{ cm}^3 = (252\,000 \text{ cm}^3)\left(\dfrac{1 \text{ m}\ell}{1 \text{ cm}^3}\right)\left(\dfrac{1 \ell}{1000 \text{ m}\ell}\right) = 252 \ell$

    b) $252 \ell = (252 \ell)\left(\dfrac{1 \text{ kg}}{1 \ell}\right) = 252 \text{ kg}$

57. $1 \text{ kg} = (1 \text{ kg})\left(\dfrac{1 \text{ lb}}{0.45 \text{ kg}}\right) = 2.\overline{2} \text{ lb}; \dfrac{\$3.50}{2.\overline{2}} = \$1.575 \approx \$1.58 \text{ per pound}$

Chapter Test

1. $1800 \ell = (1800)(0.001) = 1.8 \text{ k}\ell$

2. $46.2 \text{ cm} == (46.2)(0.0001) = 0.00462 \text{ hm}$

3. $1 \text{ km} = (1 \text{ km})\left(\dfrac{100 \text{ dam}}{1 \text{ km}}\right) = 100 \text{ dam}$ or 100 times greater

4. $400(6) = 2400 \text{ m}; (2400 \text{ m})\left(\dfrac{1 \text{ km}}{1000 \text{ m}}\right) = 2.4 \text{ km}$

5. b
6. a
7. c
8. b
9. b

10. $1 \text{ m}^2 = (1 \text{ m}^2)\left(\dfrac{100^2 \text{ cm}^2}{1 \text{ m}^2}\right) = 10\,000 \text{ cm}^2$ or 10,000 times greater

11. $1 \text{ m}^3 = (1 \text{ m}^3)\left(\dfrac{1000^2 \text{ mm}^3}{1 \text{ m}^3}\right) = 1\,000\,000 \text{ cm}^3$ or 1,000,000 times greater

12. $149 \text{ lb} = (149 \text{ lb})\left(\dfrac{0.45 \text{ kg}}{1 \text{ lb}}\right) = 67.05 \text{ kg}$

13. $169.29 \text{ m} = (169.29 \text{ m})\left(\dfrac{1 \text{ yd}}{0.9 \text{ m}}\right)\left(\dfrac{3 \text{ ft}}{1 \text{ yd}}\right) = 564.3 \text{ ft}$

14. $53{,}321 \text{ km}^2 = (53{,}321 \text{ km}^2)\left(\dfrac{1 \text{ mi}^2}{2.6 \text{ km}^2}\right)$
    $\approx 20{,}508.1 \text{ mi}^2$

15. $30° \text{ F} = \dfrac{5}{9}(30 - 32) = -1.\overline{1} \approx -1.11° \text{ C}$

16. $-5° \text{ C} = \dfrac{9}{5}(-5) + 32 = 23° \text{ F}$

17. a) $300 \text{ kg} = (300 \text{ kg})\left(\dfrac{1000 \text{ g}}{1 \text{ kg}}\right) = 300\,000 \text{ g}$

    b) $300 \text{ kg} = (300 \text{ kg})\left(\dfrac{1 \text{ lb}}{0.45 \text{ kg}}\right) = 666.\overline{6} \text{ lb} \approx 666.67 \text{ lb}$

18. a) $V = lwh = 20(20)(8) = 3200 \text{ m}^3$

   b) $3200 \text{ m}^3 = (3200 \text{ m}^3)\left(\dfrac{1000 \text{ k}\ell}{1 \text{ m}^3}\right)$

   $= 3\ 200\ 000\ell$ (or $3200\ k\ell$)

   c) $3200 \text{ k}\ell = (3200 \text{ k}\ell)\left(\dfrac{1000\ \ell}{1 \text{ k}\ell}\right)\left(\dfrac{1 \text{ kg}}{1\ \ell}\right) = 3\ 200\ 000 \text{ kg}$

19. Total surface area: $2lh + 2wh = 2(20)(6) + 2(15)(6) = 420 \text{ m}^2$

   Liters needed for first coat: $(420 \text{ m}^2)\left(\dfrac{1\ \ell}{10 \text{ m}^2}\right) = 42\ \ell$

   Liters needed for second coat: $(420 \text{ m}^2)\left(\dfrac{1\ \ell}{15 \text{ m}^2}\right) = 28\ \ell$

   Total liters needed: $42 + 28 = 70\ \ell$

   Total cost: $(70\ \ell)\left(\dfrac{\$3.50}{1\ \ell}\right) = \$245$

20. a) $\left(\dfrac{2.09 \text{ euros}}{1\ \ell}\right)\left(\dfrac{3.8\ \ell}{1 \text{ gal}}\right) = 7.94 \text{ euros per gallon}$

   b) $7.94 \text{ euros per gallon} = (7.94 \text{ euros per gallon})\left(\dfrac{\$1}{0.75 \text{ euros}}\right) \approx \$10.59 \text{ per gallon}$

CHAPTER EIGHT

GEOMETRY

Exercise Set 8.1

1. Parallel

3. Angle

5. Supplementary

7. Right

9. Acute

11. Ray, $\overrightarrow{AB}$

13. Line segment, $\overline{AB}$

15. Ray, $\overrightarrow{BA}$

17. Line, $\overleftrightarrow{AB}$

19. $\overline{AD}$

21. $\overleftrightarrow{BD}$

23. $\{B, F\}$

25. $\measuredangle CBF$ or $\measuredangle FBC$

27. $\triangle BCF$

29. $\overrightarrow{BC}$

31. DE

33. $\overline{BC}$

35. $\measuredangle ABE$

37. $\overline{BF}$ or $\overline{FB}$

39. $\overleftrightarrow{AC}$

41. $\overrightarrow{BE}$

43. Acute

45. Obtuse

47. None of these

49. $90° - 7° = 83°$

51. $90° - 32\frac{3}{4}° = 57\frac{1}{4}°$

53. $90° - 64.7° = 25.3°$

55. $180° - 13° = 167°$

57. $180° - 20.5° = 159.5°$

59. $180° - 43\frac{5}{7}° = 136\frac{2}{7}°$

61. c

63. b

65. a

173

67. Let x = measure of $\angle 1$
$4x$ = measure of $\angle 2$
$x + 4x = 90$
$5x = 90$
$x = \dfrac{90}{5} = 18°, m\angle 1$
$4x = (4)(18) = 72°, m\angle 2$

69. Let x = measure of $\angle 1$
$180 - x$ = measure of $\angle 2$
$x - (180 - x) = 102$
$x - 180 + x = 102$
$2x - 180 = 102$
$2x = 282$
$x = \dfrac{282}{2} = 141°, m\angle 1$
$180 - x = 180 - 141 = 39°, m\angle 2$

71. $m\angle 1 + 130° = 180°$
$m\angle 1 = 50°$
$m\angle 2 = m\angle 1$ (vertical angles)
$m\angle 3 = 130°$ (vertical angles)
$m\angle 5 = m\angle 2$ (alternate interior angles)
$m\angle 4 = m\angle 3$ (alternate interior angles)
$m\angle 7 = m\angle 4$ (vertical angles)
$m\angle 6 = m\angle 5$ (vertical angles)
Measures of angles 3, 4, and 7 are each 130°.
Measures of angles 1, 2, 5, and 6 are each 50°.

73. $m\angle 3 + 25° = 180°$
$m\angle 3 = 155°$
$m\angle 1 = 25°$ (vertical angles)
$m\angle 2 = m\angle 3$ (vertical angles)
$m\angle 4 = m\angle 1$ (corresponding angles)
$m\angle 7 = m\angle 4$ (vertical angles)
$m\angle 6 = m\angle 3$ (alternate interior angles)
$m\angle 5 = m\angle 6$ (vertical angles)
Measures of angles 1, 4, and 7 are each 25°.
Measures of angles 2, 3, 5, and 6 are each 155°.

75. $x + 2x + 12 = 90$
$3x + 12 = 90$
$3x = 78$
$x = \dfrac{78}{3} = 26°, m\angle 2$
$90 - x = 90 - 26 = 64°, m\angle 1$

77. $x + 2x - 9 = 90$
$3x - 9 = 90$
$3x = 99$
$x = \dfrac{99}{3} = 33°, m\angle 1$
$90 - x = 90 - 33 = 57°, m\angle 2$

79. $x + 3x - 4 = 180$
$4x - 4 = 180$
$4x = 184$
$x = \dfrac{184}{4} = 46°, m\angle 2$
$180 - x = 180 - 46 = 134°, m\angle 1$

81. $x + 5x + 6 = 180$
$6x + 6 = 180$
$6x = 174$
$x = \dfrac{174}{6} = 29°, m\angle 1$
$180 - x = 180 - 29 = 151°, m\angle 2$

For Exercises 83 - 90, the answers given are one of many possible answers.

83. $\overrightarrow{BG}$ and $\overrightarrow{DG}$

85. Plane ABG and plane JCD

87. Plane $AGB \cap$ plane $ABC \cap$ plane $BCD = \{B\}$

89. $\overleftrightarrow{BC} \cap$ plane $ABG = \{B\}$

91. Always true. If any two lines are parallel to a third line, then they must be parallel to each other.
93. Sometimes true. Vertical angles are only complementary when each is equal to 45°.

95. Sometimes true. Alternate interior angles are only complementary when each is equal to 45°.
97. a) An infinite number of lines can be drawn through a given point.
    b) An infinite number of planes can be drawn through a given point.
99. An infinite number of planes can be drawn through a given line.
101. Answers will vary.
103. No. Line l and line n may be parallel or skew.

105. a)

   Other answers are possible.
   b) Let $m\angle ABC = x$ and $m\angle CBD = y$.

   $x + y = 90°$ and $y = 2x$

   Substitute $y = 2x$ into $x + y = 90°$.

   $x + 2x = 90°$

   $3x = 90°$

   $\dfrac{3x}{3} = \dfrac{90°}{3}$

   $x = 30° = m\angle ABC$

c) $m\angle CBD = y$

   $y = 2x = 2(30°) = 60°$

d) $m\angle ABD + m\angle DBE = 180°$

   $m\angle ABD = x + y = 30° + 60° = 90°$.

   $90° + m\angle DBE = 180°$

   $m\angle DBE = 180° - 90° = 90°$.

Exercise Set 8.2
1. Polygon
3. Proportion
5. Congruent
7. a) Trapezoid
   b) Not regular
11. a) Heptagon
    b) Not regular
15. a) Isosceles
    b) Right
19. a) Scalene
    b) Acute
23. Trapezoid
27. Rhombus

9. a) Pentagon
   b) Regular
13. a) Decagon
    b) Regular
17. a) Isosceles
    b) Acute
21. a) Equilateral
    b) Acute
25. Square

29. The measures of the other two angles of the triangle are 38° and 180° − 131° (supplementary angles). The measure of the third angle of the triangle is 180° − (38°) − (180° − 131°) = 93°. Since angle x is a vertical angle with the 93° angle, the measure of angle x is 93°.
31. The measures of two angles of the triangle are 180° − 105° and 180° − 133° (supplementary angles). The measure of the third angle of the triangle is 180° − (180° − 105°) − (180° − 133°) = 58°. Since angle x is a vertical angle with the 58° angle, the measure of angle x is 58°.

33.

Angle	Measure	Reason
1	90°	$\angle$ 1 and $\angle$ 7 are vertical angles
2	50°	$\angle$ 2 and $\angle$ 4 are corresponding angles
3	130°	$\angle$ 3 and $\angle$ 4 form a straight angle
4	50°	Vertical angle with the given 50° angle
5	50°	$\angle$ 2 and $\angle$ 5 are vertical angles
6	40°	Vertical angle with the given 40° angle
7	90°	$\angle$ 2, $\angle$ 6, and $\angle$ 7 form a straight angle
8	130°	$\angle$ 3 and $\angle$ 8 are vertical angles
9	140°	$\angle$ 9 and $\angle$ 10 form a straight angle
10	40°	$\angle$ 10 and $\angle$ 12 are vertical angles
11	140°	$\angle$ 9 and $\angle$ 11 are vertical angles
12	40°	$\angle$ 6 and $\angle$ 12 are corresponding angles

35. $n = 5$
$(5 - 2) \times 180° = 3 \times 180° = 540°$

37. $n = 9$
$(9 - 2) \times 180° = 7 \times 180° = 1260$

39. $n = 12$
$(12 - 2) \times 180° = 10 \times 180° = 1800°$

41. a) The sum of the measures of the interior angles of a triangle is 180°. Dividing by 3, the number of angles, each interior angle measures 60°.
b) Each exterior angle measures $180° - 60° = 120°$.

43. a) The sum of the measures of the interior angles of a hexagon is $(6 - 2) \times 180° = 4 \times 180° = 720°$. Dividing by 6, the number of angles, each interior angle measures 120°.
b) Each exterior angle measures $180° - 120° = 60°$.

45. a) The sum of the measures of the interior angles of a decagon is $(10 - 2) \times 180° = 8 \times 180° = 1440°$. Dividing by 10, the number of angles, each interior angle measures 144°.
b) Each exterior angle measures $180° - 144° = 36°$.

47. Let $x = A'C'$

$$\frac{A'C'}{AC} = \frac{A'B'}{AB}$$

$$\frac{x}{50} = \frac{10}{25}$$

$$25x = 500$$

$$x = 20$$

Let $y = B'C'$

$$\frac{B'C'}{BC} = \frac{A'B'}{AB}$$

$$\frac{y}{40} = \frac{10}{25}$$

$$25y = 400$$

$$y = \frac{400}{25} = 16$$

49. Let $x = DC$

$$\frac{DC}{D'C'} = \frac{AB}{A'B'}$$

$$\frac{x}{10} = \frac{9}{15}$$

$$15x = 90$$

$$x = \frac{90}{15} = 6$$

Let $y = B'C'$

$$\frac{B'C'}{BC} = \frac{A'B'}{AB}$$

$$\frac{y}{6} = \frac{15}{9}$$

$$9y = 90$$

$$y = \frac{90}{9} = 10$$

51. Let $x = D'C'$

$$\frac{D'C'}{DC} = \frac{A'D'}{AD}$$

$$\frac{x}{16} = \frac{22.5}{18}$$

$$18x = 360$$

$$x = 20$$

Let $y = A'B'$

$$\frac{A'B'}{AB} = \frac{A'D'}{AD}$$

$$\frac{y}{17} = \frac{22.5}{18}$$

$$18y = 382.5$$

$$y = 21.25$$

53. Let $x = BC$

$$\frac{BC}{EC} = \frac{AB}{DE}$$

$$\frac{x}{1} = \frac{3}{1}$$

$$x = 3$$

55. $AD = AC - DC = 5 - \frac{5}{3} = \frac{15}{3} - \frac{5}{3} = 3\frac{1}{3}$

57. $AB = A'B' = 33$

59. $B'C' = BC = 18$

61.

$m\angle ACB = m\angle A'C'B' = 180° - 31° - 78° - 71°$

63. $AD = A'D' = 9$

65. $A'B' = AB = 10$

67. $m\angle DAB = m\angle D'A'B'$
$= 360° - 130° - 70° - 50° = 110°$

69. Let $x =$ height of silo

$$\frac{x}{6} = \frac{105}{9}$$

$$9x = 630$$

$$x = 70 \text{ ft}$$

71. a) $\dfrac{44 \text{ mi}}{0.875 \text{ in.}} = \dfrac{\text{SP-A}}{2.25 \text{ in.}}$

$\text{SP-A} = \dfrac{(44)(2.25)}{0.875} \text{ mi} = 113.1 \text{ mi}$

b) $\dfrac{44 \text{ mi}}{0.875 \text{ in.}} = \dfrac{\text{SP-R}}{1.5 \text{ in.}}$

$\text{SP-R} = \dfrac{(44)(1.5)}{0.875} \text{ mi} = 75.4 \text{ mi}$

73. $180° - 125° = 55°$ 75. $180° - 90° - 55° = 35°$

77. $\dfrac{DE}{D'E'} = 3$ $\dfrac{EF}{E'F'} = 3$ $\dfrac{DF}{D'F'} = 3$

       $\dfrac{12}{D'E'} = 3$ $\dfrac{15}{E'F'} = 3$ $\dfrac{9}{D'F'} = 3$

       $3D'E' = 12$ $3E'F' = 15$ $3D'F' = 9$

         $D'E' = 4$ $E'F' = 5$ $D'F' = 3$

79.   a) $m\angle\ HMF = m\angle\ TMB, \ m\angle\ HFM = m\angle\ TBM, \ m\angle\ MHF = m\angle\ MTB$

      b)   Let $x =$ height of the wall

$$\frac{x}{20} = \frac{5.5}{2.5}$$

$$2.5x = 110$$

$$x = \frac{110}{2.5} = 44 \text{ ft}$$

Exercise Set 8.3

Throughout this section, on exercises involving π, we used the π key on a scientific calculator to determine the answer. If you use 3.14 for π, your answers may vary slightly.

1.   a) Perimeter
     b) Area

3.   Circle

5. $A = \dfrac{1}{2}bh = \dfrac{1}{2}(7)(5) = 17.5 \text{ cm}^2$ 7. $A = \dfrac{1}{2}bh = \dfrac{1}{2}(100)(50) = 2500 \text{ cm}^2$

9.   a) $A = lw = 8(4) = 32 \text{ ft}^2$ 11. $3 \text{ m} = 3(100) = 300 \text{ cm}$

      b) $P = 2l + 2w = 2(8) + 2(4) = 24 \text{ ft}$ a) $A = bh = 300(20) = 6000 \text{ cm}^2$

                                                        b) $P = 2l + 2w = 2(300) + 2(27) = 654 \text{ cm}$

13. $2 \text{ ft} = 2(12) = 24 \text{ in.}$ 15.   a) $A = \pi r^2 = \pi(7)^2 = 49\pi \approx 153.94 \text{ cm}^2$

      a) $A = \dfrac{1}{2}h(b_1 + b_2) = \dfrac{1}{2}(24)(5 + 19) = 288 \text{ in.}^2$ b) $C = 2\pi r = 2\pi(7) = 14\pi \approx 43.98 \text{ cm}$

      b) $P = s_1 + s_2 + b_1 + b_2 = 25 + 25 + 5 + 19 = 74 \text{ in.}$

17. $r = \dfrac{13}{2} = 6.5 \text{ ft}$ 19.   a) $c^2 = 15^2 + 8^2$

      a) $A = \pi r^2 = \pi(6.5)^2 = 42.25\pi \approx 132.73 \text{ ft}^2$ $c^2 = 225 + 64$

      b) $C = \pi d = \pi(13) \approx 40.84 \text{ ft}$ $c^2 = 289$

                                                    $c = \sqrt{289} = 17 \text{ yd}$

                                         b) $P = s_1 + s_2 + s_3 = 8 + 15 + 17 = 40 \text{ yd}$

                                         c) $A = \dfrac{1}{2}bh = \dfrac{1}{2}(8)(15) = 60 \text{ yd}^2$

21. a) $b^2 + 5^2 = 13^2$

$b^2 + 25 = 169$

$b^2 = 144$

$b = \sqrt{144} = 12$ km

b) $P = s_1 + s_2 + s_3 = 5 + 12 + 13 = 30$ km

c) $A = \dfrac{1}{2}bh = \dfrac{1}{2}(5)(12) = 30$ km^2

23. Area of square: $(6)^2 = 36$ ft^2

Area of circle: $\pi(3)^2 = 9\pi = 28.27433388$ ft^2

Shaded area:

$36 - 28.27433388 = 7.72566612 \approx 7.73$ ft^2

25. Use the Pythagorean Theorem to find the length of a side of the shaded square.

$x^2 = 2^2 + 2^2$

$x^2 = 4 + 4$

$x^2 = 8$

$x = \sqrt{8}$

Shaded area: $\sqrt{8}\left(\sqrt{8}\right) = 8$ in.^{2}

27. Find area of trapezoid minus area of unshaded triangle.

Trapezoid: $18\left(\dfrac{9+11}{2}\right) = 180$

Triangle: $\dfrac{1}{2}(18)(10) = 90$

Shaded area: $180 - 90 = 90$ yd^2

29. Area of trapezoid:

$\dfrac{1}{2}(8)(9 + 20) = \dfrac{1}{2}(8)(29) = 116$ in.^{2}

Area of circle: $\pi(4)^2 = 16\pi = 50.26548246$ in.^{2}

Shaded area:

$116 - 50.26548246 = 65.73451754 \approx 65.73$ in.^{2}

31. Radius of larger circle: $\dfrac{12}{2} = 6$ mm

Area of large circle:

$\pi(6)^2 = 36\pi = 113.0973355$ mm^2

Radius of each smaller circle: $\dfrac{6}{2} = 3$ mm

Area of each smaller circle:

$\pi(3)^2 = 9\pi = 28.27433388$ mm^2

Shaded area:

$113.0973355 - 28.27433388 - 28.27433388$

$= 56.54866776 \approx 56.55$ mm^2

33. $\dfrac{1}{x} = \dfrac{9}{99}$

$9x = 99$

$x = \dfrac{99}{9} = 11$ yd^2

35. $\dfrac{1}{13.5} = \dfrac{9}{x}$

$x = 13.5(9) = 121.5$ ft^2

37. $\dfrac{1}{9} = \dfrac{10,000}{x}$

$x = 9(10,000) = 90,000$ cm^2

39. $\dfrac{1}{x} = \dfrac{10,000}{6586}$

$10,000x = 6586$

$x = \dfrac{6586}{10,000} = 0.6586$ m^2

41. Area of living/dining room: $25(22) = 550$ ft^2

    a) $550(11.99) = \$6594.50$

    b) $550(15.99) = \$8794.50$

43. Area of kitchen: $12(14) = 168$ ft^2

    Area of first floor bathroom: $6(10) = 60$ ft^2

    Area of second floor bathroom: $8(14) = 112$ ft^2

    Area of kitchen and both bathrooms: 340 ft^2

    Cost: $340(\$10.49) = \$3566.60$

45. Area of bedroom 1: $10(14) = 140$ ft^2

    Area of bedroom 2: $10(20) = 200$ ft^2

    Area of bedroom 3: $10(14) = 140$ ft^2

    Total area: $140 + 200 + 140 = 480$ ft^2

    Cost: $480(\$9.99) = \$4795.20$

47. Area of entire lawn if all grass:

    $400(300) = 120,000$ ft^2

    Area of house: $\frac{1}{2}(50)(100 + 150) = 6250$ ft^2

    Area of goldfish pond:

    $\pi(20)^2 = 400\pi = 1256.637061$ ft^2

    Area of privacy hedge: $200(20) = 4000$ ft^2

    Area of garage: $70(30) = 2100$ ft^2

    Area of driveway: $40(25) = 1000$ ft^2

    Area of lawn:

    $120,000 - 6250 - 1256.637061 - 4000 - 2100 - 1000$

    $= 105,393.3629$ ft$^2 = \dfrac{105,393.3629}{9}$

    $= 11,710.37366$ yd^2

    Cost:

    $11,710.37366(\$0.03) = \$351.3112098 \approx \$351.31$

49. a) Perimeter $= 2(94) + 2(50) = 288$ ft

    b) Area $= (94)(50) = 4700$; 4700 tiles

51. Let $a =$ height on the wall the ladder reaches

    $a^2 + 20^2 = 29^2$

    $a^2 + 400 = 841$

    $a^2 = 441$

    $a = \sqrt{441} = 21$ ft

53. Let d be the distance.

    $d^2 = 37^2 + 310^2$

    $d^2 = 97469$

    $d = \sqrt{97469} \approx 312$ ft

55. $s = \frac{1}{2}(a + b + c) = \frac{1}{2}(8 + 6 + 10) = 12$

    $A = \sqrt{12(12-8)(12-6)(12-10)}$

    $= \sqrt{12(4)(6)(2)} = \sqrt{576} = 24$ cm^2

57. Answers will vary.

Exercise Set 8.4

In this section, we use the π key on the calculator to determine answers in calculations involving π. If you use 3.14 for π, your answers may vary slightly.

1. Volume

3. Platonic

5. Right

7. a) $V = lwh = (12)(3)(6) = 216 \text{ ft}^3$

 b) $SA = 2lw + 2wh + 2lh$

 $SA = 2(12)(3) + 2(3)(6) + 2(12)(6) = 252 \text{ ft}^2$

9. a) $V = \pi r^2 h = \pi(2^2)(12) = 48\pi$

 $V \approx 150.80 \text{ in.}^3$

 b) $SA = 2\pi r^2 + 2\pi rh$

 $SA = 2\pi(2^2) + 2\pi(2)(12) = 56\pi$

 $SA \approx 175.93 \text{ in}^2$

11. a) $V = \frac{1}{3}\pi r^2 h = \frac{1}{3}\pi(3^2)(14) = 42\pi$

 $V \approx 131.95 \text{ cm}^3$

 b) $SA = \pi r^2 + \pi r\sqrt{r^2 + h^2}$

 $SA = \pi(3^2 + 3\sqrt{3^2 + 14^2}) = \pi(9 + 3\sqrt{205})$

 $SA \approx 163.22 \text{ cm}^2$

13. a) $V = \frac{4}{3}\pi r^3$

 $V = \frac{4}{3}\pi(7^3) = \frac{4}{3}\pi(343) \approx 1436.76 \text{ mi}^3$

 b) $SA = 4\pi r^2$

 $SA = 4\pi(7^2) = 4\pi(49) \approx 615.75 \text{ mi}^2$

15. Area of the base.

 $B = \frac{1}{2}bh = \frac{1}{2}(12)(12) = 72 \text{ m}^2$

 $V = Bh = 72(18) = 1296 \text{ m}^3$

17. Area of the base: $B = s^2 = 12^2 = 144 \text{ cm}^2$

 $V = \frac{1}{3}Bh = \frac{1}{3}(144)(15) = 720 \text{ cm}^3$

19. $V = $ vol. of large prism $-$ vol. of small prism

 $V = (10)(10)(20) - (5)(5)(20) = (100 - 25)(20)$

 $V = (75)(20) = 1500 \text{ ft}^3$

21. $V = 2(\text{volume of one small trough})$

 depth of trough $= 7$

 area of triangular ends $= \frac{1}{2}(4)(7) = 14$

 $V = 2(14)(14) = 392 \text{ ft}^3$

23. $V = $ volume of cylinder $-$ volume of 3 spheres

 $= \pi(3.5)^2(20.8) - 3\left[\frac{4}{3}\pi(3.45)^3\right]$

 $= 254.8\pi - 164.2545\pi = 90.5455\pi$

 $V \approx 284.46 \text{ cm}^3$

25. $V = $ volume of rect. solid $-$ volume of pyramid

 $= 3(3)(4) - \frac{1}{3}(3^2)(4) = 36 - 12 = 24 \text{ ft}^3$

27. $6 \text{ yd}^3 = 6(27) = 162 \text{ ft}^3$

29. $453.6 \text{ ft}^3 = \frac{453.6}{27} = 16.8 \text{ yd}^3$

31. $0.25 \text{ m}^3 = 0.25(1,000,000) = 250,000 \text{ cm}^3$

33. $500,000 \text{ cm}^3 = \frac{500,000}{1,000,000} = 0.5 \text{ m}^3$

35. a) $V = lwh = (20)(15)\left(\frac{9}{12}\right) = 225 \text{ ft}^3$

 b) Cost $= (\$11)(225) = \$2475$

37. a) $V = 80(50)(30) = 120,000 \text{ cm}^3$

 b) $120,000 \text{ m}\ell$

 c) $120,000 \text{ m}\ell = \frac{120,000}{1000} = 120 \ell$

39.   Volume of the box $-$ Volume of the ball

$$(7.5)^3 - \frac{4}{3}\pi(3.75)^3 = 421.875 - 220.893$$

$$\approx 200.98 \text{ cm}^3$$

41.   a) Circular pan:

$$A = \pi r^2 = \pi\left(\frac{9}{2}\right)^2 = 20.25\pi$$

$$= 63.61725124 \approx 63.62 \text{ in.}^2$$

Rectangular pan: $A = lw = 7(9) = 63 \text{ in.}^2$

b) Circular pan:

$$V = \pi r^2 h \approx 63.62(2) = 127.24 \text{ in.}^3$$

Rectangular pan: $V = lwh = 7(9)(2) = 126 \text{ in.}^3$

   c) Circular pan

43. $V = \frac{1}{3}\pi r^2 h = \frac{1}{3}\pi\left(\frac{3}{2}\right)^2 (6) = 4.5\pi$

$$= 14.13716694 \approx 14.14 \text{ in.}^3$$

45. $r = \dfrac{3.875}{2} = 1.9375 \text{ in.}$

Volume of each cylinder:

$$\pi r^2 h = \pi(1.9375)^2 (3)$$

$$= 11.26171875\pi = 35.37973289$$

Total volume:

$$8(35.37973289) = 283.0378631 \approx 283.04 \text{ in.}^3$$

47. $8 - 12 + x = 2$

$$-4 + x = 2$$

$$x = 6$$

$$x = 6 \text{ faces}$$

49. $x - 12 + 8 = 2$

$$x - 4 = 2$$

$$x = 6 \text{ vertices}$$

51. $12 - x + 20 = 2$

$$32 - x = 2$$

$$x = 30 \text{ edges}$$

53. $r_E = \dfrac{12,756.3}{2} = 6378.15 \text{ km}$

$$r_M = \dfrac{3474.8}{2} = 1737.4 \text{ km}$$

a) $SA_E = 4\pi(6378.15^2) \approx 5.11 \times 10^8 \text{ km}^2$

b) $SA_M = 4\pi(1737.4^2) \approx 3.79 \times 10^7 \text{ km}^2$

c) $\dfrac{SA_E}{SA_M} = \dfrac{5.11 \times 10^8}{3.79 \times 10^7} \approx 13 \text{ times larger}$

d) $V_E = \dfrac{4}{3}\pi(6378.15^3) \approx 1.09 \times 10^{12} \text{ km}^3$

e) $V_M = \dfrac{4}{3}\pi(1737.4^3) \approx 2.20 \times 10^{10} \text{ km}^3$

f) $\dfrac{V_E}{V_M} = \dfrac{1.09 \times 10^{12}}{2.20 \times 10^{10}} \approx 50 \text{ times larger}$

55.   a) Find the volume of each numbered region. Since the length of each side is $a+b$, the sum of the volumes of each region will equal $(a+b)^3$. Answers will vary.

b) $V_1 = a(a)(a) = a^3$ $V_2 = a(a)(b) = a^2 b$ $V_3 = a(a)(b) = a^2 b$ $V_4 = a(b)(b) = ab^2$

$V_5 = a(a)(b) = a^2 b$ $V_6 = a(b)(b) = ab^2$ $V_7 = b(b)(b) = b^3$

c) The volume of the piece not shown is ab^2.

57. If we double the radius of a sphere, the new volume will be eight times the original volume.

Exercise Set 8.5

1. Reflection

3. Vector

5. Rotation

7. Symmetry

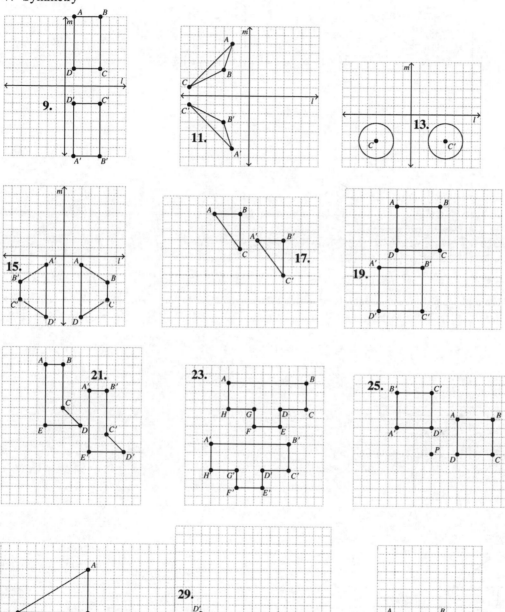

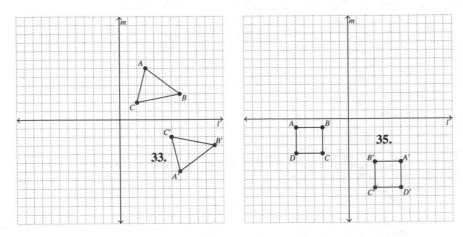

33.

35.

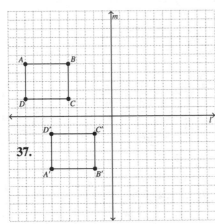

37.

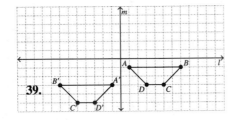

39.

41. a)

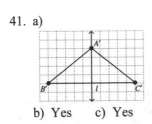

b) Yes   c) Yes

43. a)

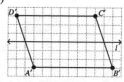

b) No   c) No

45. a)

b) No

c) No

d)

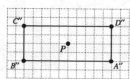

e) Yes

f) Yes

47. a) – c)

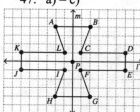

d) No. Any 90° rotation will result in the figure being in a different position than the starting position.

49. Answers will vary.

51. A regular octagon cannot be used as a tessellating shape.

53. Although answers will vary depending on the font, the following capital letters have reflective symmetry about a horizontal line drawn through the center of the letter: B, C, D, E, H, I, K, O, X.

55. Although answers will vary depending on the font, the following capital letters have 180° rotational symmetry about a point in the center of the letter: H, I, N, O, S, X, Z.

Exercise Set 8.6

1. Rubber

3. Klein

5. Jordan

7. 7 – Green; 1, 3, 5 – Yellow; 2, 4, 6 – Red. Answers will vary.

9. 1, 7 – Red; 4, 6, 8 – Green; 2, 3, 5 – Blue. Answers will vary.

11. CA, WA, MT, UT – Red
    OR, WY, AZ – Green
    ID, NM – Blue
    NV, CO – Yellow . Answers will vary.

13. YT, NU, AB, ON – Red
    NT, QC – Blue
    BC, SK – Green
    MB – Yellow. Answers will vary.

15. Outside; a straight line from point *A* to a point clearly outside the curve crosses the curve an even number of times.

17. Outside; a straight line from point *A* to a point clearly outside the curve crosses the curve an even number of times.

19. Inside; a straight line from point *C* to a point clearly outside the curve crosses the curve an odd number of times.

21. 1

23. 0

25. 5

27. 3

29. Larger than 5

31. 4

33. a) - d) Answers will vary.

35. One

37. Two

39. a) No, it has an inside and an outside.
    b) Two
    c) Two
    d) Two strips, one inside the other

41. Answers will vary.

43. Answers will vary.

Exercise Set 8.7

1. Parallel

3. Two

5. Sphere

7. Geodesic

9.

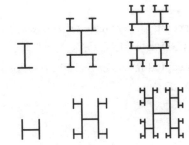

11.

13. a)

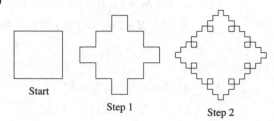

Start   Step 1   Step 2

b) Infinite.

c) Finite since it covers a finite or closed area.

15. Each type of geometry can be used in its own frame of reference.

17. Coastlines, trees, mountains, galaxies, rivers, weather patterns, brains, lungs, blood supply

Review Exercises

In the Review Exercises and Chapter Test questions, the π key on the calculator is used to determine answers in calculations involving π. If you use 3.14 for π, your answers may vary slightly.

1. $\angle CBF$

2. $\overline{BC}$

3. $\triangle BFC$

4. $\overline{BH}$

5. $\{F\}$

6. $\{\ \}$

7. $90° - 27.6° = 62.4°$

8. $180° - 100.5° = 79.5°$

9. Let $x = BC$

$$\frac{BC}{B'C} = \frac{AC}{A'C}$$
$$\frac{x}{3.4} = \frac{12}{4}$$
$$4x = 40.8$$
$$x = \frac{40.8}{4} = 10.2 \text{ in.}$$

10. Let $x = A'B'$

$$\frac{A'B'}{AB} = \frac{A'C}{AC}$$
$$\frac{x}{6} = \frac{4}{12}$$
$$12x = 24$$
$$x = \frac{24}{12} = 2 \text{ in.}$$

11. $m\angle ABC = m\angle A'B'C$
$m\angle A'B'C = 180° - 88° = 92°$
Thus, $m\angle ABC = 92°$
$m\angle BAC = 180° - 30° - 92° = 58°$

12. $m\angle ABC = m\angle A'B'C$
$m\angle A'B'C = 180° - 88° = 92°$
Thus, $m\angle ABC = 92°$

13. $m\angle 1 = 43°$
$m\angle 6 = 180° - 117° = 63°$
$m\angle 2 = m\angle 1 + m\angle 6 = 106°$
$m\angle 3 = 74°$
$m\angle 4 = m\angle 6 = 63°$
$m\angle 5 = 180° - m\angle 4 = 180° - 63° = 117°$

14. $n = 8$

$$(n-2)180° = (8-2)180° = 6(180°) = 1080°$$

15. a) $A = lw = 11(9) = 99 \text{ mi}^2$

b) $P = 2l + 2w = 2(11) + 2(9) = 40 \text{ mi}$

16. a) $A = \frac{1}{2}h(b_1 + b_2) = \frac{1}{2}(4)(6+12) = 36 \text{ m}^2$

b) $P = 5 + 6 + 5 + 12 = 28 \text{ m}$

17. a) $A = bh = 12(7) = 84 \text{ in.}^2$

b) $P = 2(9) + 2(12) = 42 \text{ in.}$

18. a) $A = \frac{1}{2}bh = \frac{1}{2}(3)(4) = 6 \text{ km}^2$

b) $P = 3 + 4 + \sqrt{3^2 + 4^2} = 7 + \sqrt{25} = 12 \text{ km}$

19. a) $A = \pi r^2 = \pi(7)^2 \approx 153.94 \text{ ft}^2$

b) $C = 2\pi r = 2\pi(7) = 14\pi \approx 43.98 \text{ ft}$

20. $A = $ area of rectangle $- 3($area of one circle$)$

Length of rectangle $= 3($diameter of circle$)$

$$= 3(10) = 30$$

Area of rectangle: $(10)(30) = 300$

Area of circle: $\pi(5^2) = 25\pi$

Shaded area: $300 - 3(25\pi) \approx 64.38 \text{ m}^2$

21. Shaded area is the area of an 12 by 12 square minus the four corner squares (each 3 by 3) and minus the area of a circle of diameter 6.

Shaded area $= (12)(12) - 4(3)(3) - \pi(3^2)$

$$= 108 - 9\pi \approx 79.73 \text{ yd}^2$$

22. $A = lw = 14(16) = 224 \text{ ft}^2$

Cost: $224(\$9.75) = \$2184$

23. a) $V = lwh = 11(3)(5) = 165 \text{ cm}^3$

b) $SA = 2lw + 2wh + 2lh$

$$SA = 2(11)(3) + 2(3)(5) + 2(11)(5) = 206 \text{ cm}^2$$

24. a) $V = \pi r^2 h = \pi(3^2)(9) = 81\pi \approx 254.47 \text{ in}^3$

b) $SA = 2\pi r^2 + 2\pi rh$

$$SA = 2\pi(3^2) + 2\pi(3)(9) = 72\pi$$

$$SA \approx 226.19 \text{ in}^2$$

25. a) $r = \frac{12}{2} = 6 \text{ mm}$

$$V = \frac{1}{3}\pi r^2 h = \frac{1}{3}\pi(6^2)(16) = 192\pi$$

$$V \approx 603.19 \text{ mm}^3$$

b) $SA = \pi r^2 + \pi r\sqrt{r^2 + h^2}$

$$SA = \pi\left(6^2 + 6\sqrt{6^2 + 16^2}\right) = \pi\left(36 + 6\sqrt{292}\right)$$

$$SA \approx 435.20 \text{ mm}^2$$

26. a) $V = \frac{4}{3}\pi r^3$

$$V = \frac{4}{3}\pi(5^3) = \frac{4}{3}\pi(125) \approx 523.60 \text{ yd}^3$$

b) $SA = 4\pi r^2$

$$SA = 4\pi(5^2) = 4\pi(25) \approx 314.16 \text{ yd}^2$$

27. $B = \frac{1}{2}bh = \frac{1}{2}(9)(12) = 54 \text{ m}^2$

   $V = Bh = 54(8) = 432 \text{ m}^3$

28. If h represents the height of the triangle which is the base of the pyramid, then

   $$h^2 + 3^2 \ = \ 5^2$$
   $$h^2 + 9 \ = \ 25$$
   $$h^2 \ = \ 16$$
   $$h \ = \ \sqrt{16} = 4 \text{ ft}$$

   $B = \frac{1}{2}bh = \frac{1}{2}(6)(4) = 12 \text{ ft}^2$

   $V = \frac{1}{3}Bh = \frac{1}{3}(12)(7) = 28 \text{ ft}^3$

29. $V = $ volume of cylinder $-$ volume of cone

   $= \pi(2)^2(9) - \frac{1}{3}\pi(2)^2(9) = 36\pi - 12\pi = 24\pi$

   $= 75.39822369 \approx 75.40 \text{ cm}^3$

30. $V = $ vol. of large sphere $-$ vol. of small sphere

   $= \frac{4}{3}\pi(7)^3 - \frac{4}{3}\pi(3.5)^3 = 457.3333\pi - 57.1667\pi$

   $= 400.1666\pi \approx 1257.16 \text{ in.}^3$

31. $$h^2 + 1^2 \ = \ 3^2$$
   $$h^2 + 1 \ = \ 9$$
   $$h^2 \ = \ 8$$
   $$h \ = \ \sqrt{8}$$

   $A = \frac{1}{2}h(h_1 + h_2) = \frac{1}{2}(\sqrt{8})(2 + 4) = 8.485281374 \text{ ft}^2$

   a) $V = Bh = 8.485281374(8)$

      $= 67.88225099 \approx 67.88 \text{ ft}^3$

31. b) Weight:
      $67.88(62.4) + 375 = 4610.7 \text{ lb}$

   Yes, it will support the trough filled with water.

   c) $(4610.7 - 375) = 4235.7 \text{ lb of water}$

      $\frac{4235.7}{8.3} = 510.3253 \approx 510.33 \text{ gal}$

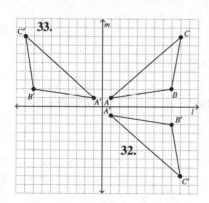

33.   **32.**

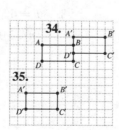

34.   **35.**

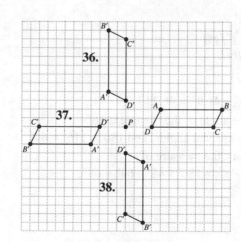

36.   **37.**   **38.**

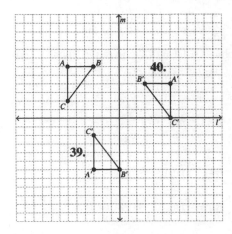

41. Yes   42. No   43. No   44. Yes

45. a) – d) Answers will vary.

46. 1, 3, 9 Red; 2, 8,10 Green; 4, 5, 6, 7 Blue Answers will vary.

47. Outside; a straight line from point A to a point clearly outside the curve crosses the curve an even number of times.

48. Euclidean: Given a line and a point not on the line, one and only one line can be drawn parallel to the given line through the given point.

    Elliptical: Given a line and a point not on the line, no line can be drawn through the given point parallel to the given line.

    Hyperbolic: Given a line and a point not on the line, two or more lines can be drawn through the given point parallel to the given line.

49.

50.

Chapter Test

1. $\angle BAE$

2. $\triangle BCD$

3. $\{D\}$

4. $\overline{AC}$

5. $90° - 41.9° = 48.2°$

6. $180° - 73.5° = 106.5°$

7. One angle of the triangle is 48° (by vertical angles) and 180° - 116° = 64°. Thus, the measure of angle x = 180° - 48° - 64° = 68°.

8. $n = 6$

    $(n-2)180° = (6-2)180° = 4(180°) = 720°$

9. Let $x = B'C'$

$$\frac{B'C'}{BC} = \frac{A'C'}{AC}$$

$$\frac{x}{7} = \frac{5}{13}$$

$$13x = 35$$

$$x = \frac{35}{13} = 2.692307692 \approx 2.69 \text{ cm}$$

10. a) $x^2 + 5^2 = 13^2$

$$x^2 + 25 = 169$$

$$x^2 = 144$$

$$x = \sqrt{144} = 12 \text{ in.}$$

  b) $P = 5 + 13 + 12 = 30$ in.

  c) $A = \frac{1}{2}bh = \frac{1}{2}(5)(12) = 30$ in.^{2}

11. $r = \frac{14}{2} = 7$ cm

  a) $V = \frac{4}{3}\pi r^3 = \frac{4}{3}\pi\left(7^3\right) \approx 1436.76$ cm^3

  b) $SA = 4\pi r^2$

$$SA = 4\pi\left(7^2\right) = 4\pi(49) \approx 615.75 \text{ cm}^2$$

12.

Shaded volume =

        volume of prism $-$ volume of cylinder

Volume of prism: $V = lwh = (6)(4)(3) = 72$ m^3

Volume of cylinder: $V = \pi r^2 h = \pi\left(1^2\right)4 = 4\pi$ m^3

Shaded volume $= 72 - 4\pi \approx 59.43$ m^3

13. $B = lw = 4(7) = 28$ ft^2

$$V = \frac{1}{3}Bh = \frac{1}{3}(28)(12) = 112 \text{ ft}^3$$

14.

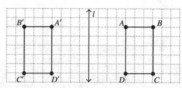

15.

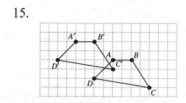

16.

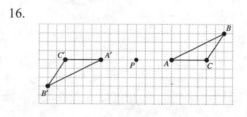

17.

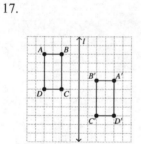

18. a) No
    b) Yes

19. **A Möbius strip** is a surface with one side and one edge.

20. a) and b) Answers will vary.

CHAPTER NINE

MATHEMATICAL SYSTEMS

Exercise Set 9.1

1. Binary
3. Closed
5. Identity
7. Inverse
9. Commutative
11. The commutative property of addition stated that $a + b = b + a$, for any elements a and b.
    Example: $3 + 4 = 4 + 3$
13. The associative property of multiplication states that $(a \times b) \times c = a \times (b \times c)$, for any real numbers a, b, and c.   EXAMPLE: $(2 \times 3) \times 4 = 2 \times (3 \times 4)$
15. $7 - 3 = 4$, BUT $3 - 7 = -4$
17. $(8 \div 4) \div 2 = 2 \div 2 = 1$ BUT $8 \div (4 \div 2) = 8 \div 2 = 4$
19. Yes. Satisfies 5 properties needed.
21. No. No identity element
23. Yes. Satisfies 4 properties needed
25. No. Not closed
27. Yes. Satisfies 5 properties needed
29. No. No identity element
31. No. Not closed; i.e., 1/0 is undefined.
33. No; the system is not closed, $\pi + (-\pi) = 0$,
    which is rational. There is no identity element.
35. ANSWERS WILL VARY.

Exercise Set 9.2

1. {1, 2, 3, 4, 5, 6, 7, 8, 9, 10, 11, 12}
3. Identity
5. Associative
7. Associative
9. Commutative
11. No. The system is not closed. It
    contains a symbol other than p,
    q, and r.

193

13. Identity element = C, Row 3 is identical to top row and column 3 is identical to left column.

15. a) The inverse of A is A because A ‡ A = C.

   b) The inverse of B is B because B ‡ B = C.

   c) The inverse of C is C because C ‡ C=C.

17. No. Not commutative, Non-symmetrical around main diagonal

19. $9 + 5 = 2$

21. $8 + 7 = 3$

23. $4 + 12 = 4$

25. $3 + (8 + 9) = 3 + 5 = 8$

27. $12 - 7 = 5$

29. $4 - 10 = 6$

31. $6 - 10 = 8$

33. $5 - 5 = 12$

35. See below.

37. $1 + 6 = 1$

39. $5 - 2 = 3$

41. $4 - 6 = 4$

35.

+	1	2	3	4	5	6
1	2	3	4	5	6	1
2	3	4	5	6	1	2
3	4	5	6	1	2	3
4	5	6	1	2	3	4
5	6	1	2	3	4	5
6	1	2	3	4	5	6

43.

+	1	2	3	4	5	6	7
1	2	3	4	5	6	7	1
2	3	4	5	6	7	1	2
3	4	5	6	7	1	2	3
4	5	6	7	1	2	3	4
5	6	7	1	2	3	4	5
6	7	1	2	3	4	5	6
7	1	2	3	4	5	6	7

43. See above.

45. $4 + 7 = 4$

47. $5 + 5 = 3$

49. $(4 - 5) - 6 = 6 - 6 = 7$

51. See above. Yes. satisfies 5 required properties

53. a) $\{0, 2, 4, 6\}$

   b) £

   c) Yes. All elements in the table are in the original set.

   d) Identity element is 0.

   e) Yes; $0 £ 0 = 0, 2 £ 6 = 0,$
      $4 £ 4 = 0, 6 £ 2 = 0$

   f) $(2 £ 4) £ 4 = 6 £ 4 = 2$
      and $2 £ (4 £ 4) = 2 £ 0 = 2$

   g) Yes; $2 £ 6 = 0 = 6 £ 2$

   h) Yes, system satisfies five properties needed.

55. a) $\{*, 5, L\}$

   b) $

   c) Yes. All elements in the table are in the original set.

   d) Yes. Identity element is L.

   e) Yes; $* \$ 5 = L, 5 \$ * = L, L \$ L = L$

   f) $(*\$5) \$5 = L \$ 5 = 5$
      and $* \$ (5 \$ 5) = * \$ * = 5$

57. a) The system is closed. All elements in the table are elements of the set.

   b) Yes; the identity element is O.

   c) G-D, O- O, L does not have an inverse, D-G

   d) $(L ☆ L) ☆ D = L; L ☆ (L ☆ D) = D$

   e) No

   f) Yes

   g) No; not every element has an inverse and the the associative property does not hold.

g) Yes; L $ * = * and * $ L = *

h) Yes, system satisfies five properties needed.

59. Not closed, not associative.

Use $(\Gamma \ddagger \Gamma) \neq \Delta$; since $\Gamma \ddagger \Gamma = $ II, and II is not an element, the next operation cannot be performed (not associative).

63. No identity element and therefore no inverses.

$(d \Leftrightarrow e) \Leftrightarrow d = d \Leftrightarrow d = e$

$d \Leftrightarrow (e \Leftrightarrow d) = d \Leftrightarrow e = d$

Not associative since $e \neq d$

$e \Leftrightarrow d = e \qquad d \Leftrightarrow e = d$

Not commutative since $e \neq d$

67. a) All elements in the table are in the set {1, 2, 3, 4, 5, 6} so the system is closed. The identity is 6. 5 and 1 are inverses of each other, and 2, 3, 4, and 6 are their own inverses. Thus, if the associative property is assumed, the system is a group.

b) $3 \infty 1 = 2$, but $1 \infty 3 = 4$

61. No inverses for ⊙ and *

(* ⊡ *) ⊡ T = ⊙ ⊡ T = *

* ⊡ (* ⊡ T) = * ⊡ ⊙ = P

Not associative since * ≠ P

65. a)

+	E	O
E	E	O
O	O	E

b) The system is closed, the identity element is E, each element is its own inverse, and the system is commutative since the table is symmetric about the main diagonal. Since the system has fewer than 6 elements and satisfies the above properties, it is a commutative group.

69. a)

*	R	S	T	U	V	I
R	V	T	U	S	I	R
S	U	I	V	R	T	S
T	S	R	I	V	U	T
U	T	V	R	I	S	U
V	I	U	S	T	R	V
I	R	S	T	U	V	I

b) Is a group: Is closed; identity = I; each element has a unique inverse; an example of associativity:

$R * (T * V) = R * U = S$

$(R * T) * V = U * V = S$

c) $R * S = T \qquad S * R = U$

Not Commutative since $T \neq U$

Exercise Set 9.3

1. $m - 1$

3. Remainder

5. Congruent

7. $20 \div 7 = 2$ remainder 6. so 20 days is 6 days later than today (Thursday), or Wednesday.

9. $102 \div 7 = 14$ remainder 4. so 102 days is 4 days later than today (Thursday), or Monday.

11. 3 years $= (3 \cdot 365)$ days $= 1095$ days;

$1095 \div 7 = 156$ remainder 3, so 3 years is 3 days later than today (Thursday), or Sunday.

13. 400 days / 7 = 57 remainder 1. So 400 days is one day after today, or Friday.

15. $9 + 18 \equiv 6 \pmod{12}$, so 18 months after October is April.

17. Since 2 years = 24 months and $24 \equiv 0 \pmod{12}$, 2 years 10 months is the same as 10 months; $9 + 10 \equiv 7 \pmod{12}$, so 10 months after October is August.

19. $83 \div 12 = 6$ remainder 11. So 83 months is 11 months after October, which is September.

21. $105 \div 12 = 8$ remainder 9. So 105 months is 9 months after October, which is July.

23. $4 + 4 = 8 \qquad 8 \equiv 3 \pmod 5$

25. $9 - 3 = 6 \qquad 6 \equiv 1 \pmod 5$

27. $8 \bullet 9 = 72 \qquad 72 \equiv 2 \pmod 5$

29. $4 - 8 \equiv (5+4) - 8 \pmod 5$
$\qquad \equiv 9 - 8 \pmod 5$
$\qquad \equiv 1 \pmod 5$

31. $(15 \bullet 4) - 8 = 60 - 8 = 52 \qquad 52 \equiv 2 \pmod 5$

33. $17 \pmod 2 \equiv 1$

35. $60 \pmod 7 \equiv 4$

37. $41 \pmod 9 \equiv 5$

39. $-1 \pmod 7 \equiv 6$

41. $-27 \pmod 8 \equiv 5$

43. $4 + 5 \equiv 3 \pmod 6$

45. $6 + 3 \equiv 2 \pmod 7$

47. $4 - 5 \equiv 5 \pmod 6$

49. $5 \bullet 5 \equiv 7 \pmod 9$

51. $2 \bullet \{ \} \equiv 1 \pmod 6$
No solution

53. $4 \bullet \{ \} \equiv 4 \pmod{10}$
$\{1, 6\}$

55. $5 - 8 \equiv 9 \pmod{12}$

57. a) 1792, 1796, 1800,
1804, 1808
b) 2024, 2028, 2032,
2036, 2040

59.   am/pm am/pm am/pm am/pm am/pm am/pm am/pm r r

$\uparrow$

today

a) $30 \equiv 3 \pmod 9$; 3 days after today he works at Longboat Key.

b) $100 \equiv 1 \pmod 9$; 1 days after today he rests.

c) $366 \equiv 6 \pmod 9$; 6 days after today he works.
at The Meadows.

61. The manager's schedule is repeated every seven weeks. If this is week two of his schedule, then this is his second weekend that he works, or week 1 in a mod 7 system. His schedule in mod 7 on any given weekend is shown in the following table:

Weekend (mod 7):

Work/off	0	1	2	3	4	5	6
	w	w	w	w	w	w	o

a) If this is weekend 1, then in 5 more weeks $(1 + 5 = 6)$ he will have the weekend off.

b) $25 \cong 7 = 3$, remainder 4. Thus $25 \cong \pmod 7$ and 4 weeks from weekend 1 will be weekend 5. He will not have off.

c) $50 \cong 7 = 7$, remainder 1. One week from weekend 1 will be weekend 2. It will be 4 more weeks before he has off. Thus, in 54 weeks he will have the weekend off.

63. The waiter's schedule in a mod 14 system is given in the following table:

Day:	0	1	2	3	4	5	6	7	8	9	10	11	12	13
shift:	d	d	d	d	d	e	e	e	d	d	d	d	e	e

$\uparrow$

today

a) $20 \equiv 6 \pmod{14}$; 6 days from today is an evening shift.

b) $52 \equiv 10 \pmod{14}$; 10 days from today is the day shift.

c) $365 \equiv 1 \pmod{14}$; 1 day from today is the day shift.

65. a)

+	0	1	2
0	0	1	2
1	1	2	0
2	2	0	1

b) Yes. All the numbers in the table are from the set {0, 1, 2}.

c) The identity element is 0.

d) Yes.   element + inverse = identity
$0 + 0 = 0$ $1 + 2 = 0$ $2 + 1 = 0$

e) $(1 + 2) + 2 = 2$ $1 + (2 + 2) = 1 + 1 = 2$
Associative since $2 = 2$.

f) Yes, the table is symmetric about the main diagonal. $1 + 2 = 0 = 2 + 1$

g) Yes. All five properties are satisfied.

h) Yes. The modulo system behaves the same no matter how many elements are in the system.

67. a)

×	0	1	2	3
0	0	0	0	0
1	0	1	2	3
2	0	2	0	2
3	0	3	2	1

b) Yes. All the elements in the table are from the set {0, 1, 2, 3}.

c) Yes. The identity element is 1.

d) elem. × inverse = identity

$0 \times$ none $= 1$ $1 \times 1 = 1$ $2 \times$ none $= 1$

$3 \times 3 = 1$ Elements 0 and 2 do not have inverses.

e) $(1 \times 3) \times 0 = 3 \times 0 = 0$

$1 \times (3 \times 0) = 1 \times 0 = 0$ Yes, Associative

f) Yes. $2 \times 3 = 2 = 3 \times 2$

g) No. Not all elements have inverses.

For the operation of division in modular systems, we define n ÷ d = n • i, where i is the multiplicative inverse of d.

69. ? ÷ 3 ≡ 4 (mod 5)

Since 4 · 3 = 12, and 12 ≡ 2 (mod 5),

2 ÷ 3 ≡ 4 (mod 5)

71. ? ÷ ? ≡ 1 (mod 4)   0 ÷ 0 is undefined.

1 ÷ 1 ≡ 1 (mod 4)   2 ÷ 2 ≡ 1 (mod 4)

3 ÷ 3 ≡ 1 (mod 4)   ? = {1, 2, 3}

73. $8k \equiv x$ (mod 8)   8(1) ≡ 0 (mod 8)

8(2) = 16 ≡ 0 (mod 8) $x = 0$

75. $4k - 2 \equiv x$ (mod 4)   4(0) − 2 = -2 ≡ 2 (mod 4)

4(1) − 2 = 2 ≡ 2 (mod 4)   4(2) − 2 = 6 ≡

2 (mod 4), $x = 2$

77. a) 365 ≡ 1 (mod 7) so his birthday will be one day
    later in the week next year and will fall on
    Tuesday.

    b) 366 ≡ 2 (mod 7) so if next year is a leap year
    his birthday will fall on Wednesday.

79. If 5 is subtracted from each number on the wheel,

16 19 20 1 17 10 9 12 10 5 14 24 5   21 20 1 10 23

becomes:

11 14 15 23 12 5 4 7 5 0 9 19 0 16 15 23 5 18

which is equivalent to   **K N O W L E D G E   I S   P O W E R**

Exercise Set 9.4

1. Matrix
3. 2
5. Dimensions
7. Columns, rows

9. $A = \begin{bmatrix} 2 & 7 \\ 5 & -8 \end{bmatrix}$ $B = \begin{bmatrix} -7 & 1 \\ -5 & 2 \end{bmatrix}$ $A + B = \begin{bmatrix} 2 + (-7) & 7 + (1) \\ 5 + -5 & -8 + 2 \end{bmatrix} = \begin{bmatrix} -5 & 8 \\ 0 & -6 \end{bmatrix}$

11. $\begin{bmatrix} 5 & 2 \\ -1 & 4 \\ 7 & 0 \end{bmatrix} + \begin{bmatrix} -3 & 3 \\ -4 & 0 \\ 1 & 6 \end{bmatrix} = \begin{bmatrix} 5 + (-3) & 2 + 3 \\ -1 + (-4) & 4 + 0 \\ 7 + 1 & 0 + 6 \end{bmatrix} = \begin{bmatrix} 2 & 5 \\ -5 & 4 \\ 8 & 6 \end{bmatrix}$

13. $A - B = \begin{bmatrix} 8 & 6 \\ 4 & -2 \end{bmatrix} - \begin{bmatrix} -2 & 5 \\ 9 & 1 \end{bmatrix} = \begin{bmatrix} 8 - (-2) & 6 - 5 \\ 4 - (9) & -2 - 1 \end{bmatrix} = \begin{bmatrix} 10 & 1 \\ -5 & -3 \end{bmatrix}$

15. $A - B = \begin{bmatrix} -5 & 1 \\ 8 & 6 \\ 1 & -5 \end{bmatrix} - \begin{bmatrix} -6 & -8 \\ -10 & -11 \\ 3 & -7 \end{bmatrix} = \begin{bmatrix} -5 + 6 & 1 + 8 \\ 8 + 10 & 6 + 11 \\ 1 - 3 & -5 + 7 \end{bmatrix} = \begin{bmatrix} 1 & 9 \\ 18 & 17 \\ -2 & 2 \end{bmatrix}$

17. $2A = 2\begin{bmatrix} 1 & 2 \\ 0 & 5 \end{bmatrix} = \begin{bmatrix} 2(1) & 2(2) \\ 2(0) & 2(5) \end{bmatrix} = \begin{bmatrix} 2 & 4 \\ 0 & 10 \end{bmatrix}$

19. $2B + 3C = 2\begin{bmatrix} 3 & 2 \\ 5 & 0 \end{bmatrix} + 3\begin{bmatrix} -2 & 3 \\ 4 & 0 \end{bmatrix} = \begin{bmatrix} 6 & 4 \\ 10 & 0 \end{bmatrix} + \begin{bmatrix} -6 & 9 \\ 12 & 0 \end{bmatrix} = \begin{bmatrix} 6-6 & 4+9 \\ 10+12 & 0+0 \end{bmatrix} = \begin{bmatrix} 0 & 13 \\ 22 & 0 \end{bmatrix}$

21. $A \times B = \begin{bmatrix} 3 & 5 \\ 0 & 6 \end{bmatrix}\begin{bmatrix} 4 & 2 \\ 8 & 3 \end{bmatrix} = \begin{bmatrix} 3(4)+5(8) & 3(2)+5(3) \\ 0(4)+6(8) & 0(2)+6(3) \end{bmatrix} = \begin{bmatrix} 52 & 21 \\ 48 & 18 \end{bmatrix}$

23. $A \times B = \begin{bmatrix} 2 & 3 & -1 \\ 0 & 4 & 6 \end{bmatrix}\begin{bmatrix} 2 \\ 4 \\ 1 \end{bmatrix} = \begin{bmatrix} 2(2)+3(4)-1(1) \\ 0(2)+4(4)+6(1) \end{bmatrix} = \begin{bmatrix} 15 \\ 22 \end{bmatrix}$

25. $A \times B = \begin{bmatrix} 2 & -5 & 0 \end{bmatrix}\begin{bmatrix} 4 & 1 \\ -1 & 0 \\ -2 & 6 \end{bmatrix} = \begin{bmatrix} 2(4)+(-5)(-1)+0(-2) & 2(1)+(-5)(0)+0(6) \end{bmatrix} = \begin{bmatrix} 13 & 2 \end{bmatrix}$

27. $A + B = \begin{bmatrix} -2 & 5 \\ 0 & 4 \end{bmatrix} + \begin{bmatrix} 1 & 0 \\ -6 & 3 \end{bmatrix} = \begin{bmatrix} (-2)+1 & 5+0 \\ 0+(-6) & 4+3 \end{bmatrix} = \begin{bmatrix} -1 & 5 \\ -6 & 7 \end{bmatrix}$

$A \times B = \begin{bmatrix} -2 & 5 \\ 0 & 4 \end{bmatrix}\begin{bmatrix} 1 & 0 \\ -6 & 3 \end{bmatrix} = \begin{bmatrix} (-2)(1)+5(-6) & (-2)(0)+5(3) \\ 0(1)+4(-6) & 0(0)+4(3) \end{bmatrix} = \begin{bmatrix} -32 & 15 \\ -24 & 12 \end{bmatrix}$

29. $A + B = \begin{bmatrix} 1 & 3 & 0 \\ 2 & 4 & -1 \end{bmatrix} + \begin{bmatrix} 7 & -2 & 3 \\ 2 & -1 & 1 \end{bmatrix} = \begin{bmatrix} 1+7 & 3+(-2) & 0+3 \\ 2+2 & 4+(-1) & -1+1 \end{bmatrix} = \begin{bmatrix} 8 & 1 & 3 \\ 4 & 3 & 0 \end{bmatrix}$

$A \times B = \begin{bmatrix} 1 & 3 & 0 \\ 2 & 4 & -1 \end{bmatrix} \times \begin{bmatrix} 7 & -2 & 3 \\ 2 & -1 & 1 \end{bmatrix}$

Operation cannot be performed because number of columns of A is not equal to number of rows of B.

31. Matrices A and B cannot be added because they do not have the same dimensions.

$A \times B = \begin{bmatrix} 2 & 5 & 1 \\ 8 & 3 & 6 \end{bmatrix} \times \begin{bmatrix} 3 & 2 \\ 4 & 6 \\ -2 & 0 \end{bmatrix} = \begin{bmatrix} 2(3)+5(4)+1(-2) & 2(2)+5(6)+1(0) \\ 8(3)+3(4)+6(-2) & 8(2)+3(6)+6(0) \end{bmatrix} = \begin{bmatrix} 24 & 34 \\ 24 & 34 \end{bmatrix}$

33. Cannot be added; cannot be multiplied

35. $A + B = \begin{bmatrix} 4 & -3 \\ 7 & 1 \end{bmatrix} + \begin{bmatrix} 2 & 3 \\ 6 & 7 \end{bmatrix} = \begin{bmatrix} 4+2 & -3+3 \\ 7+6 & 1+7 \end{bmatrix} = \begin{bmatrix} 6 & 0 \\ 13 & 8 \end{bmatrix}$

$B + A = \begin{bmatrix} 2 & 3 \\ 6 & 7 \end{bmatrix} + \begin{bmatrix} 4 & -3 \\ 7 & 1 \end{bmatrix} = \begin{bmatrix} 2+4 & 3+(-3) \\ 6+7 & 7+1 \end{bmatrix} = \begin{bmatrix} 6 & 0 \\ 13 & 8 \end{bmatrix}$ Thus $A + B = B + A$.

37. $A + B = \begin{bmatrix} 0 & -1 \\ 3 & -4 \end{bmatrix} + \begin{bmatrix} 8 & 1 \\ 3 & -4 \end{bmatrix} = \begin{bmatrix} 0+8 & -1+1 \\ 3+3 & -4+(-4) \end{bmatrix} = \begin{bmatrix} 8 & 0 \\ 6 & -8 \end{bmatrix}$

$B + A = \begin{bmatrix} 8 & 1 \\ 3 & -4 \end{bmatrix} + \begin{bmatrix} 0 & -1 \\ 3 & -4 \end{bmatrix} = \begin{bmatrix} 8+0 & 1+(-1) \\ 3+3 & -4+(-4) \end{bmatrix} = \begin{bmatrix} 8 & 0 \\ 6 & -8 \end{bmatrix}$ Thus $A + B = B + A$.

39. $(A + B) + C = \left(\begin{bmatrix} 6 & 5 \\ -1 & 3 \end{bmatrix} + \begin{bmatrix} 3 & 4 \\ -2 & 7 \end{bmatrix} \right) + \begin{bmatrix} -2 & 4 \\ 5 & 0 \end{bmatrix} = \begin{bmatrix} 9 & 9 \\ -3 & 10 \end{bmatrix} + \begin{bmatrix} -2 & 4 \\ 5 & 0 \end{bmatrix} = \begin{bmatrix} 7 & 13 \\ 2 & 10 \end{bmatrix}$

$A + (B + C) = \begin{bmatrix} 6 & 5 \\ -1 & 3 \end{bmatrix} + \left(\begin{bmatrix} 3 & 4 \\ -2 & 7 \end{bmatrix} + \begin{bmatrix} -2 & 4 \\ 5 & 0 \end{bmatrix} \right) = \begin{bmatrix} 6 & 5 \\ -1 & 3 \end{bmatrix} + \begin{bmatrix} 1 & 8 \\ 3 & 7 \end{bmatrix} = \begin{bmatrix} 7 & 13 \\ 2 & 10 \end{bmatrix}$

Thus, $(A + B) + C = A + (B + C)$.

41. $(A + B) + C = \left(\begin{bmatrix} 3 & 2 \\ 2 & 0 \\ -5 & 9 \end{bmatrix} + \begin{bmatrix} 1 & 8 \\ 0 & 6 \\ 4 & -4 \end{bmatrix} \right) + \begin{bmatrix} 0 & 4 \\ 1 & 9 \\ 9 & -6 \end{bmatrix} = \begin{bmatrix} 4 & 10 \\ 2 & 6 \\ -1 & 5 \end{bmatrix} + \begin{bmatrix} 0 & 4 \\ 1 & 9 \\ 9 & -6 \end{bmatrix} = \begin{bmatrix} 4 & 14 \\ 3 & 15 \\ 8 & -1 \end{bmatrix}$

$A + (B + C) = \begin{bmatrix} 3 & 2 \\ 2 & 0 \\ -5 & 9 \end{bmatrix} + \left(\begin{bmatrix} 1 & 8 \\ 0 & 6 \\ 4 & -4 \end{bmatrix} + \begin{bmatrix} 0 & 4 \\ 1 & 9 \\ 9 & -6 \end{bmatrix} \right) = \begin{bmatrix} 3 & 2 \\ 2 & 0 \\ -5 & 9 \end{bmatrix} + \begin{bmatrix} 1 & 12 \\ 1 & 15 \\ 13 & -10 \end{bmatrix} = \begin{bmatrix} 4 & 14 \\ 3 & 15 \\ 8 & -1 \end{bmatrix}$

Thus, $(A + B) + C = A + (B + C)$.

43. $A \times B = \begin{bmatrix} 1 & -2 \\ 4 & -3 \end{bmatrix} \begin{bmatrix} -1 & -3 \\ 2 & 4 \end{bmatrix} = \begin{bmatrix} 1(-1)-2(2) & 1(-3)-2(4) \\ 4(-1)+(-3)(2) & 4(-3)+(-3)(4) \end{bmatrix} = \begin{bmatrix} -5 & -11 \\ -10 & -24 \end{bmatrix}$

$B \times A = \begin{bmatrix} -1 & -3 \\ 2 & 4 \end{bmatrix} \begin{bmatrix} 1 & -2 \\ 4 & -3 \end{bmatrix} = \begin{bmatrix} -1(1)+(-3)4 & -1(-2)+(-3)(-3) \\ 2(1)+4(4) & 2(-2)+4(-3) \end{bmatrix} = \begin{bmatrix} -13 & 11 \\ 18 & -16 \end{bmatrix}$

Thus, $A \times B \neq B \times A$.

45. $A \times B = \begin{bmatrix} 2 & 0 \\ -4 & 1 \end{bmatrix} \begin{bmatrix} 2 & 0 \\ 8 & 4 \end{bmatrix} = \begin{bmatrix} 2(2)+0(8) & 2(0)+0(4) \\ -4(2)+1(8) & -4(0)+1(4) \end{bmatrix} = \begin{bmatrix} 4 & 0 \\ 0 & 4 \end{bmatrix}$

$B \times A = \begin{bmatrix} 2 & 0 \\ 8 & 4 \end{bmatrix} \begin{bmatrix} 2 & 0 \\ -4 & 1 \end{bmatrix} = \begin{bmatrix} 2(2)+0(-4) & 2(0)+0(1) \\ 8(2)+4(-4) & 8(0)+4(1) \end{bmatrix} = \begin{bmatrix} 4 & 0 \\ 0 & 4 \end{bmatrix}$ Thus, $A \times B = B \times A$.

47. $(A \times B) \times C = \left(\begin{bmatrix} 1 & 3 \\ 4 & 0 \end{bmatrix} \begin{bmatrix} 4 & 2 \\ 3 & 1 \end{bmatrix} \right) \begin{bmatrix} 2 & 1 \\ 3 & 0 \end{bmatrix} = \begin{bmatrix} 13 & 5 \\ 16 & 8 \end{bmatrix} \begin{bmatrix} 2 & 1 \\ 3 & 0 \end{bmatrix} = \begin{bmatrix} 41 & 13 \\ 56 & 16 \end{bmatrix}$

$A \times (B \times C) = \begin{bmatrix} 1 & 3 \\ 4 & 0 \end{bmatrix} \left(\begin{bmatrix} 4 & 2 \\ 3 & 1 \end{bmatrix} \begin{bmatrix} 2 & 1 \\ 3 & 0 \end{bmatrix} \right) = \begin{bmatrix} 1 & 3 \\ 4 & 0 \end{bmatrix} \begin{bmatrix} 14 & 4 \\ 9 & 3 \end{bmatrix} = \begin{bmatrix} 41 & 13 \\ 56 & 16 \end{bmatrix}$ Thus, $(A \times B) \times C = A \times (B \times C)$.

49. $(A \times B) \times C = \left(\begin{bmatrix} 4 & 3 \\ -6 & 2 \end{bmatrix} \begin{bmatrix} 1 & 2 \\ 0 & 1 \end{bmatrix} \right) \begin{bmatrix} 4 & 3 \\ 0 & -2 \end{bmatrix} = \begin{bmatrix} 4 & 11 \\ -6 & -10 \end{bmatrix} \begin{bmatrix} 4 & 3 \\ 0 & -2 \end{bmatrix} = \begin{bmatrix} 16 & -10 \\ -24 & 2 \end{bmatrix}$

$A \times (B \times C) = \begin{bmatrix} 4 & 3 \\ -6 & 2 \end{bmatrix} \left(\begin{bmatrix} 1 & 2 \\ 0 & 1 \end{bmatrix} \begin{bmatrix} 4 & 3 \\ 0 & -2 \end{bmatrix} \right) = \begin{bmatrix} 4 & 3 \\ -6 & 2 \end{bmatrix} \begin{bmatrix} 4 & -1 \\ 0 & -2 \end{bmatrix} = \begin{bmatrix} 16 & -10 \\ -24 & 2 \end{bmatrix}$

Thus, $(A \times B) \times C = A \times (B \times C)$.

51. $A + B = \begin{bmatrix} 85 & 150 & 50 \\ 95 & 162 & 41 \end{bmatrix} + \begin{bmatrix} 73 & 130 & 45 \\ 120 & 200 & 53 \end{bmatrix}$

$= \begin{bmatrix} 85+73 & 150+130 & 50+45 \\ 95+120 & 162+200 & 41+53 \end{bmatrix} = \begin{bmatrix} 158 & 280 & 95 \\ 215 & 362 & 94 \end{bmatrix}$

 S A SC

Total tickets : $\begin{bmatrix} 158 & 280 & 95 \\ 215 & 362 & 94 \end{bmatrix}$ Matinee
 Evening

 Large Small

53. $A \times B = \begin{bmatrix} 2 & 2 & .5 & 1 \\ 3 & 2 & 1 & 2 \\ 0 & 1 & 0 & 3 \\ .5 & 1 & 0 & 0 \end{bmatrix} \begin{bmatrix} 10 & 12 \\ 5 & 8 \\ 8 & 8 \\ 4 & 6 \end{bmatrix} = \begin{bmatrix} 2 \cdot 10 + 2 \cdot 5 + .5 \cdot 8 + 1 \cdot 4 & 2 \cdot 12 + 2 \cdot 8 + .5 \cdot 8 + 1 \cdot 6 \\ 3 \cdot 10 + 2 \cdot 5 + 1 \cdot 8 + 2 \cdot 4 & 3 \cdot 12 + 2 \cdot 8 + 1 \cdot 8 + 2 \cdot 6 \\ 0 \cdot 10 + 1 \cdot 5 + 0 \cdot 8 + 3 \cdot 4 & 0 \cdot 12 + 1 \cdot 8 + 0 \cdot 8 + 3 \cdot 6 \\ .5 \cdot 10 + 1 \cdot 5 + 0 \cdot 8 + 0 \cdot 4 & .5 \cdot 12 + 1 \cdot 8 + 0 \cdot 8 + 0 \cdot 6 \end{bmatrix} = \begin{bmatrix} 38 & 50 \\ 56 & 72 \\ 17 & 26 \\ 10 & 14 \end{bmatrix}$ Sugar
 Flour
 Milk
 Eggs

55. $C(A \times B) = [40 \quad 30 \quad 12 \quad 20] \begin{bmatrix} 38 & 50 \\ 56 & 72 \\ 17 & 26 \\ 10 & 14 \end{bmatrix} = [36.04 \quad 47.52]$ cents $\qquad$ small \$36.04, large \$47.52

57. a) Yes

b) Yes, $\begin{bmatrix} 0 & 0 \\ 0 & 0 \\ 0 & 0 \end{bmatrix}$

c) Yes
d) Answers will vary.
e) Yes
f) Yes

59. $A + B = \begin{bmatrix} 2 & 3 \\ 4 & 5 \end{bmatrix} + \begin{bmatrix} 3 & 4 \\ 5 & 6 \end{bmatrix} = \begin{bmatrix} 2+3 & 3+4 \\ 4+5 & 5+6 \end{bmatrix} = \begin{bmatrix} 5 & 7 \\ 9 & 11 \end{bmatrix}$

$B + A = \begin{bmatrix} 3 & 4 \\ 5 & 6 \end{bmatrix} + \begin{bmatrix} 2 & 3 \\ 4 & 5 \end{bmatrix} = \begin{bmatrix} 3+2 & 4+3 \\ 5+4 & 6+5 \end{bmatrix} = \begin{bmatrix} 5 & 7 \\ 9 & 11 \end{bmatrix}$ $\quad$ Thus $A + B = B + A$. Answers will vary

61. Answers will vary.

63. $A \times I = \begin{bmatrix} a & b \\ c & d \end{bmatrix} \begin{bmatrix} 1 & 0 \\ 0 & 1 \end{bmatrix} = \begin{bmatrix} a & b \\ c & d \end{bmatrix}$

$I \times A = \begin{bmatrix} 1 & 0 \\ 0 & 1 \end{bmatrix} \begin{bmatrix} a & b \\ c & d \end{bmatrix} = \begin{bmatrix} a & b \\ c & d \end{bmatrix}$

Thus, $A \times I = I \times A$

Review Exercises

1. A binary operation is an operation that can be performed on two and only two elements of a set. The result is a single element.

2. A mathematical system consists of a set of elements and at least one binary operation.

3. No. Example: $2 \div 3 = 2/3$, and $2/3$, is not a whole number.

4. Yes. The difference of any two real numbers is always a real number.

5. $10 + 7 = 14 \equiv 5$ (mod 12)   6. $11 + 9 = 8 \equiv 8$ (mod 12)   7. $3 - 6 = -3 \equiv 9$ (mod 12)

8. $7 - 4 + 6 = 9 \equiv 9$ (mod 12)

9. a) The system is closed. If the binary operation is $\boxdot$ then for any elements a and b in the set, a $\boxdot$ b is a member of the set.

   b) There exists an identity element in the set. For any element a in the set, if a $\boxdot$ i = i $\boxdot$ a = a, then i is called the identity element.

   c) Every element in the set has a unique inverse. For any element a in the set, there   exists an element b such that a $\boxdot$ b = b $\boxdot$ a = i. Then b is the inverse of a, and a is the inverse of b.

   d) The set is associative under the operation   For elements a, b, and c in the set, (a $\boxdot$ b) $\boxdot$ c = a $\boxdot$ (b $\boxdot$ c).

10. Yes.

11. The set of integers with the operation of multiplication does not form a group since not all elements have an inverse. Example: $4 \cdot \dfrac{1}{4} = 1$, but $\dfrac{1}{4}$ is not an integer. Only 1 and -1 have inverses.

12. No, there is no identity element.

13. The set of rational numbers with the operation of multiplication does not form a group since zero does not have an inverse. $0 \bullet \underline{?} = 1$

14. There is no identity element and no inverses.

15. Not every element has an inverse.
    Not Associative   Example: (P ? P) ? 4 = L ? 4 = #   P ? (P ? 4) = P ? L = 4;   # $\neq$ 4

16. Not Associative   Example: (! $\square$ p) $\square$? = p $\square$? = ! and ! $\square$ (p $\square$?) = ! $\square$! = ?;   ! $\neq$?

17. a) $\{ ☺ , ●, ♀, ♂ \}$

    b) ⌂

    c) Yes. All the elements in the table are from
       the set $\{ ☺ , ●, ♀, ♂ \}$.

    d) The identity element is ☺ .

    e) Yes.

       ☺ ⌂ ☺ = ☺   ● ⌂ ♂ = ☺
       ♀ ⌂ ♀ = ☺   ♂ ⌂ ● = ☺
       Every element has an inverse.

f) Yes, Associative

$$(\text{☉} \,\triangle\, \text{♀})\,\triangle\, \text{♂} = \text{♀} \,\triangle\, \text{♂} = \text{●}$$

$$\text{☺} \,\triangle\, (\text{♀}\,\triangle\, \text{♂}) = \text{☺} \,\triangle\, \text{●} = \text{●}$$

g) Yes. The elements in the table are symmetric around the diagonal.

$$\text{♂} \,\triangle\, \text{♀} = \text{●} = \text{♀} \,\triangle\, \text{♂}$$

h)   Yes, all five properties are satisfied.

18. $23 \div 6 = 3$, remainder 5 $23 \equiv 5 \ (\text{mod } 6)$

19. $32 \div 4 = 6$, remainder 0 $32 \equiv 0 \ (\text{mod } 8)$

20. $56 \div 9 = 6$, remainder 2 $56 \equiv 2 \ (\text{mod } 9)$

21. $68 \div 8 = 8$, remainder 4 $68 \equiv 4 \ (\text{mod } 8)$

22. $71 \div 12 = 5$, remainder 11 $71 \equiv 11 \ (\text{mod } 12)$

23. $54 \div 14 = 3$, remainder 12 $54 \equiv 12 \ (\text{mod } 14)$

24. $2 + 5 = 7 \equiv 1 \ (\text{mod } 6)$; ? = 1

25. $2 - ? \equiv 3 \ (\text{mod } 7)$

$2 - 0 \equiv 2 \ (\text{mod } 7)$ $2 - 2 \equiv 0 \ (\text{mod } 7)$

$2 - 1 \equiv 1 \ (\text{mod } 7)$ $2 - 3 \equiv 6 \ (\text{mod } 7)$

$2 - 4 \equiv 5 \ (\text{mod } 7)$ $2 - 5 \equiv 4 \ (\text{mod } 7)$

$2 - 6 \equiv 3 \ (\text{mod } 7)$;

? = 6

26. $7 \bullet ? \equiv 5 \ (\text{mod } 8)$

$7 \bullet 0 \equiv 0 \ (\text{mod } 8)$ $7 \bullet 1 \equiv 7 \ (\text{mod } 8)$

$7 \bullet 2 = 14 \equiv 6 \ (\text{mod } 8)$ $7 \bullet 3 = 21 \equiv 5 \ (\text{mod } 8)$

? = 3

27. $? \bullet 4 \equiv 0 \ (\text{mod } 8)$

$0 \bullet 4 \equiv 0 \ (\text{mod } 8)$ $1 \bullet 4 \equiv 4 \ (\text{mod } 8)$

$2 \bullet 4 = 8 \equiv 0 \ (\text{mod } 8)$ $3 \bullet 4 = 12 \equiv 4 \ (\text{mod } 8)$

$4 \bullet 4 = 16 \equiv 0 \ (\text{mod } 8)$ $5 \bullet 4 = 20 \equiv 4 \ (\text{mod } 8)$

$6 \bullet 4 = 24 \equiv 0 \ (\text{mod } 8)$ $7 \bullet 4 = 28 \equiv 4 \ (\text{mod } 8)$

Replace ? with {0, 2, 4, 6}.

28. $10 \bullet 7 \equiv ? \ (\text{mod } 11)$

$10 \bullet 7 = 70$; $70 \div 11 \equiv 6$, remainder 4

Thus, $10 \bullet 7 \equiv 4 \ (\text{mod } 11)$.

Replace ? with 4.

29. $? \bullet 7 \equiv 3 \ (\text{mod } 10)$

$0 \bullet 7 \equiv 0 \ (\text{mod } 10)$ $1 \bullet 7 \equiv 7 \ (\text{mod } 10)$

$2 \bullet 7 = 14 \equiv 4 \ (\text{mod } 10)$ $3 \bullet 7 = 21 \equiv 1 \ (\text{mod } 10)$

$4 \bullet 7 = 28 \equiv 8 \ (\text{mod } 10)$ $5 \bullet 7 = 35 \equiv 5 \ (\text{mod } 10)$

$6 \bullet 7 = 42 \equiv 2 \ (\text{mod } 10)$ $7 \bullet 7 = 49 \equiv 9 \ (\text{mod } 10)$

$8 \bullet 7 = 56 \equiv 6 \ (\text{mod } 10)$ $9 \bullet 7 = 63 \equiv 3 \ (\text{mod } 10)$

$10 \bullet 7 = 70 \equiv 0 \ (\text{mod } 10)$

Replace ? with 9.

30. $? \bullet 3 \equiv 5 \ (\text{mod } 6)$

$0 \bullet 3 \equiv 0 \ (\text{mod } 6)$ $1 \bullet 3 = 3 \equiv 3 \ (\text{mod } 6)$

$2 \bullet 3 = 6 \equiv 0 \ (\text{mod } 6)$ $3 \bullet 3 = 9 \equiv 3 \ (\text{mod } 6)$

$4 \bullet 3 = 12 \equiv 0 \ (\text{mod } 6)$ $5 \bullet 3 = 15 \equiv 3 \ (\text{mod } 6)$

NO SOLUTION

31. $7 \bullet ? \equiv 2 \ (\text{mod } 9)$

$7 \bullet 0 \equiv 0 \ (\text{mod } 9)$ $7 \bullet 1 \equiv 7 \ (\text{mod } 9)$

$7 \bullet 2 = 14 \equiv 5 \ (\text{mod } 9)$ $7 \bullet 3 = 21 \equiv 3 \ (\text{mod } 9)$

$7 \bullet 4 = 28 \equiv 1 \ (\text{mod } 9)$ $7 \bullet 5 = 35 \equiv 8 \ (\text{mod } 9)$

$7 \bullet 6 = 42 \equiv 6 \ (\text{mod } 9)$ $7 \bullet 7 = 49 \equiv 4 \ (\text{mod } 9)$

$7 \bullet 8 = 56 \equiv 2 \ (\text{mod } 9)$

Replace ? with 8.

32.

+	0	1	2	3	4	5
0	0	1	2	3	4	5
1	1	2	3	4	5	0
2	2	3	4	5	0	1
3	3	4	5	0	1	2
4	4	5	0	1	2	3
5	5	0	1	2	3	4

Yes, it is a commutative group.

33.

+	0	1	2	3
0	0	0	0	0
1	0	1	2	3
2	0	2	0	2
3	0	3	2	1

No, no inverse for 0 or 2.

34. Day (mod 10): 0 1 2 3 4 5 6 7 8 9 10 11 12 13

Work/off : w w w o o o o w w w w o o o

↑

today

a) Since $30 \equiv 2 \pmod{14}$, Julie will be working 30 days after today.

b) Since $45 \equiv 3 \pmod{14}$, Julie will have the day off in 45 days.

35. $A + B = \begin{bmatrix} 2 & -1 \\ -3 & 0 \end{bmatrix} + \begin{bmatrix} 3 & 5 \\ 1 & 2 \end{bmatrix} = \begin{bmatrix} 2+3 & (-1)+5 \\ (-3)+1 & 0+2 \end{bmatrix} = \begin{bmatrix} 5 & 4 \\ -2 & 2 \end{bmatrix}$

36. $A - B = \begin{bmatrix} 2 & -1 \\ -3 & 0 \end{bmatrix} - \begin{bmatrix} 3 & 5 \\ 1 & 2 \end{bmatrix} = \begin{bmatrix} 2-3 & (-1)-5 \\ (-3)-1 & 0-2 \end{bmatrix} = \begin{bmatrix} -1 & -6 \\ -4 & -2 \end{bmatrix}$

37. $3A - 2B = 3\begin{bmatrix} 2 & -1 \\ -3 & 0 \end{bmatrix} - 2\begin{bmatrix} 3 & 5 \\ 1 & 2 \end{bmatrix} = \begin{bmatrix} 6 & -3 \\ -9 & 0 \end{bmatrix} - \begin{bmatrix} 6 & 10 \\ 2 & 4 \end{bmatrix} = \begin{bmatrix} 6-6 & (-3)-10 \\ (-9)-2 & 0-4 \end{bmatrix} = \begin{bmatrix} 0 & -13 \\ -11 & -4 \end{bmatrix}$

38. $A \times B = \begin{bmatrix} 2 & -1 \\ -3 & 0 \end{bmatrix}\begin{bmatrix} 3 & 5 \\ 1 & 2 \end{bmatrix} = \begin{bmatrix} 2(3)+(-1)(1) & 2(5)+(-1)(2) \\ (-3)(3)+0(1) & (-3)(5)+0(2) \end{bmatrix} = \begin{bmatrix} 5 & 8 \\ -9 & -15 \end{bmatrix}$

39. $B \times A = \begin{bmatrix} 3 & 5 \\ 1 & 2 \end{bmatrix}\begin{bmatrix} 2 & -1 \\ -3 & 0 \end{bmatrix} = \begin{bmatrix} 3(2)+5(-3) & 3(-1)+5(0) \\ 1(2)+2(-3) & 1(-1)+2(0) \end{bmatrix} = \begin{bmatrix} -9 & -3 \\ -4 & -1 \end{bmatrix}$

Chapter Test

1. A mathematical system consists of a set of elements and at least one binary operation.

2. Closure, identity element, inverses, associative property, and commutative property.

3. Yes.

4. No, the set of natural numbers is not a commutative group under the operation of subtraction because it is not closed, not associative, and not commutative.

5.

+	1	2	3	4	5
1	2	3	4	5	1
2	3	4	5	1	2
3	4	5	1	2	3
4	5	1	2	3	4
5	1	2	3	4	5

6. Yes. It is closed since the only elements in the table are from the set {1, 2, 3, 4, 5}. The identity element is 5. The inverses are $1 \leftrightarrow 4, 2 \leftrightarrow 3, 3 \leftrightarrow 2$, $4 \leftrightarrow 1$, and $5 \leftrightarrow 5$. The system is associative. The system is commutative since the table is symmetric about the main diagonal. Thus, all five properties are satisfied.

7. a) The binary operation is □ .

   b) Yes. All elements in the table are from the set {W, S, T, R}.

   c) The identity element is T, since T □ x = x = x □ T, where x is any member of the set {W, S, T, R}.

   d) The inverse of R is S, since R □ S = T and S □ R = T.

   e) (T □ R) □ W = R □ W = S

8. The system is not a group. It does not have the closure property since c * c = d, and d is not a member of {a, b, c}, and it is not associative.

9. Since all the numbers in the table are elements of {1, 2, 3}, the system is closed. The commutative property holds since the elements are symmetric about the main diagonal. The identity element is 2 and the inverses are 1 – 3, 2 – 2, 3 – 1. If it is assumed the associative property holds as illustrated by the example: (1 ? 2) ? 3 = 2 = 1 ? (2 ? 3), then the system is a commutative group.

10. Since all the numbers in the table are elements of {@, $, &, %}, the system is closed. The commutative property holds since the elements are symmetric about the main diagonal. The identity element is $ and the inverses are @ – &, $ – $, & – @, % – %. It is assumed the associative property holds as illustrated by the example: (@ O $) O % = & = @ O ($ O %), then the system is a commutative group.

11. 39 ÷ 4 = 9, remainder 3   39 ≡ 3 (mod 4)

12. 107 ÷ 11 = 9, remainder 8   107 ≡ 8 (mod 11)

13. 4 + 5 = 9 ≡ 3 (mod 6); replace ? with 4

14. 2 – 3 – (5 + 2) -3 = 4 ≡ 4 (mod 5); replace ? with 2.

15. 3 – 5 ≡ 7 (mod 9)

   3 – 5 = (3 + 9) – 5 = 12 – 5 ≡ 7 (mod 9)

   12 – 5 ≡ 7 (mod 9)

   Replace ? with 5.

16. 3 • ? • ≡ 2 (mod 6)

3 • 0 ≡ 0 (mod 6)	3 • 1 ≡ 3 (mod 6)
3 • 2 ≡ 0 (mod 6)	3 • 3 ≡ 3 (mod 6)
3 • 4 ≡ 0 (mod 6)	3 • 5 ≡ 3 (mod 6)

   There is no solution for ?   The answer is { }.

17. a)

×	0	1	2	3	4
0	0	0	0	0	0
1	0	1	2	3	4
2	0	2	4	1	3
3	0	3	1	4	2
4	0	4	3	2	1

   b) The system is closed. The identity is 1. However, 0 does not have an inverse, so the system is not a commutative group.

18. $A + B = \begin{bmatrix} 1 & 0 \\ 4 & 6 \end{bmatrix} + \begin{bmatrix} -2 & 2 \\ 5 & 3 \end{bmatrix} = \begin{bmatrix} 1+(-2) & 0+2 \\ (-4)+5 & 6+3 \end{bmatrix} = \begin{bmatrix} -1 & 2 \\ 1 & 9 \end{bmatrix}$

19. $2A - 3B = 2\begin{bmatrix} 1 & 0 \\ -4 & 6 \end{bmatrix} - 3\begin{bmatrix} -2 & 2 \\ 5 & 3 \end{bmatrix} = \begin{bmatrix} 2 & 0 \\ -8 & 12 \end{bmatrix} - \begin{bmatrix} -6 & 6 \\ 15 & 9 \end{bmatrix} = \begin{bmatrix} 2-(-6) & 0-6 \\ (-8)-15 & 12-9 \end{bmatrix} = \begin{bmatrix} 8 & -6 \\ -23 & 3 \end{bmatrix}$

20. $A \times B = \begin{bmatrix} 1 & 0 \\ -4 & 6 \end{bmatrix}\begin{bmatrix} -2 & 2 \\ 5 & 3 \end{bmatrix} = \begin{bmatrix} 1(-2)+0(5) & 1(2)+0(3) \\ (-4)(-2)+6(5) & (-4)(2)+6(3) \end{bmatrix} = \begin{bmatrix} -2 & 2 \\ 38 & 10 \end{bmatrix}$

CHAPTER TEN

CONSUMER MATHEMATICS

Exercise Set 10.1

1. 100

3. 100

5. Previous

7. $\dfrac{3}{5} = 0.60 = (0.6)(100)\% = 60.0\%$

9. $\dfrac{11}{20} = 0.55 = (0.55)(100)\% = 55.0\%$

11. $0.007654 = (0.007654)(100)\% = 0.8\%$

13. $3.78 = (3.78)(100)\% = 378\%$

15. $7\% = \dfrac{7}{100} = 0.07$

17. $5.15\% = \dfrac{5.15}{100} = 0.0515$

19. $\dfrac{1}{4}\% = 0.25\% = \dfrac{0.25}{100} = 0.0025$

21. $135.9\% = \dfrac{135.9}{100} = 1.359$

23. $\dfrac{5}{20} = \dfrac{25}{100} = 25\%$

25. $8(.4125) = 3.3 \qquad 8.0 - 3.3$
    $\qquad\qquad\qquad\quad = 4.7\ g$

27. $(1743)(0.27) = 471$ comedy DVDs

29. $(1743)(0.11) = 192$ horror DVDs

31. $(25)(0.8584) = 21.46$ ml of oxygen

33. $(500)(0.0194) = 9.7$ ml of chlorine

35. $\dfrac{921}{1960} = 0.4698 \qquad 0.470 = 47.0\%$

37. $\dfrac{274}{1960} = 0.1398 \qquad 0.140 = 14.0\%$

39. $\dfrac{144\ \text{mill.} - 148\ \text{mill.}}{148\ \text{mill.}} = -0.027 = 2.7\%$ decrease

41. a) $\dfrac{1.68 - 0.86}{0.86} = 0.953 = 95.3\%$ increase

    b) $\dfrac{1.75 - 1.68}{1.68} = 0.042 = 4.2\%$ increase

    c) $\dfrac{2.94 - 1.75}{1.75} = 0.68 = 68\%$ increase

    d) $\dfrac{3.54 - 2.94}{2.94} = 0.204 = 20.4\%$ increase

43. $(0.15)(45) = \$\,6.75$

45. $(0.12)(40) = 4.8$

47. $24/96 = 0.25$; $(0.25)(100\%)$
    $= 25\%$

49. $54/600 = 0.09$; $(0.09)(100\%)$
    $= 9\%$

51. $0.05x = 15$; $x = 15/0.05 = 300$

53. $21 = 0.14x$, $x = 21/0.14 = 150$

55. a) tax = 6% of $43.50 = (0.06)(43.50) = \$2.61$
    b) tip = 15% of $43.50 = 0.15(43.50) = \$6.53$
    c) total cost = $43.50 + 6.53 + 2.61 = \$52.64$

57. $1.50(x) = 18$ $x = \dfrac{18}{1.50} = 12$

    12 students got an A on the 2nd test.

59. Mr. Browns' increase was $0.07(36,500) = \$2,555$
    His new salary = $\$36,500 + \$2,555 = \$39,055$

61. Percent change $= \left(\dfrac{407 - 430}{430}\right)(100\%) =$

    $\left(\dfrac{-23}{430}\right)(100\%) = -5.3\%$

63. Percent decrease from regular price =

    $\left(\dfrac{\$439 - 539.62}{539.62}\right)(100\%) = \left(\dfrac{-100.62}{539.62}\right)(100\%) =$

    -18.6%

    The sale price is 18.6% lower than the regular price.

65. Percent markup $= \left(\dfrac{49 - 35}{35}\right)(100\%) =$

    $(0.40)(100) = 40\%$ markup

67. $1000 increased by 10% is $1000 + 0.10(\$1000) = \$1000 + \$100 = \$1,100$.
    $1,100 decreased by 10% is $\$1,100 - 0.10(\$1,100) = \$1,100 - \$110 = \$990$.
    Therefore if he sells the car at the reduced price he will lose $10.

69. a) No, the 25% discount is greater. (see part b)
    b) $189.99 - 0.10(189.99) = 189.99 - 19.00 = 170.99$ $170.99 - 0.15(170.99) = 170.99 - 25.65 = \$145.34$
    c) $189.99 - 0.25(189.99) = 189.99 - 47.50 = \$142.49$
    d) Yes

71. a) $200 decreased by 25%
       is greater by $25.
    b) $100 increased by 50%
       is greater by $50
    c) $100 increased by 100%
       is greater by $200.

Exercise Set 10.2
1. Principal
3. Interest
5. Rate
7. United States

9. $i = prt = (425)(0.0192)(6) = \$48.96$

11. $i = (1100)(0.0875)(90/360) = \$24.06$

13. $i = prt = (587)(0.00045)(60) = \$15.85$

15. $i = (6,712)(0.02625)(59/360) = \$28.88$

17. $i = (1372.11)(0.01375)(6) = \$113.20$

19. $(1725)(r)(3) = 82.80$

    $r = \dfrac{82.80}{(1725)(3)}(100\%) = 1.6\%$

21. $175 = p(0.06)\left(\dfrac{5}{12}\right) = p(0.025)$ $p = \dfrac{175}{0.025} = \$7000$

23. $124.49 = (957.62)(0.065)(t) = 62.2453t$

    $\dfrac{124.49}{62.2453} = t$ $t = 2$ years

25. $i = (25,000)(0.015)(2) = 750$
Repayment $= p + i = 25,000 + 750 = \$25,750$

27. a) $i = prt$ $i = (3500)(0.075)(6/12) = \$131.25$
b) $A = p + i$ $A = 3500 + 131.25 = \$3,631.25$

29. a) $i = prt$ $i = (3650)(0.076)(18/12) = \$416.10$
b) $3650.00 - 416.10 = \$3233.90$, which is the amount Julie received.
c) $i = prt$ $416.10 = (3233.90)(r)(18/12) = 4850.85r$

$$\frac{416.10}{4850.85} = r = 0.0858 \text{ or } 8.58\%$$

31. $i = 280.50 - 270.00 = 10.50$
$10.50 = (270.00)(r)(7/360) = 5.25r$
$$r = \frac{10.50}{5.25} = 2.00 \text{ or } 200\%$$

33. [Apr 18–Nov 11]: $315 - 108 = 207$ days

35. [02/02–10/31]: $304 - 33 = 271$ days
Because of Leap Year, $271 + 1 = 272$ days

37. [08/24–05/15]: $(365 - 236) + 135 = 129 + 135 = 264$ days

39. [07/04] for 150 days: $185 + 150 = 335$, which is December 1

41. [11/25] for 120 days: $329 + 120 = 449$;
$449 - 365 = 84$ $84 - 1$ leap year day = day 83, which is March 24

43. [04/01 to 05/01]: $121 - 91 = 30$ days
$(2400)(0.02)(30/360) = 4.00$
$1200.00 - 4.00 = 1196.00$
$2400.00 - 1196.00 = \$1204.00$

[05/01 to 07/30]: $211 - 121 = 90$ days
$(1204.00)(0.02)(90/360) = 6.02$
$1204.00 + 6.02 = \$1210.02$

45. [02/01 to 05/01]: $= 121 - 32 = 89$ days
$(2400)(0.055)(89/360) = \$32.63$
$1000.00 - 32.63 = \$967.37$
$2400 - 967.37 = \$1432.63$

[05/01 to 08/31]: $= 243 - 121 = 122$ days
$(1432.63)(0.055)(122/360) = \$26.70$
$1432.63 + 26.70 = \$1459.33$

47. [07/15 to 12/27]: $361 - 196 = 165$ days
$(9000)(0.06)(165/360) = \$247.50$
$4000.00 - 247.50 = \$3752.50$
$9000.00 - 3752.50 = \$5247.50$

[12/27 to 02/01]: $(365 - 361) + 32 = 36$ days
$(5247.50)(0.06)(36/360) = \$31.49$
$5247.50 + 31.49 = \$5278.99$

49. [08/01 to 09/01]: $244 - 213 = 31$ days
$(1800)(0.15)(31/360) = \$ 23.25$
$500.00 - 23.25 = \$476.75$
$1800.00 - 476.75 = \$1323.25$

[09/01 to 10/01]: $274 - 244 = 30$ days
$(1323.25)(0.15)(30/360) = \$16.54$
$500.00 - 16.54 = \$483.46$
$1323.25 - 483.46 = \$839.79$

[10/01 to 11/01]: $305 - 274 = 31$ days
$(839.79)(0.15)(31/360) = \$10.85$
$839.79 + 10.85 = \$850.64$

51. [03/01 to 08/01]: $213 - 60 = 153$ days
$(11,600)(0.06)(153/360) = \$295.80$
$2000.00 - 295.80 = \$1704.20$
$11,600.00 - 1704.20 = \$9895.80$

[08/01 to 11/15]: $319 - 213 = 106$ days
$(9895.80)(0.06)(106/360) = \$174.83$
$4000.00 - 174.83 = \$3825.17$
$9895.8 - 3825.17 = \$6070.63$

[11/15 to 12/01]: $335 - 319 = 16$ days
$(6070.63)(0.06)(16/360) = \$16.19$
$6070.63 + 16.19 = \$6086.82$

53. [03/01 to 05/01]: $121 - 60 = 61$ days
    $(6500)(0.105)(61/360) = \$115.65$
    $1750.00 - 115.65 = \$1634.35$
    $6500.00 - 1634.35 = \$4865.65$

    [05/01 to 07/01]: $182 - 121 = 61$ days
    $(4865.65)(0.105)(61/360) = \$86.57$
    $2350.00 - 86.57 = \$2263.43$
    $4865.65 - 2263.43 = \$2602.22$

    Since the loan was for 180 days, the maturity
    date occurs in $180 - 61 - 61 = 58$ days
    $(2602.22)(0.105)(58/360) = \$44.02$
    $2602.22 + 44.02 = \$2646.24$

55. a) Nov. 4 is day 318 $318 + 364 = 682$
    $682 - 365 = 317$. Day 317 is Nov. 13, 2015
    b) $i = (1000)(0.0015)(364/360) = \$1.52$
    Amt. paid $= 1000 - 1.52 = \$998.48$
    c) interest $= \$1.52$
    d) $r = \dfrac{i}{pt} = \dfrac{1.52}{998.48\left(\frac{364}{360}\right)} = 0.001506$ or
    0.1506%

57. a) Amt. received $= 743.21 - 39.95 = \$703.26$
    $i = prt$
    $39.95 = (703.26)(r)(5/360)$
    $39.95 = (9.7675)(r)$
    $r = 39.95/9.7675 = 4.0901$ or 409.01%
    b) $39.95 = (703.26)(r)10/360$
    $39.95 = (19.535)(r)$
    $r = 39.95/19.535 = 2.045$ or 204.50%
    c) $39.95 = (703.26)(r)(20/360)$
    $39.95 = (39.07)r$
    $r = 39.95/39.07 = 1.0225$ or 102.25%

59. a) $(25,000)(0.03)(180/360) = \$375$
    b) $(25,000)(0.03)(180/365) = \$369.86$
    c) $375 - 369.86 = \$5.14$
    d) $5.14/369.86 = 0.0139 = 1.39\%$

Exercise Set 10.3

1. Profit
3. Variable
5. Compound

7. $n = 2$, $r = 0.014$, $t = 6$, $p = 1000$
    a) $A = 1000\left(1 + \dfrac{0.014}{2}\right)^{(2)(6)} = \$1,087.31$
    b) $i = 1087.31 - 1000 = \$87.31$

9. $n = 4$, $r = 0.03$, $t = 4$, $p = 2000$
    a) $A = 2000\left(1 + \dfrac{0.03}{4}\right)^{(4)(4)} = \$2,253.98$
    b) $i = 2253.98 - 2000 = \$253.98$

11. $n = 12$, $r = 0.055$, $t = 3$, $p = 7000$
    $A = 7000\left(1 + \dfrac{0.055}{12}\right)^{(12)(3)} = \$8,252.64$
    $i = 8252.64 - 7000 = \$1,252.64$

13. $n = 360$, $r = 0.04$, $t = 2$, $p = 8000$
    $A = 8000\left(1 + \dfrac{0.04}{360}\right)^{(360)(2)} = \$8,666.26$
    $i = 8666.26 - 8000 = \$666.26$

15. $n = 1$, $r = 0.036$, $t = 10$, $A = 45,000$
    $p = \dfrac{45,000}{\left(1 + \dfrac{0.036}{12}\right)^{(12)(10)}} = \$31,412.36$

17. $n = 4$, $r = 0.04$, $t = 4$, $A = 100,000$
    $p = \dfrac{100,000}{\left(1 + \dfrac{0.04}{4}\right)^{(4)(4)}} = \$85,282.13$

19. $A = 6500\left(1 + \dfrac{0.0186}{2}\right)^{(2)(5)} = \$7,130.44$

21. $A = 1500\left(1 + \dfrac{0.039}{12}\right)^{12 \cdot 2.5} = \$1,653.36$

23. $A = 10,000\left(1 + \dfrac{0.0175}{2}\right)^{10} = \$10,910.27$

25. $A = 3000\left(1 + \dfrac{0.0175}{4}\right)^{4 \cdot 2} = \$3,106.62$

27. $A = 2250\left(1 + \dfrac{0.02}{365}\right)^{365 \cdot 2} = \$2,341.82$

29. $A = 1\left(1 + \dfrac{0.035}{2}\right)^{2 \cdot 1} \approx 1.0353$

$i = A - 1 = 1.0353 - 1 = 0.0353$

APY $= 3.53\%$

31. a) $A = 1\left(1 + \dfrac{0.019}{4}\right)^{4 \cdot 1} \approx 1.0191$

$i = A - 1 = 1.0191 - 1 = 0.0191$

APY $= 1.91\%$

b) $A = 1\left(1 + \dfrac{0.018}{360}\right)^{360 \cdot 1} \approx 1.0182$

$i = A - 1 = 1.0182 - 1 = 0.0182$

APY $= 1.82\%$

c) Prospero Bank

33. $A = 1\left(1 + \dfrac{0.024}{12}\right)^{12 \cdot 1} \approx 1.0243$

$i = A - 1 = 1.0243 - 1 = 0.0243$

Yes, APY $= 2.43\%$, not 2.6%

35. Present value $= \dfrac{276,000}{\left(1 + \dfrac{0.036}{12}\right)^{(12)(5)}} = \$230,596.71$

37. Present value $= \dfrac{20,000}{\left(1 + \dfrac{0.0155}{4}\right)^{(4)(18)}} = \$15,138.96$

39. a) $925,000 - 370,000 = \$555,000$

b) $\dfrac{555,000}{\left(1 + \dfrac{0.075}{12}\right)^{(12)(30)}} = \$58,907.61$

c) surcharge $= \dfrac{58,907.60}{598} = \$98.51$

41. a) $A = 1000\left(1 + \dfrac{0.02}{2}\right)^{4} = \$1,040.60$

$i = \$1040.60 - \$1000 = \$40.60$

b) $A = 1000\left(1 + \dfrac{0.04}{2}\right)^{4} = \$1,082.43$

$i = \$1082.43 - \$1000 = \$82.43$

c) $A = 1000\left(1 + \dfrac{0.08}{2}\right)^{4} = \$1,169.86$

$i = \$1169.86 - \$1000 = \$169.86$

d) No predictable outcome.

43. $i = prt = (100,000)(0.05)(4) = \$20,000$

$A = 100,000\left(1 + \dfrac{0.05}{360}\right)^{360 \cdot 4} = \$122,138.58$

interest earned: $\$122,138.58 - 100,000 = \$22,138.58$

Select investing at the compounded daily interest because the compound interest is greater by $\$2,138.58$.

45. $p = 2000, A = 3586.58, n = 12, t = 5$

$3586.58 = 2000\left(1 + \dfrac{r}{12}\right)^{60}$

$\dfrac{3586.58}{2000} = \left(1 + \dfrac{r}{12}\right)^{60}$

$(1.79329)^{1/60} = 1 + \dfrac{r}{12} = 1.00978$

$0.00978 = \dfrac{r}{12}$ $r = 0.00978(12) = .1174$ or 11.74%

47. $A = 2000\,[1 + (.08/2)]^6 = 2000\,(1.04)^6 = \$2530.64$

$i = \$2530.64 - 2000 = \$530.64$

Simple interest: $i = prt = 530.64 = 2000(r)(3)$

$530.64 = 6000r$ $r = \dfrac{530.64}{6000} = 0.0884$ or $8.84\,\%$

Exercise Set 10.4

1. Installment

3. Annual

5. Installment

7. From Table 10.2 the finance charge per $100 at 6 % for 12 payments is 3.28.

Total finance charge $= (3.28)\left(\dfrac{985}{100}\right) = \$32.31$

Total amount due $= 985.00 + 32.31 = \$1017.31$

Monthly payment $= \dfrac{1017.31}{12} = \$84.78$

9. From Table 10.2 the finance charge per $100 at 6.5 % for 48 payments is 13.83.

Total finance charge $= (13.83)\left(\dfrac{11,000}{100}\right) = \$1521.30$

Total amount due $= 11,000.00 + 1521.30 = \$12,521.30$

Monthly payment $= \dfrac{12,521.30}{48} = \$260.86$

11. a) From Table 10.2 the finance charge per $100 at 4 % for 48 payments is 8.38.

Total finance charge $= (8.38)\left(\dfrac{7650}{100}\right) = \$641.07$

b) Total amount due $= 7650.00 + 641.07 = \$8291.07$

Monthly payment $= \dfrac{8291.07}{48} = \$172.73$

13. a) Down payment = 0.20(7500) = $1500; Jaime borrowed 7500 − 1500 = $6000

Total installment price = (36 · 189.40) = $6818.40

Finance charge = 6818.40 − 6000 = $818.40

b) $\left(\dfrac{\text{finance charge}}{\text{amt. financed}}\right)(100) = \left(\dfrac{818.40}{6000}\right)(100) = 13.64$

From Table 10.2 for 36 payments, the value of 13.64 corresponds with an APR of 8.5 %.

15. a) Down payment = $17,000   Amount financed = 56,214 − 17,000 = $39,214

Total installment price = (767.29)(60) = $46,037.40

Finance charge = 46,037.40 − 39,214 = $6823.40

b) $\left(\dfrac{\text{finance charge}}{\text{amt. financed}}\right)(100) = \left(\dfrac{6823.40}{39,214}\right)(100) = 17.40$

From Table 10.2, for 60 payments, the value of $17.40 corresponds with an APR of 6.5 %.

17. Down payment = $0.00   Amount financed = $12,000.00   Monthly payments = $232.00

a) Installment price = (60)(232) = 13,920

Finance charge = $13,920.00 − $12,000.00 = $1920.00

$\left(\dfrac{\text{finance charge}}{\text{amt. financed}}\right)(100) = \left(\dfrac{1920.00}{12,000}\right)(100) = 16.00$

From Table 10.2, for 60 payments, the value of $16.00 corresponds to an APR of 6.0 %.

b) From Table 10.2, the monthly payment per $100 for the remaining 36 months at 6.0% APR is $9.52.

$$u = \dfrac{nPV}{100+V} = \dfrac{(36)(232)(9.52)}{(100+9.52)} = \dfrac{79,511.04}{109.52} = 725.9956 = \$726.00$$

c) (232)(23) = 5336;   5336 + 726 = 6062;   13,920 − 6062 = $7,858.00

19. a) Amount financed = 32,000 − 10,000 = $22,000

From Table 10.2, the finance charge per 100 financed at 8 % for 36 payments is 12.81.

Total finance charge = $(12.81)\left(\dfrac{22,000}{100}\right) = 2818.20$

b) Total amt. due = 22,000 + 2818.20 = $24,818.20

Monthly payment = $\dfrac{24,818.20}{36} = \$689.39$

c) From Table 10.2, the monthly payment per $100 for the remaining 12 months at 8.0% APR is $4.39.

$$u = \dfrac{nPV}{100+V} = \dfrac{(12)(689.39)(4.39)}{100+4.39} = \dfrac{36,317.07}{104.39} = \$347.90$$

d) (23)(689.39) = 15,855.97;   15,855.97 + 347.90 = 16,203.87;   24,818.20 − 16,203.87 = $8614.33

21. a) $(0.01)(2600) = \$26$

    b) Principal on which interest is charged during April:

    $2600 - 500 = \$2100$

    Interest for April:

    $(2100)(0.00039698)(30) = \$25.01$

    1% of outstanding principal:

    $(0.01)(2100) = \$21.00$

    Minimum monthly payment:

    $25.01 + 21.00 = \$46.01$ which rounds up to \$47.

23. a) Total charges: $677 + 452 + 139 + 141 = \$1409$

    $(0.015)(1409) = \$21.14$ which rounds up to \$22

    b) Principal on which interest is paid during December:

    $1409 - 300 = \$1109$

    Interest for December:

    $(1109)(0.0005163)(31) = \$17.75$

    1.5% of outstanding principal:

    $(0.015)(1109) = \$16.64$

    Minimum monthly payment:

    $17.75 + 16.64 = \$34.39$ which rounds up to \$35.

25. a) Finance charge $= (1203.85)(0.013)(1) = \$15.65$

    b) Bal. due May 5 $= (1203.85 + 15.65 + 106.94) - 525 = \$801.44$

27. a) Finance charge $= (124.78)(0.0125)(1) = \$1.56$

    b) old balance + finance charge – payment + art supplies + flowers + music CD = new balance

    $124.78 + 1.56 - 100.00 + 25.64 + 67.23 + 13.90 = \$133.11$

29. a)

Date	Balance Due	Number of Days	(Balance)(Days)
May 12	$378.50	1	$(378.50)(\ 1) = \$\ \ 378.50$
May 13	$508.29	2	$(508.29)(\ 2) = \ \ 1{,}016.58$
May 15	$458.29	17	$(458.29)(17) = \ \ 7{,}790.93$
June 01	$594.14	7	$(594.14)(\ 7) = \ \ 4{,}158.98$
June 08	$631.77	4	$\underline{(631.77)(\ 4) = \ \ 2{,}527.08}$
		31	sum $= \$15{,}872.07$

Average daily balance $= \dfrac{15{,}872.07}{31}$

$= \$512$

b) Finance charge $= prt =$

$(512.00)(0.013)(1) = \$6.66$

c) Balance due $= 631.77 + 6.66 =$

$\$638.43$

31. a)

Date	Balance Due	Number of Days	(Balance)(Days)
Feb. 03	$124.78	5	$(124.78)(5) = \$623.90$
Feb. 08	$150.42	4	$(150.42)(4) = 601.68$
Feb. 12	$ 50.42	2	$(50.42)(2) = 100.84$
Feb. 14	$117.65	11	$(117.65)(11) = 1,294.15$
Feb. 25	$131.55	6	$(131.55)(6) = \underline{789.30}$
Mar 3		28	sum $= \$3,409.87$

Average daily balance $= \dfrac{3409.87}{28}$

$= \$121.78$

b) Finance charge $= prt =$
$(121.78)(0.0125)(1) = \$1.52$

c) Balance due $= 131.55 + 1.52$
$= \$133.07$

d) The finance charge using the avg. daily balance method is $0.04 less than the finance charge using the unpaid balance method.

33. a) $i = (875)(0.0004273)(32) = \$11.96$ b) $A = 875 + 11.96 = \$886.96$

35. $1000.00   5 %   6 payments
   a) State National Bank (SNB): $(1000)(0.05)(0.5) = \$25.00$
   b) Consumers Credit Union (CCU): $(86.30)(12) = 1,035.60$
      $1035.60 - 1000.00 = \$35.60$
   c) $\left(\dfrac{25}{1000}\right)(100) = 2.50$

   In Table 10.2, $2.49 is the closest value to $2.50, which corresponds to an APR of 8.5 %.

   d) $\left(\dfrac{35.60}{1000}\right)(100) = 3.56$

   In Table 10.2, $3.56 corresponds to an APR of 6.5 %.

37. Let $p =$ amount Ken borrowed
   $p + 2500 =$ purchase price
   Installment price: $2500 + (379.50)(36) = \$16,162$
   Interest $=$ Installment price $-$ purchase price
   $i = 16,162 - (p + 2500) = 16,162 - p - 2500 = 13,662 - p$

   Since $i = prt$ we have:
   $13,662 - p = (p)(0.06)(3) =$
   $13,662 = 0.18p + p$ $p = 11,577.97$
   purchase price $= 11,577.97 + 2500 =$
   $\$14,077.97$

39. $35,000   15 % down payment   60 month fixed loan   APR $= 8.5 %$
   $(35,000)(.15) = 5250$ $35,000 - 5250 = 29750$
   a) From Table 10.2, 60 payments at an APR of 8.5 % yields a finance charge of $23.10 per $100.
   $\left(\dfrac{29,750}{100}\right)(23.10) = \$6872.25$

   b) $29,750.00 + 6872.25 = 36622.25$ $\dfrac{36,622.25}{60} = \$610.37$

   c) In Table 10.2, 36 payments at an APR of 8.5 % yields a finance charge of $13.64 per $100.
   $u = \dfrac{(36)(610.37)(13.64)}{100 + 13.64} = \dfrac{29,9716.08}{113.64} = \$2637.42$

   d) $u = \dfrac{f \cdot k(k+1)}{n(n+1)} = \dfrac{(6872.25)(36)(37)}{60(61)} = \dfrac{9,153,837}{3660} = \$2501.05$

   e) The actuarial method

Exercise Set 10.5

1. Mortgage

3. Points

5. Adjusted

7. From Table 10.4 the monthly principal and interest payment per $1000 at 6 % for 15 years is 8.43857.

Monthly principal and interest payment $= (8.43857)\left(\dfrac{85,000}{100}\right) = \$717.28$

9. From Table 10.4 the monthly principal and interest payment per $1000 at 5.5 % for 20 years is 6.87887.

Monthly principal and interest payment $= (6.87887)\left(\dfrac{285,000}{100}\right) = \$1,960.48$

11. a) Down payment = 15% of $225,000
    (0.15)(225,000) = $33,750
  b) amt. of mortgage = 225,000 – 33,750 = $191,250
    Table 10.4 yields $5.36822 per $1000 of
    mortgage
  c) Monthly payment $= \left(\dfrac{191,250}{1000}\right)(5.36822)$
    $= \$1,026.67$

13. a) Down payment = 19% of $2,100,000
    (0.19)(2,100,000) = $399,000
  b) amt. of mortgage = 2,100,000 – 399,000
    = $1,701,000
    Table 10.4 yields $6.75207 per $1000 of
    mortgage
  c) Monthly payment $= \left(\dfrac{1,701,000}{1000}\right)(6.75207)$
    $= \$11,485.27$

15. a) Down payment = 20% of $195,000
    (0.20)(195,000) = $39,000
  b) amt. of mortgage = 195,000 – 39,000 = $156,000
  c) (156,000)(0.02) = $3,120.00

17. $3200 = monthly income
    3200 – 335 = 2865 = adjusted monthly income
  a) (0.28)(2865) = $802.20
  b) Table 10.4 yields $7.91 per $1000 of mortgage
    $\left(\dfrac{150,000}{1000}\right)(7.91) = \$1186.50$
    1186.50 + 225.00 = $1411.50
  c) No; $1,411.50 > $802.20

19. a) (490.18)(30)(12) = $176,464.80
    176,464.80 + 11,250.00 = $187,714.80
  b) 187,714.80 – 75,000 = $112,714.80
  c) $i = prt = (63,750)(0.085)(1/12) = 451.56$
    490.18 – 451.56 = $38.62

21. a) down payment = (0.20)(550,000) = $110,000
  b) amount of mortgage = 550,000 – 110,000
    = $440,000
    cost of three points = (0.03)(440,000)
    = $13,200
  c) 15,375 – 995 = $14,380.00 adjusted monthly
    income
  d) At a rate of 5.5% for 30 years, Table 10.4
    yields 5.67789.
    mortgage payment $= \left(\dfrac{440,000}{1000}\right)(5.67789)$
    $= \$2498.27$

  e) Insurance + taxes $= \dfrac{5634 + 2325}{12} = \$663.25$ /mo.
    2498.27 + 663.25 = $3,161.52 total mo. Payment
  f) Yes, Since $4,026.40 is greater than $3161.52, the
    Nejems qualify.
  g) interest on first payment $= i = prt =$
    (440,000)(0.055)(1/12) = $2016.67
    amount applied to principal = 2498.27 – 2016.67
    = $481.60

23. <u>Bank A</u>   Down payment $= (0.10)(105,000) =$ $10,500

amount of mortgage $105,000 - 10,500 = \$94,500$

At a rate of 4% for 30 years, Table 10.4 yields $4.77415.

monthly mortgage payment =

$$\left(\frac{94,500}{1000}\right)(4.77415) = \$451.16$$

cost of three points $= (0.03)(94,500) = \$2835$

Total cost of the house =

$10,500 + 2835 + (451.16)(12)(30) =$ $175,752.60

<u>Bank B</u>   Down payment $= (0.20)(105,000) =$ $21,000

amount of mortgage $105,000 - 21,000 = \$84,000$

At a rate of 5.5% for 25 years,   Table 10.4 yields $6.14087.

monthly mortgage payment =

$$\left(\frac{84,000}{1000}\right)(6.14087) = \$515.83$$

cost of the house $= 21,000 + (515.83)(12)(25) =$ $175,749.00

The Riveras should select Bank B.

25. Down payment: $= \$47,000$

Amount of mortgage: $235,000 - 47,000 = \$188,000$

a) One-year Treasury Bill Rate + Add-on Rate = Initial ARM rate

$1.5\% + 2.5\% = 4\%$

b) Monthly payment: $(6.05980)\left(\dfrac{188,000}{1000}\right) = \$1139.24$

c) New ARM rate: $3.0\% + 2.5\% = 5.5\%$

27. Down payment: $(0.20)(450,000) = \$90,000$

Amount of mortgage: $450,000 - 90,000 = \$360,000$

1 point: $(0.01)(360,000) = \$3,600$

a) Monthly payment: $(13.21507)\left(\dfrac{360,000}{1000}\right) = \$4757.43$

Total payments: $90,000 + 3600 + (10)(12)(4757.43) = \$664,491.60$

b) Monthly payment: $(9.65022)\left(\dfrac{360,000}{1000}\right) = \$3474.08$

Total payments: $90,000 + 3600 + (20)(12)(3474.08) = \$927,379.20$

c) Monthly payment: $(8.77572)\left(\dfrac{360,000}{1000}\right) = \$3159.26$

Total payments: $90,000 + 3600 + (30)(12)(3159.26) = \$1,230,933.60$

29. a) The variable rate mortgage would be the cheapest.

b) To find the total payments for the variable rate mortgage, add the 6 monthly payments and multiply by 12, giving $52,337.64. The payments over 6 years for the fixed rate mortgage are 6 times 12 times 90 times the table value 8.40854 for a 30-year mortgage at 9.5%. This yields $54,487.34. The variable rate saves $2149.70. Using the payment formula for the fixed rate mortgage will yield $54,487.44 & $2149.80.

Exercise Set 10.6

1. Annuity

3. Sinking

5. Immediate

7. 401k

9. $A = \dfrac{1200\left[\left(1+\dfrac{0.035}{1}\right)^{(1)(30)} - 1\right]}{\dfrac{0.035}{1}} = \$61,947.21$

11. $A = \dfrac{400\left[\left(1+\dfrac{0.08}{4}\right)^{(4)(35)} - 1\right]}{\dfrac{0.08}{4}} = \$299,929.32$

13. $p = \dfrac{20,000\left(\dfrac{0.045}{2}\right)}{\left(1+\dfrac{0.045}{2}\right)^{(2)(15)} - 1} = \$473.99$

15. $p = \dfrac{250,000\left(\dfrac{0.06}{12}\right)}{\left(1+\dfrac{0.06}{12}\right)^{(12)(35)} - 1} = \$175.48$

17. $A = \dfrac{150\left[\left(1+\dfrac{0.033}{12}\right)^{(12)(40)} - 1\right]}{\dfrac{0.033}{12}} = \$149,271.58$

19. $A = \dfrac{2000\left[\left(1+\dfrac{0.07}{2}\right)^{(2)(10)} - 1\right]}{\dfrac{0.07}{2}} = \$56,559.36$

21. $p = \dfrac{24,000\left(\dfrac{0.075}{12}\right)}{\left(1+\dfrac{0.075}{12}\right)^{(12)(6)} - 1} = \$264.97$

23. $p = \dfrac{1000000\left(\dfrac{0.08}{4}\right)}{\left(1+\dfrac{0.08}{4}\right)^{(4)(20)} - 1} = \$5,160.71$

25. a) $A = \dfrac{100\left[\left(1+\dfrac{0.12}{12}\right)^{(12)(10)} - 1\right]}{\dfrac{0.12}{12}} = \$23,003.87$

b) $A = 23,003.87\left(1+\dfrac{0.12}{12}\right)^{(12)(30)} = \$826,980.88$

c) $A = \dfrac{100\left[\left(1+\dfrac{0.12}{12}\right)^{(12)(30)} - 1\right]}{\dfrac{0.12}{12}} = \$349,496.41$

d) $(100)(12)(10) = \$12,000$

e) $(100)(12)(30) = \$36,000$

f) Alberto

Review Exercises

1. $3/20 = 0.15 = 15.0\%$

2. $2/3 \approx 0.667 = 66.7\%$

3. $5/8 = 0.625 = 62.5\%$

4. $0.041 = 4.1\%$

5. $0.0098 = 0.98\% \approx 1.0\%$

6. $3.141 = 314.1\%$

7. 2% $\dfrac{2}{100} = 0.02$

8. 21.7% $\dfrac{21.7}{100} = 0.217$

9. 123% $\dfrac{123}{100} = 1.23$

10. $\dfrac{1}{4}\% = 0.25\%$ $\dfrac{0.25}{100} = 0.0025$

11. $\dfrac{5}{6}\% = 0.8\overline{3}\%$ $\dfrac{0.8\overline{3}}{100} = 0.008\overline{3}$

12. 0.00045% $\dfrac{0.00045}{100} = 0.0000045$

13. $\dfrac{42{,}745 - 41{,}500}{41{,}500} = 0.03 = 3.0\%$

14. $\dfrac{2800 - 3200}{3200} = -0.125 = -12.5\%$

15. $(x\%)(60) = 12$ $x\% = 12/60 = 0.20$
    $0.20 = 20\%$
    12 is 20% of 60.

16. $0.55x = 44$ $x = 44/0.55 = 80$
    44 is 55% of 80.

17. $(0.17)(540) = x$ $91.8 = x$
    17% of 540 is 91.8.

18. Tip = 15% of $42.79 = (0.15)(42.79) = \$6.42$

19. $0.20(x) = 8$ $x = 8/0.20 = 40$
    The original number was 40 people.

20. $\dfrac{126}{1050} = 0.12 = 12\%$

21. $i = (2500)(0.015)(3) = \$112.50$

22. $37.50 = (2700)(r)(100/360)$
    $37.50 = \left(\dfrac{270{,}000}{360}\right)(r)$ $r = 0.05$ or 5%

23. $114.75 = (p)(0.085)(3)$
    $114.75 = (p)(0.255)$ $\$450 = p$

24. $555.75 = (3600)(0.0325)(t)$
    $555.75 = (117.00)(t)$ $t = 4.75$ yrs.

25. $i = (7500)(0.025)(24/12) = \$375$
    Total amount due at maturity $= 7500 + 375 =$
    $\$7875$

26. a) $i = (6000)(0.115)(24/120) = \$1380.00$
    b) amount received: $6000.00 - 1380.00$
       $= \$4620.00$
    c) $i = prt$ $1380 = (4620)(r)(24/12)$ $1380 = 9240r$
       $r = (1380)/(9240) = 0.1494$ $(0.1494)(100)$
       $= 14.9\%$

27. Partial payment on June 1 (61 days)
    $i = (8{,}400)(0.045)(61/360) = \$64.05$
    $\$2900.00 - 64.05 = \$2835.95$
    $8400.00 - 2835.95 = \$5564.045$

    $i = (5{,}564.05)(0.045)(119/360) = \$82.77$
    $5564.05 + 82.77 = \$5646.82$

28. Partial payment on Aug 31 (108 days)
    $i = (6{,}700)(0.061)(108/360) = \$122.61$
    $\$3250.00 - 122.61 = \$3127.39$
    $6700.00 - 3127.39 = \$3572.61$

    $i = (3{,}572.61)(0.061)(102/360) = \$61.75$
    $3572.61 + 61.75 = \$3634.36$

29. a) $A = 5,000\left(1 + \dfrac{0.06}{1}\right)^{(1)(5)} = \$6691.13$ $\$6691.13 - \$5000 = \$1691.13$

   b) $A = 5,000\left(1 + \dfrac{0.06}{2}\right)^{(2)(5)} = \$6719.58$ $\$6719.58 - \$5000 = \$1719.58$

   c) $A = 5,000\left(1 + \dfrac{0.06}{4}\right)^{(4)(5)} = \$6734.28$ $\$6734.28 - \$5000 = \$1734.28$

   d) $A = 5,000\left(1 + \dfrac{0.06}{12}\right)^{(12)(5)} = \$6744.25$ $\$6744.25 - \$5000 = \$1744.25$

   e) $A = 5,000\left(1 + \dfrac{0.06}{360}\right)^{(360)(5)} = \$6749.13$ $\$6749.13 - \$5000 = \$1749.13$

30. $A = p\left(1 + \dfrac{r}{n}\right)^{nt}$

   $A = 3400\left(1 + \dfrac{0.015}{12}\right)^{12 \cdot 2} = \$3503.48$

31. Let $p = 1.00$. Then $A = 1\left(1 + \dfrac{0.056}{360}\right)^{360} = 1.05759$

   $i = 1.05759 - 1.00 = 0.05759$
   The effective annual yield is 5.76% .

32. $p\left(1 + \dfrac{0.026}{4}\right)^{20} = 4750$ $p = \dfrac{4750}{(1.0065)^{20}} = 4172.71$

   You need to invest $4172.71

33. a) Down payment: $(26,000)(0.30) = \$7800$
   b) Amount Financed: $26,000 - 7,800 = \$18,200$
      From Table 10.2, $13.23 indicates an APR of 5.0%

   c) $\left(\dfrac{18,200}{100}\right)(13.23) = \$2407.86$

   d) Monthly payment: $\dfrac{18,200 + 2,407.86}{60} = \$343.46$

34. a) Down payment: $(12,850)(0.20) = \$2570$
   b) Amount Financed: $12,850 - 2,570 = \$10,280$
      From Table 10.2, $14.48 indicates an APR of 9.0%

   c) $\left(\dfrac{10,280}{100}\right)(14.48) = \$1488.54$

   d) Monthly payment: $\dfrac{10,280 + 1488.53}{36} = \$326.90$

35. a) $(\$140,000)(0.40) = \$56,000$
   b) $140,000 - 56,000 = \$84,000$
   c) $(1624)(60) = \$97,440$;
      $97,440 - 84,000 = \$13,440$

d) $\left(\dfrac{\text{finance charge}}{\text{amount financed}}\right)(100)$

$= \left(\dfrac{13,440}{84,000}\right)(100) = 16;$

Using Table 10.2, where no. of
payments is 60, APR = 6%

36. 48 mo. $176.14/mo. $7,500   24 payments

a) $(176.14)(48) = 8454.72$ $8454.72 - 7500 = \$954.72$ $\left(\dfrac{954.72}{7500}\right)(100) = \$12.73$

From Table 10.2, $12.73 indicates an APR of 6.0%

b) $n = 24$, $p = 176.14$, $v = 6.37$ $u = \dfrac{(24)(176.14)(6.37)}{100 + 6.37} = \dfrac{26,928.28}{106.37} = \$253.16$

c) $(176.14)(48) = 8454.72$ $(176.14)(23) = 4051.22$ $8454.72 - 4051.22 = \$4403.50$
   $4403.50 - 253.16 = \$4150.34$

37. 24 mo. $111.73/mo.   Down payment = $860   24 payments

a) $3,420 - 860 = \$2560.00$ $(111.73)(24) = 2681.52$ $2681.52 - 2560.00 = \$121.52$

   $\left(\dfrac{121.52}{2560}\right)(100) = \$4.75$

From Table 10.2, $4.75 indicates an APR of 4.5%

b) $n = 12$, $p = 111.73$, $v = 2.45$ $u = \dfrac{(12)(111.73)(2.45)}{100 + 2.45} = \dfrac{3,284.86}{102.45} = \$32.06$

c) $(111.73)(11) = 1229.03$ $2681.52 - 1229.03 = 1452.49$ $1452.49 - 32.06 = \$1420.43$

38. Balance = $485.75 as of June 01 $i = 1.3\%$
   June 04: $485.75 - 375.00 = \$110.75$ June 08: $110.75 + 370.00 = \$480.75$
   June 21: $480.75 + 175.80 = \$656.55$ June 28: $656.55 + 184.75 = \$841.30$
   a) $(485.75)(0.013)(1) = \$6.31$ b) $841.30 + 6.31 = \$847.61$
   c) $(485.75)(3) + (110.75)(4) + (480.75)(13) + (656.55)(7) + (841.30)(3) = \$15,269.75$
      $15,269.75/30 = \$508.99$
   d) $(508.99)(0.013)(1) = \$6.62$ e) $841.30 + 6.62 = \$847.92$

39. a) $i = (185.72)(0.14)(1) = \$2.60$
    b) Aug. 05: 185.72
       Aug. 08: $185.72 + 85.75 = \$271.47$
       Aug. 10: $271.47 - 75.00 = \$196.47$
       Aug. 15: $196.47 + 72.85 = \$269.32$
       Aug. 21: $269.32 + 275.00 = \$544.32$
       As of Sep 5, the new account balance
       is $544.32 + $2.60 = $546.92.
    d) $i = (\$382.68)(0.014)(1) = \$5.36$

c)

Date	Balance	# of Days	(Balance)(Days)
Aug. 05	185.72	3	(185.72)(3) = 557.16
Aug. 08	271.47	2	(271.47)(2) = 542.94
Aug. 10	196.47	5	(196.47)(5) = 982.35
Aug. 15	269.32	6	(269.32)(6) = 1615.92
Aug. 21	544.32	15	(544.32)(15) = 8164.80
Sep 05		31	sum = $11,863.17

average daily balance $= \dfrac{11,863.17}{31} = \$382.68$

e) $544.32 + 5.36 = \$549.68$ is the amount due on
   Sep 5

40. a) down payment = $(0.25)(135,700) = \$33,925$

    b) gross monthly income = $64,000/12 = \$5,333.33$
       adjusted monthly income:
       $5333.33 - 528.00 = \$4,805.33$
       28% of adjusted monthly income:
       $(0.28)(4,805.33) = \$1345.49$

    c) $\left(\dfrac{101,775}{1000}\right)(8.40854) = \$855.78$

    d) total monthly payment:
       $855.78 + 316.67 = \$1172.45$

    e) Yes, $1345.49 is greater than $1172.45.

41. a) down payment = $(0.15)(89,900) = \$13,485$

    b) amount of mortgage = $89,900 - 13,485 =$ $76,415
       At 11.5% for 30 years, the monthly payment formula yields a monthly mortgage payment of $756.73.

    c) $i = prt = (76,415)(0.115)(1/12) = \$732.31$
       amount applied to principal:
       $756.73 - 732.31 = \$24.42$

    d) total cost of house: $13,485 + (756.73)(12)(30) =$ $285,907.80

    e) total interest paid: $285,907.80 - 89,900 =$ $196,007.80

42. Down payment: = $140,000
    Amount of mortgage: $375,000 - 140,000 = \$235,000$

    a) One-year Treasury Bill Rate + Add-on Rate = Initial ARM rate
       $1.0\% + 3.5\% = 4.5\%$

    b) Monthly payment: $(7.64993)\left(\dfrac{235,000}{1000}\right) = \$1797.73$

    c) New ARM rate: $2.5\% + 3.5\% = 6.0\%$

43. $A = \dfrac{250\left[\left(1+\dfrac{0.09}{12}\right)^{(12)(10)} - 1\right]}{\dfrac{0.09}{12}} = \$48,378.57$

44. $p = \dfrac{80,000\left(\dfrac{0.036}{4}\right)}{\left(1+\dfrac{0.036}{4}\right)^{(4)(5)} - 1} = \$3668.72$

Chapter Test

1. a) $i = (1150)(0.011)(42/12) = \$44.28$

   b) $351 = (2600)(0.045)(t); \; 351 = 117t; \; t = 3$ years

2. $i = prt = (1,640)(0.0315)(18/12) = \$77.49$

3. Total amount paid to the bank,
   $1640 + 77.49 = \$1717.49$

4. Partial payment on Sept. 15 (45 days)
   $i = (5,400)(0.125)(45/360) = \$84.375$
   $\$3000.00 - 84.375 = \$2915.62$
   $5000.00 - 2915.625 = \$2484.375$

   $i = (2,484.375)(0.125)(45/360) = \$38.82$
   $2484.38 + 38.82 = \$2523.20$

5. $84.38 + 38.82 = \$123.20$

6. a) $A = 7500\left(1+\dfrac{0.03}{4}\right)^8 = \$7961.99$
      interest $= 7961.99 - 7500.00 = \$461.99$

   b) $A = 2,500\left(1+\dfrac{0.065}{12}\right)^{36} = \$3036.68$
      interest $= 3036.68 - 2500.00 = \$536.68$

7. $(2350)0(0.15) = 352.50$
   $2350 - 352.50 = \$1997.50$

8. $(90.79)(24) = 2178.96$
   $2178.96 - 1997.50 = \$181.46$

9. $\left(\dfrac{181.46}{1,997.50}\right)(100) = \$9.08$ /$100   In Table 10.2, \$9.08 is closest to \$9.09 which yields an APR of 8.5% .

10. \$7500   36 mo.   \$223.10 per mo.   (223.10)(36) = 8031.60   8031.60 − 7500.00 = 531.60

   a) $\left(\dfrac{531.60}{7500}\right)(100) = 7.09$ In Table 10.2, \$7.09 yields an APR of 4.5% .

   b) $u = \dfrac{n \cdot P \cdot V}{100 + V} = \dfrac{(12)(223.10)(2.45)}{100 + 2.45} = \dfrac{6559.14}{102.45} = \$64.02$

   c) $(223.10)(23) = 5{,}131.30$ $8{,}031.60 − 5{,}131.30 = \$2{,}900.30$ $2{,}900.30 − 64.02 = \$2{,}836.28$

11. Mar. 23: \$878.25
    Mar. 26: 878.25 + 95.89 = \$974.14
    Mar. 30: 974.14 + 68.76 = \$1042.90
    Apr. 03: 1042.90 − 450.00 = \$592.90
    Apr. 15: 592.90 + 90.52 = \$683.42
    Apr. 22: 683.42 + 450.85 = \$1134.27

   a) $i = (878.25)(0.014)(1) = \$12.30$
   b) 1134.27 + 12.30 = \$1146.57

   c)
Date	Balance	# of Days	Balance-Days
Mar. 23	878.25	3	(878.25)(3) = 2634.75
Mar. 26	974.14	4	(974.14)(4) = 3896.56
Mar. 30	1042.90	4	(1042.90)(4) = 4171.60
Apr. 03	592.90	12	(592.90)(12) = 7114.80
Apr. 15	683.42	7	(683.42)(7) = 4783.94
Apr. 22	1134.27	1	(1134.27)(1) = 1134.27
Apr 23		31	sum = \$23,735.92

   avg. daily balance $= \dfrac{23{,}735.92}{31} = \$765.67$

   d) $(765.67)(0.014)(1) = \$10.72$
   e) 1134.27 + 10.72 = \$1144.99

12. down payment = (0.15)(144,500) = \$21,675.00

13. amount of loan = 144,500 − 21,675 − \$122,825
    points = (0.02)(122,825) = \$2456.50

14. gross monthly income = 86,500 ÷ 12 = \$7208.33
    7208.33 − 605.00 = \$6603.33   adj. mo. income
    maximum monthly payment = (0.28)(6603.33) = \$1848.93

15. At 10.5% interest for 30 years, Table 10.4 yields \$9.15.

    monthly payments $= \left(\dfrac{122{,}825}{1000}\right)(9.15) = \$1123.85$

16. 1123.53 + 304.17 = \$1427.70 total mo. payment

17. Yes.

18. a) Total cost of the house:
       21,675 + 2456.50 + (1123.53)(12)(30)
       = \$428,602.30
    b) interest = 428,602.30 − 144,500 = \$284,102.30

19. $A = \dfrac{400\left[\left(1 + \dfrac{0.03}{12}\right)^{(12)(5)} - 1\right]}{\dfrac{0.03}{12}} = \$25{,}858.69$

20. $p = \dfrac{15{,}000\left(\dfrac{0.06}{12}\right)}{\left(1 + \dfrac{0.06}{12}\right)^{(12)(4)} - 1} = \$277.28$

CHAPTER ELEVEN

PROBABILITY

Exercise Set 11.1

1. Experiment

3. Event

5. Empirical

7. a) 0 b) 1 c) 0,1

9. 1

11. a) – c Answers will vary.   13. a) – c) Answers will vary.

15. a) P(correct) = 1/5   b) P(correct) = 1/4

17. Of 30 boats: 14 sunfish   10 kayaks   6 rowboats

   a) P(s) = 14/30 = 7/15   b) P(k) = 10/30 = 1/3   c) P(r) = 6/30 = 1/5

19. Of 105 animals: 45 are dogs.   40 are cats   15 are birds   5 are rabbits

   a) P(dog) = 45/105 = 3/7   b) P(cat) = 8/21   c) P(rabbit) = 5/105 = 1/21

21. a) P(Call of Duty: Ghosts) =   b) P(Tomb Raider) =   c) P(Pokemon X&Y) =

   $\dfrac{12,710,000}{75,480,000} \approx 0.1684$ $\dfrac{3,180,000}{75,480,000} \approx 0.0421$ $\dfrac{7,640,000}{75,480,000} \approx 0.1012$

23. a) P(affecting circular) = $\dfrac{0}{150} = 0$ 25. P(5) = $\dfrac{4}{52} = \dfrac{1}{13}$

   b) P(affecting elliptical) = $\dfrac{50}{250} = 0.2$

   c) P(affecting irregular) = $\dfrac{100}{100} = 1$

27. P(not 5) = $\dfrac{48}{52} = \dfrac{12}{13}$

29. P(black) = $\dfrac{13+13}{52} = \dfrac{26}{52} = \dfrac{1}{2}$

31. P(red or black) = $\dfrac{26+26}{52} = \dfrac{52}{52} = \dfrac{1}{1} = 1$

33. $P(> 3 \text{ and } < 8) = P(4,5,6,7) = \dfrac{16}{52} = \dfrac{4}{13}$

35. a) $P(\text{red}) = \dfrac{1}{4}$ b) $P(\text{green}) = \dfrac{1}{2}$

    c) $P(\text{yellow}) = \dfrac{1}{4}$ d) $P(\text{not yellow}) = \dfrac{3}{4}$

37. a) $P(\text{red}) = \dfrac{4}{8} = \dfrac{1}{2}$ b) $P(\text{green}) = \dfrac{3}{8}$

    c) $P(\text{yellow}) = \dfrac{1}{8}$ d) $P(\text{not yellow}) = \dfrac{7}{8}$

Of 100 bars: 40 are Vanilla (v)   20 are Mint Chocolate Chip (mc)   30 are Krunch (K)
10 are Rocky Road (rr)

39. $P(\text{mc}) = \dfrac{20}{100} = \dfrac{1}{5}$ 41. $P(\text{v, mc, k}) = \dfrac{90}{100} = \dfrac{9}{10}$ 43. $P(600) = \dfrac{1}{12}$ 45. $P(\text{lose/bankrupt}) = \dfrac{2}{12} = \dfrac{1}{6}$

Of 11 letters: 1 = T   4 = E   2 = N   2 = S

47. $P(S) = \dfrac{2}{9}$ 49. $P(\text{consonant}) = \dfrac{5}{9}$ 51. $P(W) = 0$

53. $P(CA) = \dfrac{1}{10}$ 55. $P(\text{not CA}) = 1 - \dfrac{1}{10} = \dfrac{9}{10}$ 57. $P(AZ) = \dfrac{1}{10}$

59. $P(\text{car}) = \dfrac{85}{200} = \dfrac{17}{40}$ 61. $P(\text{not a car}) = 1 - \dfrac{17}{40} = \dfrac{23}{40}$ 63. $P(\text{GM vehicle}) = \dfrac{123}{200}$

65. $P(\text{GM car}) = \dfrac{55}{200} = \dfrac{11}{40}$ 67. $P(\text{Ort}) = \dfrac{10}{42} = \dfrac{5}{21}$ 69. $P(\text{not Ort}) = 1 - \dfrac{5}{21} = \dfrac{16}{21}$

71. $P(\text{mild}) = \dfrac{11}{42}$ 73. $P(\text{TB medium}) = \dfrac{8}{42} = \dfrac{4}{21}$

75. Not necessarily, but it does mean that if a coin was flipped many times, about one-half of the tosses would land heads up.

77. a) Roll a die 100 times and determine the number of times that a 5 occurs out of 100.
    b) Answers will vary.   c) Answers will vary.

79. $P(\text{red}) = \dfrac{2}{18} + \dfrac{1}{12} + \dfrac{1}{6} = \dfrac{4}{36} + \dfrac{3}{36} + \dfrac{6}{36} = \dfrac{13}{36}$ 81. $P(\text{not red}) = 1 - \dfrac{13}{36} = \dfrac{23}{36}$

83. $P(\text{yellow}) = \dfrac{1}{6} + \dfrac{1}{12} + \dfrac{1}{12} = \dfrac{2}{12} + \dfrac{2}{12} = \dfrac{4}{12} = \dfrac{1}{3}$ 85. $P(\text{yellow or green}) = \dfrac{1}{3} + \dfrac{11}{36} = \dfrac{12}{36} + \dfrac{11}{36} = \dfrac{23}{36}$

87. a) $P(R/R) = \dfrac{2}{4} \cdot \dfrac{2}{4} = \dfrac{4}{16} = \dfrac{1}{4}$

　b) $P(G/G) = \dfrac{2}{4} \cdot \dfrac{2}{4} = \dfrac{4}{16} = \dfrac{1}{4}$

　c) $P(R/G) = \dfrac{2}{4} \cdot \dfrac{2}{4} = \dfrac{4}{16} = \dfrac{1}{4}$

89. a) – c) Answers will vary.　　　90. $4 \cdot 7 + 1 = 29$

Exercise Set 11.2

1. Against

3. 4 : 1

5. 2 : 1

7. $\dfrac{1}{4}$

9. a) $P(\text{liberal arts}) = \dfrac{18}{30} = \dfrac{3}{5}$

　b) $P(\text{not liberal arts}) = 1 - \dfrac{3}{5} = \dfrac{2}{5}$

　c) odds against liberal arts =

　　$\dfrac{P(\text{not liberal arts})}{P(\text{liberal arts})} = \dfrac{\frac{2}{5}}{\frac{3}{5}} = \dfrac{2}{3}$ or 2 : 3

　d) odds in favor of liberal arts are 3 : 2

11. Since there is only one 2, the odds against a 2 are 5 : 1.

13. odds against rolling less than 5 $= \dfrac{P(5 \text{ or greater})}{P(\text{less than } 5)} =$

$\dfrac{2/6}{4/6} = \dfrac{2}{6} \cdot \dfrac{6}{4} = \dfrac{2}{4} = \dfrac{1}{2}$ or 1:2

15. odds against a king $= \dfrac{P(\text{failure to pick a king})}{P(\text{pick a king})} =$

$\dfrac{48/52}{4/52} = \dfrac{48}{52} \cdot \dfrac{52}{4} = \dfrac{48}{4} = \dfrac{12}{1}$ or 12:1

Therefore, odds in favor of picking a king are 1:12.

17. odds against a picture card =

$\dfrac{P(\text{failure to pick a picture})}{P(\text{pick a picture})} = \dfrac{40/52}{12/52} = \dfrac{40}{12} = \dfrac{10}{3}$

or 10:3

Therefore, odds in favor of picking a picture card are 3:10.

19. odds against red =

$\dfrac{P(\text{not red})}{P(\text{red})} = \dfrac{1/2}{1/2} = \dfrac{1}{2} \cdot \dfrac{2}{1} = \dfrac{2}{2} = \dfrac{1}{1}$ or 1:1

21. odds against red $= \dfrac{P(\text{not red})}{P(\text{red})} = \dfrac{5/8}{3/8} = \dfrac{5}{8} \cdot \dfrac{8}{3} = \dfrac{5}{3}$

or 5:3

23. a) odds against selecting a red marble =

$\dfrac{P(\text{failure to select red})}{P(\text{select red})} = \dfrac{12/28}{16/28} = \dfrac{12}{16} = \dfrac{3}{4}$

or 3 : 4 .

　b) odds against selecting a blue marble =

$\dfrac{P(\text{failure to select blue})}{P(\text{select blue})} = \dfrac{16/28}{12/28} = \dfrac{16}{12} = \dfrac{4}{3}$

or 4 : 3.

25. odds against a stripe = $\dfrac{P(\text{not a stripe})}{P(\text{stripe})} =$

$\dfrac{8/15}{7/15} = \dfrac{8}{15} \cdot \dfrac{15}{7} = \dfrac{8}{7}$ or 8:7

27. odds in favor of even are

$\dfrac{P(\text{even})}{P(\text{not even})} = \dfrac{7/15}{8/15} = \dfrac{7}{15} \cdot \dfrac{15}{8} = \dfrac{7}{8}$ or 7:8

29. odds against a ball with 9 or greater are

$\dfrac{P(\text{less than 9})}{P(\text{9 or greater})} = \dfrac{8/15}{7/15} = \dfrac{8}{15} \cdot \dfrac{15}{7} = \dfrac{8}{7}$ or 8:7

31. a) $\dfrac{3}{10}$ b) 7 : 3

33. The odds against testing negative =

$\dfrac{P(\text{test positive})}{P(\text{test negative})} = \dfrac{5/85}{80/85} = \dfrac{5}{80} = \dfrac{1}{16}$ or 1 : 16

35. a) P(Wendy wins) = $\dfrac{7}{7+4} = \dfrac{7}{11}$

b) P(Wendy loses) = $\dfrac{4}{7+4} = \dfrac{4}{11}$

37. a) Odds against 2 : 9   P(sells out) = $\dfrac{9}{9+2} = \dfrac{9}{11}$

b) P(not sold out) = $\dfrac{2}{9+2} = \dfrac{2}{11}$

39. P(B) = $\dfrac{15}{75} = \dfrac{1}{5}$

41. Odds in favor of B = $\dfrac{P(B)}{P(\text{not B})} = \dfrac{1/5}{4/5} = \dfrac{1}{4}$ or 1:4

43. Odds against G50 =

$\dfrac{P(\text{not G50})}{P(\text{G50})} = \dfrac{74/75}{1/75} = \left(\dfrac{74}{75}\right)\left(\dfrac{75}{1}\right) = \dfrac{74}{1}$ or 74:1

45. P(A+) = $\dfrac{34}{100} = 0.34$

47. $\dfrac{66}{34} = \dfrac{33}{17}$ or 33 : 17

49. P(O or O−) = $\dfrac{43}{100} = \dfrac{43}{43+57}$ or 43 : 57

51. If P(high blood pressure) = $0.31 = \dfrac{31}{100}$, then

P(not high blood presure) = $1 - \dfrac{31}{100} = \dfrac{69}{100}$.

The odds in favor of high blood pressure

$= \dfrac{31/100}{69/100} = \dfrac{31}{69}$ or 31:69.

53. a) P(audited) = $\dfrac{1}{104}$

b) P(not audited) =

$1 - \dfrac{1}{104} = \dfrac{103}{104}$.

The odds against
being audited =

$\dfrac{103/104}{1/104} = \dfrac{103}{1}$

or 103:1.

55. a) P(birth defect) = $\dfrac{1}{33}$

b) Number without
birth defect:

$33 - 1 = 32$

Odds against birth
defect: 32 : 1

57. $P(\# 2 \text{ wins}) = \dfrac{1}{(1+15)} = \dfrac{1}{16}$

   $P(\# 3 \text{ wins}) = \dfrac{1}{(1+1)} = \dfrac{1}{2}$

   $P(\# 4 \text{ wins}) = \dfrac{1}{(1+3)} = \dfrac{1}{4}$

   $P(\# 5 \text{ wins}) = \dfrac{3}{(3+13)} = \dfrac{3}{16}$

59. If multiple births are 3% of births, then single births are 97% of births, and the odds against a multiple birth are 97 : 3.

Exercise Set 11.3

1. Expected

3. Positive

5. $E = P_1 A_1 + P_2 A_2 = 0.20(20) + 0.80(60) = 52$ hot dogs

7. $E = P_1 A_1 + P_2 A_2 = 0.60(80,000) + 0.40(-20,000) = 48,000 - 8000 = \$40,000$

9. $E = P_1 A_1 + P_2 A_2 = 0.40(1.2\,M) + 0.60(1.6\,M) = 0.48\,M + 0.96\,M = 1.44\,M$ viewers

11. $E = P_1 A_1 + P_2 A_2 = (0.70)(15,000) + (0.10)(0) + (0.20)(-8500) = 10,500 + 0 + -7500 = \$8800$

13. $E = P_1 A_1 + P_2 A_2 + P_3 A_3 = P(\$1\text{ off})(\$1) + P(\$2\text{ off})(\$2) + P(\$5\text{ off})(\$5)$

   $E = (7/10)(1) + (2/10)(2) + (1/10)(5) = 7/10 + 4/10 + 5/10 = 16/10 = \$1.60$ off

15. a) $(3/6)(3) + (2/6)(2) + (1/6)(-14) \approx -\$0.17$

   b) Gabriel's expectation is the negative of Alyssa's, or $0.17.

17. a) $(1/5)(5) + (0)(0) + (4/5)(-1) = 1 - 4/5 = 1/5$

   Yes, positive expectation $= 1/5$

   b) $(1/4)(5) + (0)(0) + (3/4)(-1) = 5/4 - 3/4 = 1/2$

   Yes, positive expectation $= 1/2$

19. Fair Price = expected value + cost to play

   Fair Price $= -\$4.50 + \$10.00 = \$5.50$

21. a) $\left(\dfrac{1}{500}\right)(497) + \left(\dfrac{499}{500}\right)(-3) = \dfrac{497 - 1497}{500} =$

   $\dfrac{-1000}{500} = -\$2.00$

   b) Fair price $= -2.00 + 3.00 = \$1.00$

23. a)

   $\left(\dfrac{1}{2000}\right)(997) + \left(\dfrac{2}{2000}\right)(497) + \left(\dfrac{1997}{2000}\right)(-3) = \dfrac{997 + 994 - 5991}{2000} = \dfrac{-4000}{2000} = -\$2.00$

   b) Fair price $= -2.00 + 3.00 = \$1.00$

25. $\frac{1}{2}(5) + \frac{1}{2}(10) = \frac{5}{2} + \frac{10}{2} = 7.5 = \$7.50$

27. $\frac{1}{2}(5) + \frac{1}{4}(-2) + \frac{1}{4}(-15) =$

$2.50 - 0.50 - 3.75 = -\$1.75$

29. $(500)\left(\frac{3}{6}\right) + (1000)\left(\frac{3}{6}\right) = \$750$

31. $(100 + 200 + 300 + 400 + 500 + 1000)\left(\frac{1}{6}\right) = \$416.67$

33. a) $(8)\left(\frac{1}{2}\right) + (-1)\left(\frac{1}{2}\right) = \$3.50$

   b) Fair price = $\$3.50 + 2.00 = \$5.50$

35. a) $(8)\left(\frac{1}{4}\right) + (3)\left(\frac{1}{4}\right) + (-1)\left(\frac{1}{2}\right) = \$2.25$

   b) Fair price = $2.25 + 2.00 = \$4.25$

37. $(10 - 15)\left(\frac{2}{4}\right) + (25 - 15)\left(\frac{2}{4}\right) = \$2.50$

39. $(0 - 15)\left(\frac{1}{4}\right) + (2 - 15)\left(\frac{1}{4}\right) + (5 - 15)\left(\frac{1}{4}\right)$

$+ (20 - 15)\left(\frac{1}{4}\right) = -\$8.25$

41. $E = P_1A_1 + P_2A_2 + P_3A_3 =$

   $0.20(300) + 0.70(400) + 0.10(-500) =$
   Gain of 290 employees

43. $E = (0.65)(75) + (0.35)(20) = 55.75$
   Expected number of new employees is 56.

45. $E = (0.85)(200,000) + (0.05)(0)$

   $+ (0.10)(-30,000) = \$167,000$

47. $E = P(1)(1) + P(2)(2) + P(3)(3) + P(4)(4) + P(5)(5)$
   $+ P(6)(6)$

$= \frac{1}{6}(1) + \frac{1}{6}(2) + \frac{1}{6}(3) + \frac{1}{6}(4) + \frac{1}{6}(5) + \frac{1}{6}(6)$

$= \frac{21}{6} = 3.5$ points

49. $E = P_1A_1 + P_2A_2 + P_3A_3$

$= \frac{200}{365}(110) + \frac{100}{365}(160) + \frac{65}{365}(210)$

$= 60.27 + 43.84 + 37.40 = 141.51$ calls/day

51. a) $P(1) = \frac{1}{2} + \frac{1}{16} = \frac{8}{16} + \frac{1}{16} = \frac{9}{16}$,

   $P(10) = \frac{1}{4} = \frac{4}{16}$,

   $P(\$20) = \frac{1}{8} = \frac{2}{16}$, $P(\$100) = \frac{1}{16}$

   b) $E = P_1A_1 + P_2A_2 + P_3A_3 + P_4A_4$

$= \frac{9}{16}(\$1) + \frac{4}{16}(\$10) + \frac{2}{16}(\$20) + \frac{1}{16}(\$100)$

$= \frac{9}{16} + \frac{40}{16} + \frac{40}{16} + \frac{100}{16} = \frac{189}{16} = \$11.81$

53. $E = P(\text{insured lives})(\text{cost}) + P(\text{insured dies})(\text{cost} - \$40,000)$

   $= 0.97(\text{cost}) + 0.03(\text{cost} - 40,000)$

   $= 0.97(\text{cost}) + 0.03(\text{cost}) - 1200$

   $= 1.00(\text{cost}) - 1200$

   Thus, in order for the company to make a profit,
   the cost must exceed $1200

55. $E = P(\text{win})(\text{amount won}) + P(\text{lose})(\text{amount lost})$

$= \left(\frac{1}{38}\right)(35) + \left(\frac{37}{38}\right)(-1) = \frac{35}{38} - \frac{37}{38} = -\frac{2}{38}$

$= -\$0.053$

57. a) $E = \frac{1}{12}(100) + \frac{1}{12}(200) + \frac{1}{12}(300) + \frac{1}{12}(400) + \frac{1}{12}(500) + \frac{1}{12}(600) + \frac{1}{12}(700) + \frac{1}{12}(800) + \frac{1}{12}(900)$

$\frac{1}{12}(1000) = \left(\frac{5500}{12}\right) = \$458.33$

b) $E = \frac{1}{12}(5500) + \frac{1}{12}(-1800) = \frac{3700}{12} = \$308.33$

Exercise Set 11.4
1. Sample
3. 24
5. a) $(26)(26) = 676$ b) $(26)(25) = 650$
7. a) $(5)(5)(5) = 125$ b) $(5)(4)(3) = 60$
9. a) $(2)(2) = 4$ points
   b)

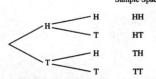

   c) P(no tails) = 1/4
   d) P(exactly one tail) = 2/4 = 1/2
   e) P(two tails) = 1/4
   f) P(at least one tail) = 3/4

11. a) $(3)(3) = 9$ points
    b)

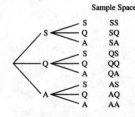

    c) P(two apples) = 1/9
    d) P(sun and then question mark) = 1/9
    e) P(at least one apple) = 5/9

13. a) (4)(3) = 12 points

b)

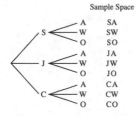

Sample Space

Y	R	YR
	B	YB
	G	YG
R	Y	RY
	B	RB
	G	RG
B	Y	BY
	R	BR
	G	BG
G	Y	GY
	R	GR
	B	GB

c) P(exactly one red) = 6/12 = 1/2

d) P(at least one is not red) = 12/12 = 1

e) P(no green) = 6/12 = 1/2

15. a) (3)(3) = 9 points

b)

Sample Space

S	A	SA
	W	SW
	O	SO
J	A	JA
	W	JW
	O	JO
C	A	CA
	W	CW
	O	CO

c) P(Java) = 1/3

d) P(Java and Oyster) = 1/9

e) P(paint other than Java) = 2/3

17. a) (6)(6) = 36 points
    b)

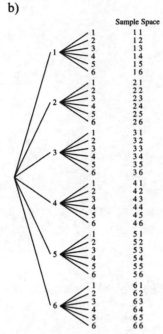

    c) P(double) = 6/36 = 1/6
    d) P(sum of 8) = 5/36
    e) P(sum of 2) = 1/36

19. a) (2)(2)(3) = 12
    b)

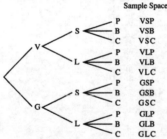

    c) P(guacamole) = 6/12 = 1/2
    d) P(lemonade and brownies) = 2/12 = 1/6
    e) P(not pie) = 8/12 = 2/3

21. a) (4)(2)(2) = 16 points

b)

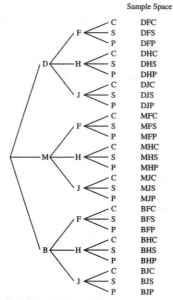

	Sample Space
	BCA
	BCS
	BLA
	BLS
	GCA
	GCS
	GLA
	GLS
	PCA
	PCS
	PLA
	PLS
	OCA
	OCS
	OLA
	OLS

c) P(physics) = 4/16 = 1/4

d) P(geology and literature) = 2/16 = 1/8

e) P(not oceanography) = 12/16 = 3/4

23. a) (3)(3)(3) = 27

b)

c) P(M) = 9/27 = 1/3

d) P(D and H) = 3/27 = 1/9

e) P(L other than S) = 18/27 = 2/3

25. a) (2)(4)(3) = 24 sample points

b)

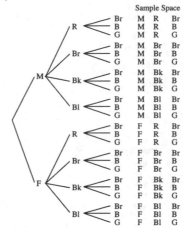

c) P(M, black, blue) = 1/24

d) P(F, blonde) = 3/24 = 1/8

27. a) P(white) = 1/3
    b) Odds against white = 2:1
    c) P(red) = 2/3
    d) No; P(white) < P(red)
    e)

Sample Space

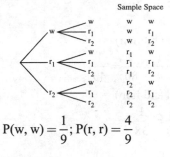

$$P(w, w) = \frac{1}{9}; P(r, r) = \frac{4}{9}$$

29. 3; 1 red, 1 blue, and 1 brown

31. a) $5 \cdot 4 \cdot 3 \cdot 2 \cdot 1 = 120$ b) STORE

Exercise Set 11.5
1. Compound
3. And
5. Independent
7. Independent
9. P(A) + P(B) – P(A and B)

11. P(A or B) = P(A) + P(B) – P(A and B)
    = 0.6 + 0.3 – 0.5 = 0.9 – 0.5 = 0.4

13. P(A or B) = P(A) + P(B) – P(A and B)
    0.9 = 0.8 + 0.2 – P(A and B)
    0.9 = 1.0 – P(A and B)
    – 0.1 = – P(A and B)
    P(A and B) = 0.1

15. P(A or B) = P(A) + P(B) – P(A and B)
    0.7 = 0.6 + P(B) – 0.3
    0.7 = P(B) + 0.3
    P(B) = 0.4

17. P(M and E) = 0.55
    P(M or E) = P(M) + P(E) – P(M and E)
    = 0.7 + 0.6 – 0.55 = 1.3 – 0.55 = 0.75

19. P(5 or 6) = 1/6 + 1/6 = 2/6 = 1/3

21. P(greater than 4 or less than 2) = P(5, 6, or 1) =
     2/6 + 1/6 = 3/6 = 1/2

23. Since these events are mutually exclusive,
     P(10 or 4) = P(10) + P(4)
     $$= \frac{4}{52} + \frac{4}{52} = \frac{8}{52} = \frac{2}{13}$$

25. Since it is possible to obtain a card that is a picture
     card and a red card, these events are not mutually
     exclusive.
     P(picture or red) = P(pict.) + P(red) – P(pict. & red)
     $$= \frac{12}{52} + \frac{26}{52} - \frac{6}{52} = \frac{32}{52} = \frac{8}{13}$$

27. Since it is possible to obtain a card less than 6 that is a club, these events are not mutually exclusive.

$$P(<6 \text{ or club}) = \frac{20}{52} + \frac{13}{52} - \frac{5}{52} = \frac{28}{52} = \frac{7}{13}$$

29. a) P(bird and bird) =
$$\frac{5}{20} \cdot \frac{5}{20} = \frac{1}{4} \cdot \frac{1}{4} = \frac{1}{16}$$
b) P(bird and bird) =
$$\frac{5}{20} \cdot \frac{4}{19} = \frac{1}{4} \cdot \frac{4}{19} = \frac{1}{19}$$

31. a) P(bird and monkey) = $\dfrac{5}{20} \cdot \dfrac{5}{20} = \dfrac{1}{4} \cdot \dfrac{1}{4} = \dfrac{1}{16}$

b) P(bird and monkey) = $\dfrac{5}{20} \cdot \dfrac{5}{19} = \dfrac{1}{4} \cdot \dfrac{5}{19} = \dfrac{5}{76}$

33. a) P(yellow bird and lion) =
$$\frac{2}{20} \cdot \frac{5}{20} = \frac{2}{20} \cdot \frac{1}{4} = \frac{2}{80} = \frac{1}{40}$$
b) P(yellow bird and lion) =
$$\frac{2}{20} \cdot \frac{5}{19} = \frac{10}{380} = \frac{1}{38}$$

35. a) P(odd and odd) = $\dfrac{12}{20} \cdot \dfrac{12}{20} = \dfrac{3}{5} \cdot \dfrac{3}{5} = \dfrac{9}{25}$

b) P(odd and odd) = $\dfrac{12}{20} \cdot \dfrac{11}{19} = \dfrac{3}{5} \cdot \dfrac{11}{19} = \dfrac{33}{95}$

37. P(lion or even) = $\dfrac{5}{20} + \dfrac{8}{20} - \dfrac{2}{20} = \dfrac{11}{20}$

39. P(lion or a 3) = $\dfrac{5}{20} + \dfrac{4}{20} - \dfrac{1}{20} = \dfrac{8}{20} = \dfrac{2}{5}$

41. P(red and green) = $\dfrac{1}{4} \cdot \dfrac{1}{2} = \dfrac{1}{8}$

43. P(2 yellows) = P(yellow and yellow)
$$= \frac{3}{8} \cdot \frac{3}{8} = \frac{9}{64}$$

45. P(2 reds) = $\dfrac{1}{4} \cdot \dfrac{3}{8} = \dfrac{3}{32}$

47. P(both not red) = $\dfrac{3}{4} \cdot \dfrac{5}{8} = \dfrac{15}{32}$

49. P(yellow or blue) = 3/7

51. P(none are red) = 1 – P(None red) = (4/7)(4/7)
= 16/49

53. P(at least one is red) = 1 – 16/49 = 33/49

55. P(all red) = (3/7)(2/6)(1/5) = 1/35

57. P(at least one is red) = 1 – P(None red) =
1 – 4/35 = 31/35

59. P(3 girls) = P(1st girl) · P(2nd girl) · P(3rd girl)
$$= \frac{1}{2} \cdot \frac{1}{2} \cdot \frac{1}{2} = \frac{1}{8}$$

61. P(at least one girl) = 1 – P(All boys) =
1 – 1/8 = 7/8

63. a) P(5 boys) = P(b) · P(b) · P(b) · P(b) · P(b)
$$= \frac{1}{2} \cdot \frac{1}{2} \cdot \frac{1}{2} \cdot \frac{1}{2} \cdot \frac{1}{2} = \frac{1}{32}$$
b) P(next child is a boy) = $\dfrac{1}{2}$

65. a) P(drama/comedy) = $\dfrac{4}{7} \cdot \dfrac{1}{7} = \dfrac{4}{49}$

b) P(drama/comedy) = $\dfrac{4}{7} \cdot \dfrac{1}{6} = \dfrac{4}{42} = \dfrac{2}{21}$

67. a) P(at least 1 science fiction) =

$$\frac{2}{7}\cdot\frac{5}{7}+\frac{5}{7}\cdot\frac{2}{7}+\frac{2}{7}\cdot\frac{2}{7}=\frac{24}{49}$$

   b) P(at least 1 science fiction)=

$$\frac{2}{7}\cdot\frac{5}{6}+\frac{5}{7}\cdot\frac{2}{6}+\frac{2}{7}\cdot\frac{1}{6}=\frac{11}{21}$$

69. P(neither had trad. ins.) $=\frac{48}{75}\cdot\frac{47}{74}=\frac{2256}{5550}=\frac{376}{925}$

71. P(at least one trad.) = 1 – P(neither trad.) =

$1-\frac{376}{925}=\frac{549}{925}$ (see Exercise 69)

73. P(all recommended) $=\frac{23}{30}\cdot\frac{22}{29}\cdot\frac{21}{28}=\frac{253}{580}$

75. P(no/no/not sure) $=\frac{3}{30}\cdot\frac{2}{29}\cdot\frac{4}{28}=\frac{1}{1015}$

77. The probability that any individual reacts
favorably is 70/100 or 0.7.

   P(Mrs. Rivera reacts favorably) = 7/10

79. P(all 3 react favorably) = (7/10)(7/10)(7/10)
= 343/1000

81. P(blue/blue) $=\left(\frac{2}{8}\right)\left(\frac{2}{12}\right)=\frac{4}{96}=\frac{1}{24}$

83. P(not red on outer and not red on inner) =

$\frac{8}{12}\cdot\frac{5}{8}=\frac{5}{12}$

85. P(no hit/no hit) = (0.6)(0.6) = 0.36

87. P(both hit) = (0.4)(0.9) = 0.36

89. a) No; The probability of the 2nd depends on the
outcome of the first.

   b) P(one afflicted) = 0.001

   c) P(both afflicted) = (0.001)(0.04) = 0.00004

   d) P(afflicted/not afflicted) = (0.001)(0.96) =
0.00096

   e) P(not afflicted/afflicted) = (0.999)(0.001) =
0.000999

   f) P(not affl/not affl) = (0.999)(0.999) = 0.998001

91. P(audit this year) $=\frac{36}{1000}=\frac{9}{250}$

93. P(audit/no audit) $=\frac{9}{250}\cdot\frac{241}{250}=\frac{2169}{62,500}$

95. P(2 - same color) = P(2 r) + P(2 b) + P(2 y)

$=\left(\frac{5}{10}\right)\left(\frac{4}{9}\right)+\left(\frac{3}{10}\right)\left(\frac{2}{9}\right)+\left(\frac{2}{10}\right)\left(\frac{1}{9}\right)$

$=\frac{20}{90}+\frac{6}{90}+\frac{2}{90}=\frac{28}{90}=\frac{14}{45}$

97. P(no diamonds) $=\left(\frac{39}{52}\right)\left(\frac{38}{51}\right)=\frac{1482}{2652}=0.56$

   The game favors the dealer since the probability
of no diamonds is greater than 1/2.

99.   P(two 2's) = (2/6)(2/6) = 4/36 = 1/9
101.   P(even or < 3) = 2/6 + 3/6 – 2/6 = 3/6 = 1/2

Exercise Set 11.6

1. Conditional

3. $\dfrac{3}{7}$

5. P(5 | yellow) = 0

7. P(even | not orange) = 2/3

9. P(red | orange) = 2/3

11. P(circle | odd) = 3/4

13. P(odd | circle) = 1

15. P(circle or square | < 4) = 2/3

17. P(even # | purple) = 3/5

19. P(purple | even) = 3/6 = 1/2

21. P(> 4 | purple) = 3/5

23. P(gold | > 5) = 1/7

25. P(1 and 1) = (1/4)(1/4) = 1/16

27. P(5 | both at least a 5) = 1/7

29. P(sum = 3) = 1/18

31. P(3 | 3) = 0

33. P(> 7 | 2nd die = 5) = 4/6 = 2/3

35. P(prefers Famous Coffee) = 115/225 = 23/45

37. P(Famous Coffee | man) = 40/100 = 2/5

39. P(man | Manhattan) = 60/110 = 6/11

41. $P(car) = \dfrac{1462}{2461} = 0.5941$

43. $P(E\text{-}Z \mid car) = \dfrac{527}{1462} = 0.3605$

45. $P(car \mid E\text{-}Z) = \dfrac{527}{843} = 0.6251$

47. $P(agg) = \dfrac{350}{650} = \dfrac{7}{13}$

49. $P(no\ sale \mid pass) = \dfrac{80}{300} = \dfrac{4}{15}$

51. $P(sale \mid pass) = \dfrac{220}{300} = \dfrac{11}{15}$

53. $P(male) = \dfrac{151}{309}$

55. $P(15 \text{ - } 65 \text{ years old} \mid female) = \dfrac{105}{158}$

57. $P(male \mid 15 \text{ - } 64 \text{ years old}) = \dfrac{102}{207} = \dfrac{34}{69}$

59. $P(good) = \dfrac{300}{330} = \dfrac{10}{11}$

61. $P(defective \mid 20 \text{ watts}) = \dfrac{15}{95} = \dfrac{3}{19}$

63. $P(good \mid 50 \text{ or } 100 \text{ watts}) = \dfrac{220}{235} = \dfrac{44}{47}$

65. $P(ABC \text{ or } NBC) = \dfrac{110}{270} = \dfrac{11}{27}$

67. $P(ABC \text{ or } NBC \mid man) = \dfrac{50}{145} = \dfrac{10}{29}$

69. $P(ABC, NBC, \text{ or } CBS \mid man) = \dfrac{55}{145} = \dfrac{11}{29}$

71. P(large company stock) = 93/200

73. P(blend | medium co. stock) = 15/52

75. a) $n(A) = 140$ b) $n(B) = 120$
c) $P(A) = 140/200 = 7/10$
d) $P(B) = 120/200 = 6/10 = 3/5$

e) $P(A \mid B) = \dfrac{n(B \text{ and } A)}{n(B)} = \dfrac{80}{120} = \dfrac{2}{3}$

f) $P(B \mid A) = \dfrac{n(A \text{ and } B)}{n(B)} = \dfrac{80}{140} = \dfrac{4}{7}$

g) $P(A) \cdot P(B) = \left(\dfrac{7}{10}\right)\left(\dfrac{3}{5}\right) = \dfrac{21}{50}$

$P(A \mid B) \neq P(A) \cdot P(B)$

A and B are not independent events.

77. a) $P(A|B) = \dfrac{P(A \text{ and } B)}{P(B)} = \dfrac{0.15}{0.5} = 0.3$

b) $P(B|A) = \dfrac{P(A \text{ and } B)}{P(A)} = \dfrac{0.15}{0.3} = 0.5$

c) Yes, $P(A) = P(A \mid B)$ and $P(B) = P(B \mid A)$.

79. $P(+ \mid \text{orange circle}) = 1/2$

81. $P(\text{green} + \mid +) = 1/3$

83. $P(\text{orange circle w/green} + \mid +) = 1/3$

Exercise Set 11.7

1. Permutation

3. n!

5. $\dfrac{n!}{(n-r)!}$

7. $_5P_3$

9. $4! = 24$

11. $0! = 1$

13. $_5P_2 = \dfrac{5!}{3!} = 5 \cdot 4 = 20$

15. $_8P_0 = \dfrac{8!}{8!} = 1$

17. $_9P_5 = \dfrac{9!}{4!} = 9 \cdot 8 \cdot 7 \cdot 6 \cdot 5 = 15,120$

19. $_8P_4 = \dfrac{8!}{4!} = 8 \cdot 7 \cdot 6 \cdot 5 = 1680$

21. $(10)(10)(10)(10) = 10,000$

23. $(3)(4)(2) = 24$

25. $(5)(4)(7)(2) = 280$ systems

27. a) $(36)(36)(36)(36) = 1,679,616$
b) $(62)(62)(62)(62) = 14,776,336$

29. $10^9 = 1,000,000,000$

31. a) $5! = 120$ b) $5! = 120$
c) $4! = 24$ d) $3! = 6$

33. a) There are 12 individuals and they can be arranged in $12! = 479,001,600$ ways
b) $10! = 3,628,800$ different ways
c) $5! \cdot 5! = 14,400$ different ways

35. a) $(10)(10)(10)(26)(26) = 676,000$
b) $(10)(9)(8)(26)(25) = 468,000$
c) $(5)(4)(8)(26)(25) = 104,000$
d) $(9)(9)(8)(26)(25) = 421,200$

37. $_{10}P_{10} = \dfrac{10!}{(10-10)!} = \dfrac{10!}{0!}$

$= (10)(9)(8)(7)(6)(5)(4)(3)(2)(1) = 3,628,800$

39.

$_{10}P_3 = \dfrac{10!}{(10-3)!} = \dfrac{10!}{7!} = \dfrac{10 \cdot 9 \cdot 8 \cdot 7!}{7!} = 720$

41. $_{10}P_3 = \dfrac{10!}{(10-3)!} = \dfrac{10!}{7!} = \dfrac{10 \cdot 9 \cdot 8 \cdot 7!}{7!} = 720$

43. The order of the flags is important. Thus, it Is a permutation problem.

$_9P_5 =$

$\dfrac{9!}{(9-5)!} = \dfrac{9!}{4!} = (9)(8)(7)(6)(5) = 15,120$

45. $_{20}P_5 = (20)(19)(18)(17)(16) = 1,860,480$

47. a) $9! = 362,880$
   b) $5 \cdot 7! \cdot 4 = 100,800$
   c) $4 \cdot 3 \cdot 7! = 60,480$

49. $_9P_9 = \dfrac{9!}{0!} = 9! = 362,880$

51. $\dfrac{10!}{3!3!2!} = 50,400$

53. $\dfrac{7!}{2!2!2!} = 630$

55. a) Since the pitcher must bat last, there is Only one possibility for the last position.

$\underline{\ \ \ \ \ \ \ \ \ \ }\underline{1}$

There are 8 possible batters left for the 1st position. Once the 1st batter has been selected, there are 7 batters left for the 2nd position, 6 for the third, etc.

$\underline{(8)}\ \underline{(7)}\ \underline{(6)}\ \underline{(5)}\ \underline{(4)}\ \underline{(3)}\ \underline{(2)}\ \underline{(1)}\ \underline{(1)} = 40,320$

   b) $9! = (9)(8)(7)(6)(5)(4)(3)(2)(1)$

   $= 362,880$

57. a) $5^5 = 3125$ different keys
   b) $400,000 \div 3,125 = 128$ cars

   c) $\dfrac{1}{3125} = 0.00032$

59. $_7P_5 = \dfrac{7!}{2!} = \dfrac{7 \cdot 6 \cdot 5 \cdot 4 \cdot 3 \cdot 2!}{2!} = 2,520$ different

letter permutations;

$2520 \times 5 \text{ sec} = 12,6000 \text{ sec or } 3\dfrac{1}{2}$ hours

61. No, for example, $_3P_2 \neq {}_3P_{(3-2)}$

$\dfrac{3!}{1!} \neq \dfrac{3!}{2!}$ because $6 \neq 3$

63. $(25)(24) = 600$ tickets

65. a) $\dfrac{7!}{2!} = 2520$ b) STUDENT

Exercise Set 11.8

1. Combination

5. $_{10}C_6$

3. Permutations

7. $_5C_3 = \dfrac{5!}{(5-3)!3!} = \dfrac{(5)(4)(3!)}{(2)(1)(3!)} = 10$

9. a) $_6C_4 = \dfrac{6!}{2!4!} = \dfrac{(6)(5)(4!)}{(2)(1)(4!)} = 15$

    b) $_6P_4 = \dfrac{6!}{(6-4)!} = \dfrac{6!}{2!} = 360$

11. a) $_8C_0 = \dfrac{8!}{8!0!} = 1$

    b) $_8P_0 = \dfrac{8!}{(8-0)!} = \dfrac{8!}{8!} = 1$

13. a) $_{10}C_3 = \dfrac{10!}{7!3!} = \dfrac{(10)(9)(8)(7!)}{(7!)(3)(2)(1)} = 120$

    b) $_{10}P_3 = \dfrac{10!}{(10-3)!} = \dfrac{(10)(9)(8)(7!)}{7!} = 720$

15. $\dfrac{_3C_2}{_{13}C_2} = \dfrac{\dfrac{3!}{1!2!}}{\dfrac{13!}{11!2!}} = \left(\dfrac{3!}{1!2!}\right)\left(\dfrac{11!2!}{13!}\right) = \dfrac{3}{78} = \dfrac{1}{26}$

17. $\dfrac{_8C_5}{_{14}C_5} = \dfrac{\dfrac{8!}{3!5!}}{\dfrac{14!}{9!5!}} = \left(\dfrac{8!}{3!5!}\right)\left(\dfrac{9!5!}{14!}\right) = \dfrac{4}{143}$

19. $\dfrac{_4C_3}{_{10}C_3} = \dfrac{\dfrac{4!}{1!3!}}{\dfrac{10!}{7!3!}} = \left(\dfrac{4!}{1!3!}\right)\left(\dfrac{7!3!}{10!}\right) = \dfrac{1}{30}$

21. $_{10}C_3 = \dfrac{10!}{7!3!} = \dfrac{(10)(9)(8)(7!)}{(3)(2)(1)(7!)} = 120$ ways

23. $_6C_4 = \dfrac{6!}{2!4!} = 15$

25. $_8C_3 = \dfrac{8!}{5!3!} = \dfrac{(8)(7)(6)}{(3)(2)(1)} = 56$

27. $_7C_4 = \dfrac{7!}{3!4!} = \dfrac{(7)(6)(5)}{(3)(2)(1)} = 35$

29. $_{53}C_6 = \dfrac{53!}{47!6!} = 22{,}957{,}480$

31. $_{12}C_4 = \dfrac{12!}{8!4!} = \dfrac{(12)(11)(10)(9)}{(4)(3)(2)(1)} = 495$

33. $_{10}C_3 \cdot {_6C_2} =$

$\left(\dfrac{10!}{7!3!}\right)\left(\dfrac{6!}{4!2!}\right) = \left(\dfrac{(10)(9)(8)}{(3)(2)(1)}\right)\left(\dfrac{(6)(5)}{(2)(1)}\right) = 1800$

35. $_9C_3 \cdot {_8C_2} =$

$\left(\dfrac{9!}{6!3!}\right)\left(\dfrac{8!}{6!2!}\right)$

$= \left(\dfrac{(9)(8)(7)}{(3)(2)(1)}\right)\left(\dfrac{(8)(7)}{(2)(1)}\right) - 2352$

37. Red: $_{10}C_4 = \dfrac{10!}{6!4!} = \dfrac{(10)(9)(8)(7)}{(4)(3)(2)(1)} = 210$

    White. $_8C_2 = \dfrac{8!}{6!2!} = \dfrac{(8)(7)}{(2)(1)} = 28$

    $(210)(28) = 5880$ different choices

39. Regular: $_{10}C_5 =$

    $\dfrac{10!}{5!5!} = \dfrac{(10)(9)(8)(7)(6)}{(5)(4)(3)(2)(1)} = 252$

    Diet: $_7C_3 =$

    $\dfrac{7!}{3!4!} = \dfrac{(7)(6)(5)}{(3)(2)(1)} = 35$

    $(252)(35) = 8820$ ways to select the sodas

41. $_8C_4 \cdot {_5C_2} =$

$\left(\dfrac{8!}{4!4!}\right)\left(\dfrac{5!}{3!2!}\right) = \left(\dfrac{(8)(7)(6)(4)}{(4)(3)(2)(1)}\right)\left(\dfrac{(5)(4)}{(2)(1)}\right) = 700$

43. $_5C_2 \cdot {_4C_3} \cdot {_6C_2} =$

$\left(\dfrac{5!}{3!2!}\right)\left(\dfrac{4!}{1!3!}\right)\left(\dfrac{6!}{4!2!}\right) =$

$\left(\dfrac{(5)(4)}{(2)(1)}\right)\left(\dfrac{(4)}{(1)}\right)\left(\dfrac{(6)(5)}{(2)(1)}\right) = 600$

45. a) $_{10}C_8 = \dfrac{10!}{2!8!} = \dfrac{(10)(9)}{(2)(1)} = 45$

    b) $_{10}C_9 = \dfrac{10!}{1!9!} = \dfrac{(10)(9!)}{(1)(9!)} = 10$

    $_{10}C_{10} = \dfrac{10!}{10!} = 1$

    $_{10}C_8 + {}_{10}C_9 + {}_{10}C_{10} = 45 + 10 + 1 = 56$

49. a) $4! = 24$ b) $4! = 24$

47. a)

$$
\begin{array}{ccccccccc}
& & & & 1 & & & & \\
& & & 1 & & 1 & & & \\
& & 1 & & 2 & & 1 & & \\
& 1 & & 3 & & 3 & & 1 & \\
1 & & 4 & & 6 & & 4 & & 1
\end{array}
$$

    b) $\quad 1 \qquad 5 \qquad 10 \qquad 10 \qquad 5 \qquad 1$

51. $(15)(14) \cdot {}_{13}C_3 = (15)(14)\left(\dfrac{13!}{10!\,3!}\right)$

    $= (15)(14)\left(\dfrac{(13)(12)(11)}{(3)(2)(1)}\right) = 60{,}060$

Exercise Set 11.9

1. $P(4 \text{ red chips}) = \dfrac{\text{no. of 4 red chip comb.}}{\text{no. of 4 chip comb.}} = \dfrac{{}_{12}C_4}{{}_{20}C_4}$

3. $P(3 \text{ kings}) = \dfrac{\text{no. of 3 king comb.}}{\text{no. of 3 card comb.}} = \dfrac{{}_4C_3}{{}_{52}C_3}$

5. $P(\text{all 5 yellow Labs}) =$

    $\dfrac{\text{no. of 5 yellow Lab comb.}}{\text{no. of 5 puppy comb.}} = \dfrac{{}_8C_5}{{}_{15}C_5}$

7. $P(\text{each of the 5 has a criminal justice degree}) =$

    $\dfrac{\text{no. of 5 crim. just. combo.}}{\text{no. of student comb.}} = \dfrac{{}_{23}C_5}{{}_{120}C_5}$

9. $_3C_2 = \dfrac{3!}{1!2!} = 3$

    $_{13}C_2 = \dfrac{13!}{11!2!} = \dfrac{(13)(12)}{(2)(1)} = 78$

    $P(2 \text{ diet soda}) = \dfrac{3}{78} = \dfrac{1}{26}$

11. $_8C_5 = \dfrac{8!}{3!5!} = \dfrac{(8)(7)(6)}{(3)(2)(1)} = 56$

    $_{14}C_5 = \dfrac{14!}{5!9!} = \dfrac{(14)(13)(12)(11)(10)}{(5)(4)(3)(2)(1)} = 2002$

    $P(5 \text{ men's names}) = \dfrac{56}{2002} = \dfrac{4}{143}$

13. $_4C_3 = \dfrac{4!}{1!3!} = \dfrac{(4)}{(1)} = 4$

    $_{10}C_3 = \dfrac{10!}{7!3!} = \dfrac{(10)(9)(8)}{(3)(2)(1)} = 120$

    $P(3 \text{ greater than 5}) = \dfrac{4}{120} = \dfrac{1}{30}$

15. $6C_2 = \dfrac{6!}{4!2!} = \dfrac{(6)(5)}{(2)(1)} = 15$

    $2C_1 = 2$

    $11C_3 = \dfrac{11!}{8!3!} = \dfrac{(11)(10)(9)}{(3)(2)(1)} = 165$

    $P(2 \text{ from mfg, 1 from acct.}) = \dfrac{(15)(2)}{165} = \dfrac{2}{11}$

17. $_4C_3 = \dfrac{4!}{1!3!} = \dfrac{(4)}{(1)} = 4$

$_4C_2 = \dfrac{4!}{2!2!} = \dfrac{(4)(3)}{(2)(1)} = 6$

$_{52}C_5 = \dfrac{52!}{47!5!} = \dfrac{(52)(51)(50)(49)(48)}{(5)(4)(3)(2)(1)} = 2,598,960$

P(3 kings, 2 queens) $= \dfrac{(4)(6)}{2,598,960} = \dfrac{1}{108,290}$

19. $_3C_2 = \dfrac{3!}{1!2!} = 3 \qquad _5C_2 = \dfrac{5!}{3!2!} = \dfrac{(5)(4)}{(2)(1)} = 10$

P(no cars) $= \dfrac{3}{10}$

21. P(at least 1 car) $= 1 - P(\text{no cars}) =$

$1 - \dfrac{3}{10} = \dfrac{7}{10}$

23. $_6C_3 = \dfrac{6!}{3!3!} = \dfrac{(6)(5)(4)}{(3)(2)(1)} = 20$

$_{25}C_3 = \dfrac{25!}{3!22!} = \dfrac{(25)(24(23)}{(3)(2)(1)} = 2300$

P(3 infielders) $= \dfrac{20}{2300} = \dfrac{1}{115}$

25. $_{10}C_2 = \dfrac{10!}{8!2!} = 45 \qquad _6C_1 = \dfrac{6!}{5!1!} = 6$

P(2 pitchers and 1 infielder) $= \dfrac{(45)(6)}{2300} = \dfrac{27}{230}$

For problems 27 and 29, use the fact that $_{25}C_6 = \dfrac{25!}{19!6!} = 177,100$

27. $_{10}C_6 = \dfrac{10!}{4!6!} = 210$

P(all mid) $= \dfrac{210}{177,100} = 0.0012$

29. $_{10}C_3 = \dfrac{10!}{7!3!} = 120$

$_{15}C_3 = \dfrac{15!}{12!3!} = 455$

P(2 mid/4 compact) $= \dfrac{(120)(455)}{117,100} = 0.3083$

For problems 31 and 33, use the fact that $_{12}C_4 =$
$\dfrac{12!}{8!4!} = 495$

33. $_4C_1 = 4$
$_5C_1 = 5$
$_3C_2 = 3$
P(2 are oat, 1 is wheat, 1 is rice) =
$\dfrac{(4)(5)(3)}{495} = \dfrac{4}{33}$

31. $_5C_2 = \dfrac{5!}{3!2!} = 10 \qquad _3C_2 = \dfrac{3!}{1!2!} = 3$

P(2 are oat, 2 are rice) = $\dfrac{(10)(3)}{495} = \dfrac{2}{33}$

35. $_{15}C_3 = 455$
$_{26}C_3 = 2600$

P(3 cashiers) = $\dfrac{455}{2600} = \dfrac{91}{520} = \dfrac{7}{40}$

37. a) P(royal spade flush) = $\dfrac{_{47}C_2}{_{52}C_7} = \dfrac{1}{123,760}$

b) P(any royal flush) = $\dfrac{4}{123,760} = \dfrac{1}{30,940}$

$\dfrac{7!}{3!4!} = \dfrac{(7)(6)(5)}{(3)(2)(1)} = 35$

$(252)(35) = 8820$ ways to select the sodas

39. a) $\left(\dfrac{\left(_4C_2\right)\left(_4C_2\right)\left(_{44}C_1\right)}{_{52}C_5} \right) = \dfrac{1584}{2,598,960} = \dfrac{33}{54,145}$

P(2 aces/2 8's/other card/ace or 8) = $\dfrac{33}{54,145}$

b) P(aces of spades and clubs/8's of spades and clubs/9 of diamonds) =
$\dfrac{1}{_{52}C_5} = \dfrac{1}{2,598,960}$

41. a) $\left(\dfrac{1}{15}\right)\left(\dfrac{1}{14}\right)\left(\dfrac{1}{13}\right)\left(\dfrac{5}{12}\right)\left(\dfrac{4}{11}\right)\left(\dfrac{3}{10}\right)\left(\dfrac{2}{9}\right)\left(\dfrac{1}{8}\right)$

$= \dfrac{120}{259,459,200} = \dfrac{1}{2,162,160}$

b) P(any 3 of 8 for officers) = $\dfrac{(8)(7)(6)}{2,162,160} = \dfrac{1}{6435}$

43. 1; Since there are more people than hairs, 2 or more people must have the same number of hairs.

Exercise Set 11.10

1. Trials

3. Success

5. $P(3) = {}_6C_3 (0.2)^3 (0.8)^{6-3}$
$= \dfrac{6!}{3!3!}(0.008)(0.512) = 0.08192$

7. $P(2) = {}_5C_2 (0.4)^2 (0.6)^{5-2}$
$= \dfrac{5!}{2!3!}(0.16)(0.216) = 0.3456$

9. $P(0) = {}_6C_0(0.5)^0(0.5)^{6-0}$

$= \dfrac{6!}{0!6!}(1)(0.0156252) = 0.015625$

11. $p = 0.14$, $q = 1 - p = 1 - 0.14 = 0.86$

a) $P(x) = {}_nC_x(0.14)^x(0.86)^{n-x}$

b) $n = 12$, $x = 2$, $p = 0.14$, $q = 0.86$

$P(2) = {}_{12}C_2(0.14)^2(0.86)^{10}$

13. $P(5) = {}_8C_5(0.6)^5(0.4)^{8-5}$

$= \dfrac{8!}{3!5!}(0.07776)(0.064) = 0.27869$

15. $P(6) = {}_{10}C_6(0.53)^6(0.47)^{10-6}$

$= \dfrac{10!}{4!6!}(0.02216436)(0.04879681) = 0.22713$

17. $P(4) = {}_6C_4(0.92)^4(0.08)^{6-4}$

$= \dfrac{6!}{4!2!}(0.7164)(0.0064) = 0.06877$

19. $P(4) = {}_5C_4(0.8)^4(0.2)^{5-4}$

$= \dfrac{5!}{1!4!}(0.4096)(0.2) = 0.4096$

21. a) $P(\text{all five}) = {}_5C_5(0.25)^5(0.75)^{5-5}$

$= \dfrac{5!}{5!}(0.0009765625)(1) = 0.00098$

b) $P(\text{exactly three}) = {}_5C_3(0.25)^3(0.75)^{5-3}$

$= \dfrac{5!}{2!3!}(0.015625)(0.5625) = 0.08789$

c) $P(\text{at least } 3) = P(3) + P(4) + P(5)$

$P(4) = {}_5C_4(0.25)^4(0.75)^{5-1}$

$- \dfrac{5!}{1!4!}(0.00390625)(0.75) - 0.01465$

$P(\text{at least } 3) = 0.08789 + 0.01465 + 0.00098$

$= 0.10352$

23. a) $P(3) = {}_6C_3\left(\dfrac{12}{52}\right)^3\left(\dfrac{40}{52}\right)^3$

$\dfrac{6!}{3!3!}(0.01229)(0.45517) = 0.1119$

b) $P(2) = {}_6C_2\left(\dfrac{13}{52}\right)^2\left(\dfrac{39}{52}\right)^4$

$= \dfrac{6!}{2!4!}(0.0625)(0.3164) = 0.2966$

25. The probability that the sun would be shining would equal 0 because 72 hours later would occur at midnight.

Review Exercises

1. Relative frequency over the long run can accurately be predicted, not individual events or totals.
2. Roll the die many times then compute the relative frequency of each outcome and compare with the expected probability of 1/6.

3. $P(\text{smart phone}) = \dfrac{40}{45} = \dfrac{8}{9}$

4. Answers will vary.

5. $P(\text{watches ABC}) = \dfrac{80}{250} = \dfrac{8}{25}$

6. $P(\text{even}) = \dfrac{5}{10} = \dfrac{1}{2}$

7. $P(\text{odd or} > 3) = \dfrac{5}{10} + \dfrac{6}{10} - \dfrac{3}{10} = \dfrac{8}{10} = \dfrac{4}{5}$

8. $P(> 3 \text{ or} < 6) = \dfrac{6}{10} + \dfrac{6}{10} - \dfrac{2}{10} = \dfrac{10}{10} = 1$

9. $P(\text{even and} > 4) = \dfrac{2}{10} = \dfrac{1}{5}$

10. $P(\text{Cheddar}) = \dfrac{21}{60} = \dfrac{7}{20}$

11. $P(\text{Gouda}) = \dfrac{13}{60}$

12. $P(\text{Cheddar or Monterey Jack}) =$
    $\dfrac{21}{60} + \dfrac{17}{60} = \dfrac{38}{60} = \dfrac{19}{30}$

13. $P(\text{not Mozzarella}) = 1 - P(\text{mozzarella}) =$
    $1 - \dfrac{9}{60} = \dfrac{17}{20}$

14. a) 69:31   b) 31:69

15. 5:3

16. $P(\text{wins Triple Crown}) = \dfrac{3}{(3+82)} = \dfrac{3}{85}$

17. $(3/4)(1/4) = 3/1 \text{ or } 3:1$

18. a) $E = P(\text{win } \$200) \cdot \$198 + P(\text{win } \$100) \cdot \$98$
    $+ P(\text{lose}) \cdot (-\$2)$
    $= (0.003)(198) + (0.002)(98) - (0.995)(2)$
    $= 0.594 + 0.196 - 1.990 = -1.200 \;\rightarrow\; -\$1.20$

    b) The expectation of a person who purchases
    three tickets would be $3(-1.20) = -\$3.60$.

19. a) $E_{Cameron} = P(\text{pic. card})(\$9) +$
    $P(\text{not pic. card})(-\$3)$
    $= \left(\dfrac{12}{52}\right)(9) - \left(\dfrac{40}{52}\right)(3) = -\$0.23$

    b) $E_{Lindsey} = P(\text{pic. card})(-\$9) + P(\text{not pic. card})(\$3)$
    $= \dfrac{-27}{13} + \dfrac{30}{13} = \dfrac{3}{13} = \$0.23$

    c) Cameron can expect to lose $(100)\left(\dfrac{3}{13}\right) \approx \$23.08$

20. $E = P(\text{sunny})(1000) + P(\text{cloudy})(500) + P(\text{rain})(100) = 0.4(1000) + 0.5(500) + 0.1(100) =$
    $400 + 250 + 10 = 660$ people

21. Fair price = expected value + cost to play
    Fair price = –$2.50 + $6.50 = $4.00

22. Fair price = expected value + cost to play
    FP = –$1.50 + $5.00 = $3.50

23. a)
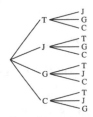

    b) Sample space:
    {TJ,TG,TC,JT,JG,JC,GT,GJ,GC,CT,CJ,CG}
    c) P(Gina is Pres. and Jake V.P.) = 1/12

24. a)

    b) Sample space:
    {H1,H2,H3,H4,T1,T2,T3,T4}
    c) P(heads and odd) = (1/2)(2/4) = 2/8 = ¼
    d) P(heads or odd) = (1/2)(2/4) + (1/2)(2/4)
    $= 4/8 + 2/8 = 6/8 = 3/4$

25. P(even and even) = (4/8)(4/8) = 16/64 = 1/4

26. P(outer is greater than 5 and inner is greater than 5)
    $= P(\text{outer is} > 5) \cdot P(\text{inner is} > 5) = \dfrac{3}{8} \cdot \dfrac{3}{8} = \dfrac{9}{64}$

27. P(outer odd and inner < 6)

$$= \text{P(outer odd) P(inner} < 6) = \frac{4}{8} \cdot \frac{5}{8} = \frac{1}{2} \cdot \frac{5}{8} = \frac{5}{16}$$

28. P(outer is even or less than 6)

$$= \text{P(even)} + \text{P}(< 6) - \text{P(even and} < 6)$$

$$= \frac{4}{8} + \frac{5}{8} - \frac{2}{8} = \frac{7}{8}$$

29. P(inner even and not green) =

$$\frac{1}{2} + \frac{6}{8} - \frac{2}{8} = \frac{1}{2} + \frac{4}{8} = 1$$

30. P(outer gold and inner not gold)

$$= \left(\frac{2}{8}\right)\left(\frac{6}{8}\right) = \left(\frac{1}{4}\right)\left(\frac{3}{4}\right) = \frac{3}{16}$$

31. P(all 3 are COLA) = $\frac{5}{12} \cdot \frac{4}{11} \cdot \frac{3}{10} = \frac{60}{1320} = \frac{1}{22}$

32. P(none are ROOT BEER) =

$$\frac{8}{12} \cdot \frac{7}{11} \cdot \frac{6}{10} = \frac{336}{1320} = \frac{14}{55}$$

33. P(at least one is P) = 1 – P(none are root beer)

$$= 1 - \frac{14}{55} = \frac{55}{55} - \frac{14}{55} = \frac{41}{55}$$

34. P(C, RB, G)

$$= \frac{5}{12} \cdot \frac{4}{11} \cdot \frac{3}{10} = \frac{60}{1320} = \frac{1}{22}$$

35. P(not green) = 1/4 + 1/4 + 1/8 = 5/8

36. Odds in favor of green   3:5
    Odds against green   5:3

37. E = P(green)($10) + P(red)($5) +
    P(yellow)(–$20)

$$= (3/8)(10) + (1/2)(5) - (1/8)(20)$$

$$= (15/4) + (10/4) - (10/4) = 15/4 \text{ ot } \$3.75$$

38. P(at least one red) = 1 – P(none are red)

$$= 1 - (1/2)(1/2)(1/2) = 1 - 1/8 = 7/8$$

39. P(rated poor) = 30/200 = 3/20

40. P(good | lunch) = 75/90 = 5/6

41. P(poor | breakfast) = 15/110 = 3/22

42. P(breakfast | poor) = 15/30 = 1/2

43. P(right handed) = $\frac{230}{400} = \frac{23}{40}$

44. P(left brained | left handed) = $\frac{30}{170} = \frac{3}{17}$

45. P(right handed | no predominance) = $\frac{60}{80} = \frac{3}{4}$

46. P(right brained | left handed) = $\frac{120}{170} = \frac{12}{17}$

47. a) 4! = (4)(3)(2)(1) = 24
    b) E = (1/4)(10K) + (1/4)(5K) + (1/4)(2K)
       + (1/4)(1K) = (1/4)(18K) = $4500

48. # of possible arrangements = $(_5C_2)(_3C_2)(_1C_1)$

$$= \left(\frac{5!}{3!2!}\right)\left(\frac{3!}{1!2!}\right)\left(\frac{1!}{1!}\right) = \frac{(5)(4)(3)}{(2)(1)} = 30$$

49. $_{10}P_3 = \frac{10!}{7!} = (10)(9)(8) = 720$

50. $_9P_3 = \frac{9!}{6!} = \frac{(9)(8)(7)(6!)}{6!} = (9)(8)(7) = 504$

51. $_6C_3 = \frac{6!}{3!3!} = \frac{(6)(5)(4)}{(3)(2)(1)} = 20$

52. Number of arrangements = 10! = 3,628,800

53. a) $P(\text{match 5 numbers}) = \dfrac{1}{_{75}C_5}$

$$= \dfrac{1}{\dfrac{75!}{70!5!}} = \dfrac{70!5!}{75!} = \dfrac{1}{17,259,390}$$

    b) $P(\text{Big game win}) = P(\text{match 5 \#s and Big \#})$

         $= P(\text{match 5 \#s}) \cdot P(\text{match Big \#})$

$$= \left(\dfrac{1}{17,259,390}\right)\left(\dfrac{1}{15}\right) = \dfrac{1}{258,890,850}$$

54. $(_{10}C_2)(_{12}C_5) =$

$$\left(\dfrac{10!}{8!2!}\right)\left(\dfrac{12!}{7!5!}\right)$$

$$= \dfrac{(10)(9)(12)(11)(10)(9)(8)}{(2)(1)(5)(4)(3)(2)(1)}$$

$$= 35,640 \text{ possible committees}$$

55. $(_8C_3)(_5C_2) =$

$$\left(\dfrac{8!}{5!3!}\right)\left(\dfrac{5!}{2!3!}\right) = \dfrac{(8)(7)(6)(5)(4)}{(3)(2)(1)(2)(1)} = 560$$

56. $P(\text{two aces}) = \dfrac{_4C_2}{_{52}C_2} = \dfrac{\dfrac{4!}{2!2!}}{\dfrac{52!}{50!2!}}$

$$= \left(\dfrac{4!}{2!2!}\right)\left(\dfrac{50!2!}{52!}\right) = \dfrac{1}{221}$$

58. $P(\text{1st 2 are red/3}^{\text{rd}}\text{ is blue}) = \left(\dfrac{5}{10}\right)\left(\dfrac{4}{9}\right)\left(\dfrac{2}{8}\right) = \dfrac{1}{18}$

57. $P(\text{all three are red}) = \left(\dfrac{5}{10}\right)\left(\dfrac{4}{9}\right)\left(\dfrac{3}{8}\right) = \dfrac{1}{12}$

59. $P(\text{1}^{\text{st}}\text{ red, 2}^{\text{nd}}\text{ white, 3}^{\text{rd}}\text{ blue})$

$$= \left(\dfrac{5}{10}\right)\left(\dfrac{3}{9}\right)\left(\dfrac{2}{8}\right) = \dfrac{1}{24}$$

60. $P(\text{at least one red}) = 1 - P(\text{none are red})$

$$= 1 - \left(\dfrac{5}{10}\right)\left(\dfrac{4}{9}\right)\left(\dfrac{3}{8}\right) = 1 - \dfrac{1}{12} = \dfrac{11}{12}$$

61. $P(\text{3 MT}) =$

$$\dfrac{_5C_3}{_{14}C_3} = \dfrac{\dfrac{5!}{3!2!}}{\dfrac{14!}{3!11!}} = \dfrac{5!3!11!}{3!2!14!} = \dfrac{(5)(4)(3)}{(14)(13)(12)} = \dfrac{5}{182}$$

62. $P(\text{2 NWs \& 1 Time}) =$

$$\dfrac{(_6C_2)(_3C_1)}{_{14}C_3} = \dfrac{\left(\dfrac{6!}{2!4!}\right)\left(\dfrac{3!}{1!2!}\right)}{\dfrac{14!}{3!11!}}$$

$$= \dfrac{(6)(5)(3)(3)(2)(1)}{(2)(1)(14)(13)(12)} = \dfrac{45}{364}$$

63. $\dfrac{_8C_3}{_{14}C_3} = \dfrac{\dfrac{8!}{3!5!}}{\dfrac{14!}{3!11!}} = \dfrac{8!3!11!}{3!5!14!}$

$$= \dfrac{(8)(7)(6)}{(14)(13)(12)} = \dfrac{336}{2184} = \dfrac{2}{13}$$

64. $P(\text{At least one parenting}) = 1 - \dfrac{2}{13} = \dfrac{11}{13}$

65. a) $P(x) = {}_nC_x (0.6)^x (0.4)^{n-x}$

   b) $P(75) = {}_{100}C_{75}(0.6)^{75}(0.4)^{25}$

66. $n = 5, x = 3, p = 1/5, q = 4/5$

$$P(3) = {}_5C_3\left(\frac{1}{5}\right)^3\left(\frac{4}{5}\right)^2 = 10 \cdot \left(\frac{1}{5}\right)^3\left(\frac{4}{5}\right)^2 = 0.0512$$

67. a) $n = 5, p = 0.6, q = 0.4$

$$P(0) = {}_5C_0 (0.6)^0 (0.4)^5$$

$$= (1)(1)(0.4)^5 = 0.01024$$

   b) P(at least 1) = 1 − P(0) = 1 − 0.01024 = 0.98976

Chapter Test

1. $P(\text{glazed}) = \dfrac{11}{25}$

2. $(P > 5) = \dfrac{4}{9}$

3. $P(\text{odd}) = \dfrac{5}{9}$

4. $P(\text{even or} < 4) = \dfrac{6}{9} = \dfrac{2}{3}$

5. $P(\text{both} < 3) = \dfrac{2}{9} \cdot \dfrac{1}{8} = \dfrac{2}{72} = \dfrac{1}{36}$

6. $P(\text{1st odd, 2nd even}) = \dfrac{5}{9} \cdot \dfrac{4}{8} = \dfrac{5}{9} \cdot \dfrac{1}{2} = \dfrac{5}{18}$

7. $P(\text{neither} > 6) = \dfrac{6}{9} \cdot \dfrac{5}{8} = \dfrac{1 \cdot 5}{3 \cdot 4} = \dfrac{5}{12}$

8. P(red or picture)
   = P(red) + P(picture) − P(red and picture)

$$= \dfrac{26}{52} + \dfrac{12}{52} - \dfrac{6}{52} = \dfrac{32}{52} = \dfrac{8}{13}$$

9. $(6)(3) = 18$

10.

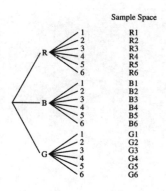

11. $P(\text{green and 2}) = \dfrac{1}{18}$

12. $P(\text{red or 1}) = \dfrac{6}{18} + \dfrac{3}{18} - \dfrac{1}{18} = \dfrac{8}{18} = \dfrac{4}{9}$

13. $P(\text{not red or even}) = \dfrac{12}{18} + \dfrac{9}{18} - \dfrac{6}{18} = \dfrac{15}{18} = \dfrac{5}{6}$

14. Number of codes = (26)(9)(10)(26)(26)
    = 1,581,840

15. a) 5:4   b) 5:4

16. E = P(club) ($8) + P(heart) ($4)
    + P(spade or diamond) (−$6)

$$= \left(\dfrac{1}{4}\right)(8) + \left(\dfrac{1}{4}\right)(4) + \left(\dfrac{2}{4}\right)(-6)$$

$$= \dfrac{8}{4} + \dfrac{4}{4} - \dfrac{12}{4} = \$0$$

17. a) $P(SUV) = \dfrac{242}{456} = \dfrac{121}{228}$

    b) $P(\text{George Washington}) = \dfrac{226}{456} = \dfrac{113}{228}$

    c) $P(SUV \mid \text{George Washington}) = \dfrac{106}{226} = \dfrac{53}{113}$

    d) $P(\text{GG Bridge} \mid \text{car}) = \dfrac{94}{214} = \dfrac{47}{107}$

19. a) $P(\text{neither is good}) = \dfrac{6}{20} \cdot \dfrac{5}{19} = \dfrac{3}{38}$

    b) $P(\geq 1 \text{ good}) = 1 - P(\text{neither -good}) =$

    $1 - \dfrac{3}{38} = \dfrac{35}{38}$

18. $_6P_3 = \dfrac{6!}{(6-3)!} = \dfrac{6!}{3!} = 6 \cdot 5 \cdot 4 = 120$

20. $(0.6)(0.6)(0.4)(0.4)(0.4) = 0.02304$

    $_5C_3 = \dfrac{5!}{3!2!} = \dfrac{(5)(4)}{(2)(1)} = 10$

    $(10)(0.02304) = 0.2304$

CHAPTER TWELVE

STATISTICS

Exercise Set 12.1
1. Statistics
3. Descriptive
5. Sample
7. Random
9. Stratified
11. Stratified sample
13. Cluster sample

15. Systematic sample
19. Random sample

17. Convenience sample

21. The patients may have improved on their own without taking honey.
23. Half the students in a population are expected to be below average.
25. A recommended toothpaste may not be better than all other types of toothpaste.
27. Most driving is done close to home. Thus, one might expect more accidents close to home.
29. We don't know how many of each professor's students were surveyed. Perhaps more of Professor Fogal's students than Professor Bond's students were surveyed. Also, because more students prefer a teacher does not mean that he or she is a better teacher. For example, a particular teacher may be an easier grader and that may be why that teacher is preferred.
31. Just because they are more expensive does not mean they will last longer.
33. There may be deep sections in the pond, so it may not be safe to go wading.
35. a)   b)

37. a)   b)

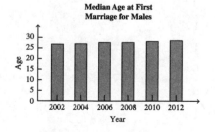

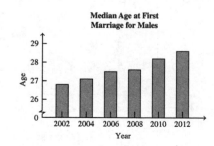

251

39. a)

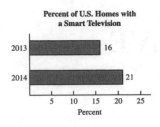

Percent of U.S. Homes with a Smart Television

b) Answers will vary.

41. Yes, the sum of its parts is 121%. The sum of the parts of a circle graph should be 100%. When the total percent of responses is more than 100%, a circle graph is not an appropriate graph to display the data. A bar graph is more appropriate in this situation.

43. Biased because the subscribers of *Consumer Reports* are not necessarily representative of the entire population.

45. President; 4 out of 43 U.S. presidents have been assassinated (Lincoln, Garfield, McKinley, Kennedy).

Exercise Set 12.2

1. Frequency

3. Mark

5. Histogram

7. a) Stem b) Leaf

9. a) Number of observations = sum of frequencies = 20

b) Width = $16 - 9 = 7$

c) $\dfrac{16 + 22}{2} = \dfrac{38}{2} = 19$

d) The modal class is the class with the greatest frequency. Thus, the modal class is $16 - 22$.

e) Since the class widths are 7, the next class would be $51 - 57$.

11.

Number of Visits	Number of Students
0	3
1	8
2	3
3	5
4	2
5	7
6	2
7	3
8	4
9	1
10	2

13.

Circulation (thousands)	Number of Newspapers
100 - 299	30
300 - 499	13
500 - 699	4
700 - 899	0
900 - 1099	0
1100 − 1299	0
1300 − 1499	0
1500 − 1699	1
1700 − 1899	1
1900 − 2099	0
2100 − 2299	0
2300 - 2499	1

15.

Circ. (thous.)	No. of News.
138 - 587	46
588 - 1037	1
1038 - 1487	0
1488 - 1937	2
1938 - 2387	1

17.

Population (hund. thous)	No. Cities
6.5 – 19.4	16
19.5 – 32.4	2
32.5 – 45.4	1
45.5 – 58.4	0
58.5 – 71.4	0
71.5 – 84.4	1

19.

Population (hund. thous.)	Number of Cities
6.0 – 20.9	16
21.0 – 35.9	2
36.0 – 50.9	1
51.0 – 65.9	0
66.0 – 80.9	0
81.0 – 95.9	1

21.

Cost of living	Number of States
85.0 – 94.9	21
95.0 – 104.9	15
105.0 – 114.9	3
115.0 – 124.9	4
125.0 – 134.9	6
135.0 – 144.9	0
145.0 – 154.9	0
155.0 – 164.9	1

23.

Cost of living	Number of States
88.0 – 99.9	32
100.0 – 111.9	7
112.0 – 123.9	4
124.0 – 135.9	6
136.0 – 147.9	0
148.0 – 159.9	0
160.0 – 171.9	1

25. 1 | 2 represents 12

```
0 | 4 6 7 8
1 | 2 2 3 5 6 7 8 9
2 | 1 2 3 5 7
3 | 3 4
4 | 0
```

27.a)

Sales (thousands)	Number of Vehicles
18 - 24	7
25 - 31	4
32 - 38	1
39 - 45	1
46 - 52	1
53 - 59	0
60 - 66	1

b) and c)

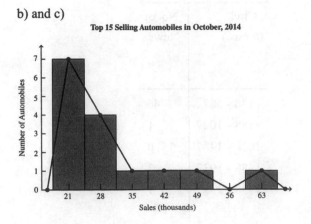

Top 15 Selling Automobiles in October, 2014

29. a)

Age	Number of People
20 – 30	10
31 – 41	14
42 – 52	9
53 – 63	4
64 – 74	3

b) and c)

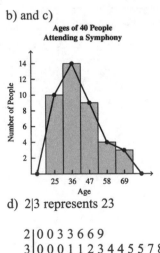

Ages of 40 People Attending a Symphony

d) 2|3 represents 23

```
2|0 0 3 3 6 6 9
3|0 0 0 1 1 2 3 4 4 5 5 7 8 9
4|0 0 0 2 5 7 9 9 9
5|0 1 1 4 7
6|2 3 6 9
7|2
```

31. a) $2+4+8+6+4+3+1=28$

b) 4

c) 2

d) $2(0)+1(4)+2(8)+6(3)+4(4)+3(5)+1(6)=75$

e)

Number of TVs	Number of homes
0	2
1	4
2	8
3	6
4	4
5	3
6	1

33. a) 7 people

b) Adding the number of people who sent 6, 5, 4, or 3 messages gives: $4 + 7 + 3 + 2 = 16$ people

c) The total number of people in the survey: $2 + 3 + 7 + 4 + 3 + 8 + 6 + 3 = 36$

d)

Number of Messages	Number of People
3	2
4	3
5	7
6	4
7	3
8	8
9	6
10	3

e)

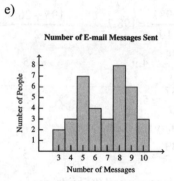

35.   Tuition: $0.48(20,986) = \$10,073.28$

Room: $0.25(20,986) = \$5246.50$

Board: $0.21(20,986) = \$4407.06$

Books/Supplies: $0.06(20,986) = \$1259.16$

37. a) – e) Answers will vary.

39. a) There are 6 F's.   b) Answers will vary.

<u>**Exercise Set 12.3**</u>

1. Average

3. Mean

5. Mode

7. Quartiles

9. a) $\overline{x}$

  b) μ

	mean	median	mode	midrange
11.	$\dfrac{135}{9} = 15$	14	14	$\dfrac{9+27}{2} = 18$
13.	$\dfrac{590}{7} \approx 84.3$	87	none	$\dfrac{57+105}{2} = 81$
15.	$\dfrac{64}{8} = 8$	$\dfrac{7+9}{2} = 8$	none	$\dfrac{1+15}{2} = 8$
17.	$\dfrac{118}{9} \approx 13.1$	11	1	$\dfrac{1+36}{2} = 18.5$
19.	$\dfrac{95}{8} \approx 11.9$	$\dfrac{12+13}{2} = 12.5$	13	$\dfrac{6+17}{2} = 11.5$
21.	$\dfrac{1987}{10} = 198.7$	$\dfrac{195+200}{2} = 197.5$	none	$\dfrac{132+285}{2} = 208.5$
23. a)	$\dfrac{34}{7} \approx 4.9$	5	5	$\dfrac{1+11}{2} = 6$
b)	$\dfrac{37}{7} \approx 5.3$	5	5	$\dfrac{1+11}{2} = 6$
c)	Only the mean			
d)	$\dfrac{33}{7} \approx 4.7$	5	5	$\dfrac{1+10}{2} = 5.5$

  The mean and the midrange

25. A 79 mean average on 10 quizzes gives a total of 790 points. An 80 mean average on 10 quizzes requires a total of 800 points. Thus, Jim missed a B by 10 points not 1 point.

27. a) Mean: $\dfrac{84.8}{10} \approx 8.5 \text{ kg}$ b) Median: $\dfrac{8.2+8.3}{2} \approx 8.3 \text{ kg}$

   c) Mode: none

   d) Midrange: $\dfrac{6.3+11.9}{2} = 9.1 \text{ kg}$

29. a) Mean: $\dfrac{145.5 \text{ million}}{10} \approx 14.6 \text{ million}$ b) Median: $\dfrac{10+12}{2} \approx 11.0 \text{ million}$

   c) Mode: none

   d) Midrange: $\dfrac{4.3+36}{2} \approx 20.2 \text{ million}$

31. Let $x =$ the sum of his scores

   $\dfrac{x}{6} = 92$

   $x = 92(6) = 552$

33. One example is 72, 73, 74, 76, 77, 78.

    Mean: $\dfrac{450}{6} = 75$, Median: $\dfrac{74+76}{2} = 75$, Midrange: $\dfrac{72+78}{2} = 75$

35. a) Yes
    b) Cannot be found since we do not know the middle two numbers in the ranked list
    c) Cannot be found without knowing all of the numbers
    d) Yes

37. a) For a mean average of 60 on 7 exams, she must have a total of $60 \times 7 = 420$ points. Sheryl presently has
       $52 + 72 + 80 + 65 + 57 + 69 = 395$ points. Thus, to pass the course, her last exam must be $420 - 395 = 25$
       or greater.
    b) A C average requires a total of $70 \times 7 = 490$ points. Sheryl has 395. Therefore, she would need
       $490 - 395 = 95$ or greater on her last exam.
    c) For a mean average of 60 on 6 exams, she must have a total of $60 \times 6 = 360$ points. If the lowest score on
       an exam she has already taken is dropped, she will have a total of $72 + 80 + 65 + 57 + 69 = 343$ points.
       Thus, to pass the course, her last exam must be $360 - 343 = 17$ or greater.
    d) For a mean average of 70 on 6 exams, she must have a total of $70 \times 6 = 420$ points. If the lowest score on
       an exam she has already taken is dropped, she will have a total of 343 points. Thus, to obtain a C, her last
       exam must be $420 - 343 = 77$ or greater.

39. One example is 1, 2, 3, 3, 4, 5 changed to 1, 2, 3, 4, 4, 5.

    First set of data: Mean: $\dfrac{18}{6} = 3$, Median: $\dfrac{3+3}{2} = 3$, Mode: 3

    Second set of data: Mean: $\dfrac{19}{6} = 3.1\overline{6}$, Median: $\dfrac{3+4}{2} = 3.5$, Mode: 4

41. No, by changing only one piece of the six pieces of data you cannot alter both the median and the midrange.

43. The data must be ranked.

45. He is taller than approximately 35% of all kindergarten children.

47. a) $Q_2 = $ Median $= \$150$

    b) $Q_1 = $ Median of the first 9 data values $= \$100$

    c) $Q_3 = $ Median of the last 9 data values $= \$180$

49. Second quartile, median

51. a) $600   b) $610   c) 25%   d) 25%   e) 17%   f) $100 \times \$620 = \$62,000$

53. a) Ruth: $\approx$ 0.290, 0.359, 0.301, 0.272, 0.315
       Mantle: $\approx$ 0.300, 0.365, 0.304, 0.275, 0.321
    b) Mantle's is greater in every case.
    c) Ruth: $\dfrac{593}{1878} \approx 0.316$; Mantle: $\dfrac{760}{2440} \approx 0.311$; Ruth's is greater.
    d) Answers will vary.
    e) Ruth: $\dfrac{1.537}{5} \approx 0.307$; Mantle: $\dfrac{1.565}{5} = 0.313$; Mantle's is greater.
    f) and g) Answers will vary.

55. $\Sigma xw = 84(0.40) + 94(0.60) = 33.6 + 56.4 = 90$

    $\Sigma w = 0.40 + 0.60 = 1.00$

    weighted average $= \dfrac{\Sigma xw}{\Sigma w} = \dfrac{90}{1.00} = 90$

57. a) – c) Answers will vary.

Exercise Set 12.4

1. Variation

3. Standard deviation

5. Sample

7. Range = $15 - 4 = 11$

$$\bar{x} = \frac{45}{5} = 9$$

x	$x - \bar{x}$	$(x - \bar{x})^2$
9	0	0
7	–2	4
4	–5	25
10	1	1
15	6	36
	0	66

$$\frac{66}{4} = 16.5, s = \sqrt{16.5} \approx 4.06$$

9. Range = $126 - 120 = 6$

$$\bar{x} = \frac{861}{7} = 123$$

x	$x - \bar{x}$	$(x - \bar{x})^2$
120	–3	9
121	–2	4
122	–1	1
123	0	0
124	1	1
125	2	4
126	3	9
	0	28

$$\frac{28}{6} \approx 4.67, s = \sqrt{4.67} \approx 2.16$$

11. Range = $15 - 4 = 11$

$$\bar{x} = \frac{60}{6} = 10$$

x	$x - \bar{x}$	$(x - \bar{x})^2$
4	–6	36
8	–2	4
9	–1	1
11	1	1
13	3	9
15	5	25
	0	76

$$\frac{76}{5} = 15.2, \ s = \sqrt{15.2} \approx 3.90$$

13. Range = 12 − 7 = 5 15.

$$\overline{x} = \frac{63}{7} = 9$$

x	$x - \overline{x}$	$(x - \overline{x})^2$
7	−2	4
9	0	0
7	−2	4
9	0	0
9	0	0
10	1	1
12	3	9
18		0

$$\frac{18}{6} = 3, s = \sqrt{3} \approx 1.73$$

Range = 160 − 70 = $90

$$\overline{x} = \frac{770}{7} = \$110$$

x	$x - \overline{x}$	$(x - \overline{x})^2$
90	−20	400
70	−40	1600
100	−10	100
150	40	1600
160	50	2500
80	−30	900
120	10	100
300	0	7200

$$\frac{7200}{6} = 1200, s = \sqrt{1200} \approx \$34.64$$

17. Range = 250 − 60 = $190

$$\overline{x} = \frac{1044}{9} = \$116$$

x	$x - \overline{x}$	$(x - \overline{x})^2$
109	−7	49
60	−56	3136
80	−36	1296
60	−56	3136
210	94	8836
60	−56	3136
100	−16	256
115	−1	1
	0	37,802

$$\frac{37,802}{8} = 4725.25, \ s = \sqrt{4725.25} \approx \$68.74$$

19.a) Range = 68 − 5 = $63

$$\overline{x} = \frac{204}{6} = \$34$$

x	$x - \overline{x}$	$(x - \overline{x})^2$
32	−2	4
60	26	676
14	−20	400
25	−9	81
5	−29	841
68	34	1156
	0	3158

$$\frac{3158}{5} = 631.6, s = \sqrt{631.6} \approx \$25.13$$

b) Answers will vary.

19. c) Range = 78 - 15 = $63

$$\bar{x} = \frac{264}{6} = \$44$$

x	$x - \bar{x}$	$(x - \bar{x})^2$
42	−2	4
70	26	676
24	−20	400
35	−9	81
15	−29	841
78	34	1156
	0	3158

$$\frac{3158}{5} = 631.6, s = \sqrt{631.6} \approx \$25.13$$

The answers remain the same.

21. a) - c) Answers will vary.

d) If each number in a distribution is multiplied by n, both the mean and standard deviation of the new distribution will be n times that of the original distribution.

e) The mean of the second set is $4 \times 5 = 20$, and the standard deviation of the second set is $2 \times 5 = 10$.

23. a) The standard deviation increases. There is a greater spread from the mean as they get older.

b) The mean weight is about 100 pounds and the normal range is about 60 to 140 pounds.

c) The mean height is about 62 inches and the normal range is about 53 to 68 inches.

d) ≈ 140 lb

e) ≈ 20 lb

f) 100% - 95% = 5%

25. Answers will vary.

27. They would be the same since the spread of data about each mean is the same.

29. a) The grades will be centered about the same number since the mean, 75.2, is the same for both classes.

b) The spread of the data about the mean is greater for the evening class since the standard deviation is greater for the evening class.

31. a)

East

Number of oil changes made	Number of days
15-20	2
21-26	2
27-32	5
33-38	4
39-44	7
45-50	1
51-56	1
57-62	2
63-68	1

West

Number of oil changes made	Number of days
15-20	0
21-26	0
27-32	6
33-38	9
39-44	4
45-50	6
51-56	0
57-62	0
63-68	0

b)

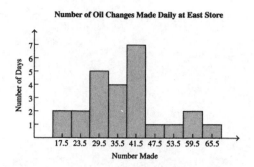

Number of Oil Changes Made Daily at East Store

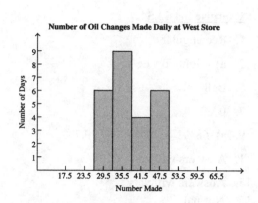

Number of Oil Changes Made Daily at West Store

c) They appear to have about the same mean since they are both centered around 38.

d) The distribution for East is more spread out. Therefore, East has a greater standard deviation.

e) East: $\dfrac{950}{25} = 38$, West: $\dfrac{950}{25} = 38$

31.. f)

East				West		
x	$x-\bar{x}$	$(x-\bar{x})^2$		x	$x-\bar{x}$	$(x-\bar{x})^2$
33	−5	25		38	0	0
30	−8	64		38	0	0
25	−13	169		37	−1	1
27	−11	121		36	−2	4
40	2	4		30	−8	64
44	6	36		45	7	49
49	11	121		28	−10	100
52	14	196		47	9	81
42	4	16		30	−8	64
59	21	441		46	8	64
19	−19	361		38	0	0
22	−16	256		39	1	1
57	19	361		40	2	4
67	29	841		34	−4	16
15	−23	529		31	−7	49
41	3	9		45	7	49
43	5	25		29	−9	81
27	−11	121		38	0	0
42	4	16		38	0	0
43	5	25		39	1	1
37	−1	1		37	−1	1
38	0	0		42	4	16
31	−7	49		46	8	64
32	-6	36		31	−7	49
35	−3	9		48	10	100
	0	3832			0	858

$\dfrac{3832}{24} \approx 159.67$, $s = \sqrt{159.67} \approx 12.64$

$\dfrac{858}{24} = 35.75$, $s = \sqrt{35.75} \approx 5.98$

33. 6, 6, 6, 6, 6

Exercise Set 12.5

1. Rectangular

3. a) Right   b) Left

5. Bell

7. 0

9. a) 68%   b) 95%   c) 99.7%

11. Answers will vary.

13. Answers will vary.

15. Normal

17. Skewed right

19. 0.5000

21. (area to the left of 2) − (area to the left of −1)
    $= 0.9772 - 0.1587 = 0.8185$

23. area to the right of 1.26
    $= 1 - (\text{area to the left of } 1.26)$
    $= 1 - 0.8962 = 0.1038$

25. area to the left of −1.78
    0.0375

27. area to the left of −2.13
    0.0166

29. area between −1.32 and −1.64
    $0.0934 - 0.0505 = 0.0429$

31. $0.7582 = 78.52\%$

33. area greater than −1.90
    $= \text{area less than } 1.90 = 0.9713 = 97.13\%$

35.
    (area to the left of 2.24) − (area to the left of −1.34)
    $= 0.9875 - 0.0901 = 0.8974 = 89.74\%$

37. (area to the left of 2.14)
    − (area to the left of 1.96)
    $= 0.9838 - 0.9750 = 0.0088 = 0.88\%$

39.
    (area to the left of 2.14) − (area to the left of 0.72)
    $= 0.9838 - 0.7642 = 0.2196 = 21.96\%$

41. a) Emily, Sarah, and Carol are taller than the mean because their z-scores are positive.
    b) Jenny and Shenice are at the mean because their z-scores are zero.
    c) Sadaf, Heather, and Kim are shorter than the mean because their z-scores are negative.

43. $0.5000 = 50\%$

45. $z_4 = \dfrac{4-3}{0.8} = \dfrac{1}{0.8} = 1.25$

    $1.000 - 0.8944 = 0.1056 = 10.56\%$

47. $z_{550} = \dfrac{550-500}{100} = \dfrac{50}{100} = 0.50$
    area less than $0.5 = 0.6915 = 69.15\%$

49. $z_{550} = \dfrac{550-500}{100} = \dfrac{50}{100} = 0.50$
    $z_{650} = \dfrac{650-500}{100} = \dfrac{150}{100} = 1.50$
    area between 1.5 and 0.5
    $= 0.9332 - 0.6915 = 0.2417 = 24.17\%$

51. $z_{575} = \dfrac{575-500}{100} = \dfrac{75}{100} = 0.75$

$z_{400} = \dfrac{400-500}{100} = \dfrac{-100}{100} = -1.00$

area between -1.00 and 0.75

$= 0.7734 - 0.1587 = 0.6147 = 61.47\%$

53. $z_{7.4} = \dfrac{7.4-7.6}{0.4} = \dfrac{-0.2}{0.4} = -0.50$

$z_{7.7} = \dfrac{7.7-7.6}{0.4} = \dfrac{0.1}{0.4} = 0.25$

$0.5987 - 0.3085 = 0.2902 = 29.02\%$

55. $z_{7.7} = \dfrac{7.7-7.6}{0.4} = \dfrac{0.1}{0.4} = 0.25$

$0.5987 = 59.87\%$

57. $0.5000 = 50.00\%$

59. $z_{56} = \dfrac{56-62}{5} = \dfrac{-6}{5} = -1.20$

$0.1151 = 11.51\%$

61. 11.51% of cars are traveling slower than 56 mph. (See Exercise 59.)

$(0.1151)(200) \approx 23$ cars

63. $z_{15.24} = \dfrac{15.24-20}{2.8} = -1.70$

$z_{28.12} = \dfrac{28.12-20}{2.8} = 2.9$

$0.9981 - 0.0446 = 0.9535 = 95.35\%$

65. $z_{15.24} = \dfrac{15.24-20}{2.8} = -1.70$

$1 - 0.9554 = 0.0446$

$(0.0446)(500) \approx 22$ watermelons

67. $z_{7100} = \dfrac{7100-8000}{1500} = \dfrac{-900}{1500} = -0.60$

area greater than $-0.60 = 1 -$ area less than -0.60

$1 - 0.2743 = 72.57\%$

69. $z_{11,750} = \dfrac{11,750-8000}{1500} = \dfrac{3750}{1500} = 2.5$

$1 - 0.9938 = 0.0062 = 0.62\%$

71. 69.15% of the families pay more than $7250 annually. (See Exercise 68.)

$(0.6915)(120) \approx 83$ families

73. We need the percentage of customers with a weight loss of less than 5 lb.

$z_5 = \dfrac{5-6.7}{0.81} = \dfrac{-1.7}{0.81} = -2.10$

$1 - 0.9821 = 0.0179 = 1.79\%$

75. The standard deviation is too large. There is too much variation.

77. a) B   b) C   c) A

79. The mean is the greatest value. The median is lower than the mean. The mode is the lowest value. The greatest frequency appears on the left side of the curve. Since the mode is the value with the greatest frequency, the mode would appear on the left side of the curve (where the lowest values are). Every value in the set of data is considered in determining the mean. The values on the far right of the curve would increase the value of the mean. Thus, the value of the mean would be farther to the right than the mode. The median would be between the mode and the mean.

81. Answers will vary.

83. a) Katie: $z_{28,408} = \dfrac{28,408-23,200}{2170} = \dfrac{5208}{2170} = 2.4$

Stella: $z_{29,510} = \dfrac{29,510-25,600}{2300} = \dfrac{3910}{2300} = 1.7$

b) Katie. Her z-score is higher than Stella's z-score. This means her sales are further above the mean than Stella's sales.

85. Answers will vary.

87. Using Table 12.8, the answer is −1.18.

89. $\dfrac{0.77}{2} = 0.385$

Using the table in Section 12.8, an area of 0.5 + 0.385 has a z-score of 1.20.

$$z = \frac{x - \overline{x}}{s}$$

$$1.20 = \frac{14.4 - 12}{s}$$

$$1.20 = \frac{2.4}{s}$$

$$\frac{1.20s}{1.20} = \frac{2.4}{1.20}$$

$$s = 2$$

Exercise Set 12.6
1. Coefficient
3. a) 1 b) - 1 c) 0
5. Positive
7. No correlation
9. Strong positive correlation
11. Yes, $\left| \ 0.81 \ \right| > 0.765$

13. No, $\left| -0.49 \right| < 0.602$

15. Yes, $\left| -0.32 \right| > 0.254$

17. Yes, $\left| 0.75 \right| < 0.917$

19. a)

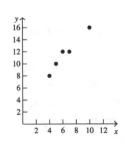

b)

x	y	x^2	y^2	xy
5	8	25	64	40
6	10	36	100	60
7	12	49	144	84
8	12	64	144	96
11	16	121	256	176
37	58	295	708	456

$$r = \frac{5(456) - 37(58)}{\sqrt{5(295) - 1369}\sqrt{5(708) - 3364}} \approx 0.981$$

c) Yes, $|0.981| > 0.878$

d) Yes, $|0.981| > 0.959$

21. a)

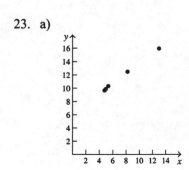

b)

x	y	x^2	y^2	xy
23	30	529	900	690
35	38	1225	1444	1330
31	27	961	729	837
43	21	1849	441	903
49	40	2401	1600	1960
181	156	6965	5114	5720

$$r = \frac{5(5720) - 181(156)}{\sqrt{5(6965) - 32,761}\sqrt{5(5114) - 24,336}} = \frac{364}{\sqrt{2064}\sqrt{1234}} \approx 0.228$$

c) No, $|0.228| < 0.878$

d) No, $|0.228| < 0.959$

23. a)

b)

x	y	x^2	y^2	xy
5.3	10.3	28.09	106.09	54.59
4.7	9.6	22.09	92.16	45.12
8.4	12.5	70.56	156.25	105
12.7	16.2	161.29	262.44	205.74
4.9	9.8	24.01	96.04	48.02
36	58.4	306.04	712.98	458.47

$$r = \frac{5(458.47) - 36(58.4)}{\sqrt{5(306.04) - 1296}\sqrt{5(712.98) - 3410.56}} = \frac{189.95}{\sqrt{234.2}\sqrt{154.34}} \approx 0.999$$

c) Yes, $|0.999| > 0.878$ \hspace{3cm} d) Yes, $|0.999| > 0.959$

25. a)

b)

x	y	x^2	y^2	xy
100	2	10,000	4	200
80	3	6400	9	240
60	5	3600	25	300
60	6	3600	36	360
40	6	1600	36	240
20	8	400	64	160
360	30	25,600	174	1500

$$r = \frac{6(1500) - 360(30)}{\sqrt{6(25,600) - 129,600}\sqrt{6(174) - 900}} = \frac{-1800}{\sqrt{24,000}\sqrt{144}} \approx -0.968$$

c) Yes, $|-0.968| > 0.811$ \hspace{2cm} d) Yes, $|-0.968| > 0.917$

27. From # 19: $\quad m = \dfrac{5(398) - 32(58)}{5(226) - 1024} = \dfrac{134}{106} \approx 1.26$

$$b = \frac{58 - \left(\dfrac{134}{106}\right)(32)}{5} \approx 3.51, \quad y = 1.26x + 3.51$$

29. From # 21: $\quad m = \dfrac{5(5720) - 181(156)}{5(6965) - 32,761} = \dfrac{364}{2064} \approx 0.18$

$$b = \frac{156 - \dfrac{364}{2064}(181)}{5} \approx 24.82, \quad y = 0.18x + 24.82$$

31. From # 23: $m = \dfrac{5(458.47) - 36(58.4)}{5(306.04) - 1296} = \dfrac{189.95}{234.2} \approx 0.81$

$b = \dfrac{58.4 - \dfrac{189.95}{234.2}(36)}{5} \approx 5.84, \quad y = 0.81x + 5.84$

33. From # 25: $m = \dfrac{6(1500) - 360(30)}{6(25,600) - 129,600} = \dfrac{-1800}{24,000} \approx -0.08$

$b = \dfrac{30 - \dfrac{-1800}{24,000}(360)}{6} \approx 9.50, \quad y = -0.08x + 9.50$

35. a)

x	y	x^2	y^2	xy
2	5	4	25	10
15	25	225	625	375
16	30	256	900	480
9	20	81	400	180
21	35	441	1225	735
5	10	25	100	50
68	125	1032	3275	1830

$r = \dfrac{6(1830) - (68)(125)}{\sqrt{6(1032) - 4624}\sqrt{6(3275) - 15625}} \approx 0.987$

b) Yes, $|\,0.987\,| > 0.811$

c) $m = \dfrac{6(1830) - (68)(125)}{(6)(1032) - 4624} = \dfrac{2480}{1568} \approx 1.58,$

$b = \dfrac{125 - \dfrac{2480}{1568}(68)}{6} \approx 2.91, \quad y = 1.58x + 2.91$

37. a)

x	y	x^2	y^2	xy
42	35	1764	1225	1470
38	45	1444	2025	1710
20	60	400	3600	1200
30	52	900	2704	1560
28	55	784	3025	1540
35	47	1225	2209	1645
17	65	289	4225	1105
210	359	6806	19,013	10,230

$r = \dfrac{7(10,230) - 210(359)}{\sqrt{7(6806) - 44,100}\sqrt{7(19,013) - 128,881}} = \dfrac{-3780}{\sqrt{3542}\sqrt{4210}} \approx -0.979$

b) Yes, $|0.979| > 0.875$

c) $m = \dfrac{7(10,230) - 210(359)}{7(6806) - 44,100} = \dfrac{-3780}{3542} \approx -1.07$,

$b = \dfrac{359 - \left(-\dfrac{3780}{3542}\right)(210)}{7} \approx 83.30, \quad y = -1.07x + 83.30$

d) $y = -1.07(33) + 83.30 \approx 48$ cups

39. a)

x	y	x^2	y^2	xy
139	120	19,321	14,400	16,680
165	165	27,225	27,225	27,225
210	208	44,100	43,264	43,680
287	275	82,369	75,625	78,925
189	160	35,721	25,600	30,240
115	100	13,225	10,000	11,500
125	118	15,625	13,924	14,750
95	90	9025	8,100	8,550
1325	1236	246,611	218,138	231,550

$r = \dfrac{8(231,550) - 1325(1236)}{\sqrt{8(246,611) - 1,755,625}\sqrt{8(218,138) - 1,527,696}} = \dfrac{214,700}{\sqrt{217,263}\sqrt{217,408}} \approx 0.988$

b) Yes, $|\ 0.988\ | > 0.834$

c) $m = \dfrac{8(231,550) - 1325(1236)}{8(246,611) - 1,755,625} = \dfrac{214,700}{217,263} \approx 0.99$,

$b = \dfrac{1236 - \dfrac{214,700}{217,263}(1325)}{8} \approx -9.17, \quad y = 0.99x - 9.17$

d) $y = 0.99(130) - 9.17 = 119.53 \approx \$120,000$

41. a)

x	y	x^2	y^2	xy
89	22	7921	484	1958
110	28	12,100	784	3080
125	30	15,625	900	3750
92	26	8464	676	2392
100	22	10,000	484	2200
95	21	9025	441	1995
108	28	11,664	784	3024
97	25	9409	625	2425
816	202	84,208	5178	20,824

$r = \dfrac{8(20,824) - 816(202)}{\sqrt{8(84,208) - 665,856}\sqrt{8(5178) - 40,804}} = \dfrac{1760}{\sqrt{7808}\sqrt{620}} \approx 0.800$

b) Yes, $|0.800| > 0.707$

c) $m = \dfrac{8(20,824) - 816(202)}{8(84,208) - 665,856} = \dfrac{1760}{7808} \approx 0.23$,

$b = \dfrac{202 - \dfrac{1760}{7808}(816)}{8} \approx 2.26$, $y = 0.23x + 2.26$

d) $y = 0.23(115) + 2.26 = 28.71 \approx 29$ units

43. a) and b) Answers will vary.

c)

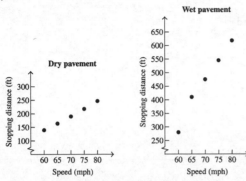

d) The values in the last row of the calculation table are:

$\Sigma x = 350,\ \Sigma y = 959,\ \Sigma x^2 = 24,750,\ \Sigma y^2 = 191,129,\ \Sigma xy = 68,470$

$r = \dfrac{5(68,470) - (350)(959)}{\sqrt{5(24,750) - (350)^2}\ \sqrt{5(191,129) - (959)^2}} \approx 0.999$

e) The values in the last row of the calculation table are:

$\Sigma x = 350,\ \Sigma y = 2328,\ \Sigma x^2 = 24,750,\ \Sigma y^2 = 1,151,074,\ \Sigma xy = 167,015$

$r = \dfrac{5(167,015) - (350)(2328)}{\sqrt{5(24,750) - (350)^2}\ \sqrt{5(1,151,074) - (2328)^2}} \approx 0.990$

f) Answers will vary.

g) $m = \dfrac{5(68,470) - (350)(959)}{5(24,750) - (350)^2} = 5.36$

$b = \dfrac{959 - (5.36)(350)}{5} = -183.40$

$y = 5.36x - 183.40$

h) $m = \dfrac{5(167,015) - (350)(2328)}{5(24,750) - (350)^2} = 16.22$

$b = \dfrac{2328 - (16.22)(350)}{5} = -669.80$

$y = 16.22x - 699.80$

i) Dry: $y = 5.36(77) - 183.40 \approx 229.3$ ft

Wet: $y = 16.22(77) - 669.80 = 579.1$ ft

45. a) The correlation coefficient will not change because $\sum xy = \sum yx$, $\left(\sum x\right)\left(\sum y\right) = \left(\sum y\right)\left(\sum x\right)$, and the square roots in the denominator will be the same.

   b) Answers will vary.

47. Answers will vary.

49. a) $SS(xy) = \sum xy - \dfrac{\left(\sum x\right)\left(\sum y\right)}{n} = 387 - \dfrac{17(106)}{6} = 86.67$

   $SS(x) = \sum x^2 - \dfrac{\left(\sum x\right)^2}{n} = 75 - \dfrac{289}{6} = 26.83$

   $SS(y) = \sum y^2 - \dfrac{\left(\sum y\right)^2}{n} = 2202 - \dfrac{11236}{6} = 329.33$

   $r = \dfrac{86.67}{\sqrt{(26.83)(329.33)}} \approx 0.922$

Review Exercises

1. a) A **population** consists of all items or people of interest.

   b) A **sample** is a subset of the population.

2. A **random sample** is one where every item in the population has the same chance of being selected.

3. The candy bars may have lots of calories, or fat, or sodium. Therefore, it may not be healthy to eat them.

4. Sales may not necessarily be a good indicator of profit. Expenses must also be considered.

5. a)   **b)**

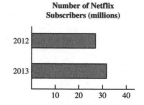

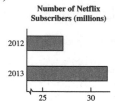

6. a)

Class	Frequency
35	1
36	3
37	6
38	2
39	3
40	0
41	4
42	1
43	3
44	1
45	1

b) and c)

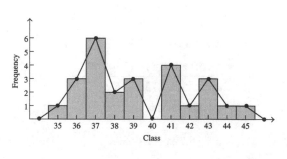

7. a)

High Temperature	Number of Cities
58 - 62	1
63 - 67	4
68 - 72	9
73 - 77	10
78 - 82	11
83 - 87	4
88 - 92	1

b) and c)

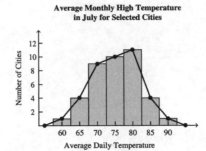

d) 6 | 5 represents 65

```
5│8
6│3 6 6 7 8 8 9
7│0 1 1 1 2 2 3 3 3 4 5 5 5 6 6 7 9 9 9
8│0 0 0 0 1 2 2 2 3 4 4 7
9│1
```

8. $\bar{x} = \dfrac{492}{6} = 82$

9. $\dfrac{81 + 85}{2} = 83$

10. None

11. $\dfrac{69 + 95}{2} = 82$

12. $93 - 67 = 26$

13.

x	$x - \bar{x}$	$(x - \bar{x})^2$
69	−13	169
76	−6	36
81	−1	1
85	3	9
86	4	16
95	13	169
	0	400

$\dfrac{400}{5} = 80,\ s = \sqrt{80} \approx 8.94$

14. $\bar{x} = \dfrac{216}{12} = 18$

15. $\dfrac{17 + 19}{2} = 18$

16. 12 and 17

17. $\dfrac{9 + 28}{2} = 18.5$

18. $28 - 9 = 19$

19.

x	$x - \bar{x}$	$(x - \bar{x})^2$
9	-9	81
10	-8	64
17	-1	1
19	1	1
24	6	36
12	-6	36
17	-1	1
28	10	100
12	-6	36
22	4	16
20	2	4
26	8	64
	0	440

$$\frac{440}{11} = 40, \quad s = \sqrt{40} \approx 6.32$$

20. $z_7 = \dfrac{7-9}{2} = \dfrac{-2}{2} = -1.00$

$z_{11} = \dfrac{11-9}{2} = \dfrac{2}{2} = 1.00$

$0.8413 - 0.1587 = 0.6826 = 68.26\%$

21. $z_5 = \dfrac{5-9}{2} = \dfrac{-4}{2} = -2.00$

$z_{13} = \dfrac{13-9}{2} = \dfrac{4}{2} = 2.00$

$0.9772 - 0.0228 = 0.9544 = 95.44\%$

22. $z_{12.2} = \dfrac{12.2-9}{2} = \dfrac{3.2}{2} = 1.6$

$0.9452 = 94.52\%$

23. Subtract the answer for Exercise 22 from 1:
$1 - 0.9452 = 0.0548 = 5.48\%$

24. $z_{7.8} = \dfrac{7.8-9}{2} = -\dfrac{1.2}{2} = -0.6$

$1 - 0.2743 = 0.7257 = 72.57\%$

25. $z_{20} = \dfrac{20-20}{5} = \dfrac{0}{5} = 0$

$z_{25} = \dfrac{25-20}{5} = \dfrac{5}{5} = 1.00$

$0.3413 = 34.13\%$

26. $z_{18} = \dfrac{18-20}{5} = \dfrac{-2}{5} = -0.40$

$0.5000 - 0.1554 = 0.3446 = 34.46\%$

27. $z_{22} = \dfrac{22-20}{5} = \dfrac{2}{5} = 0.40$

$z_{28} = \dfrac{28-20}{5} = \dfrac{8}{5} = 1.60$

$0.9452 - 0.6554 = 0.2898 = 28.98\%$

28. $z_{30} = \dfrac{30-20}{5} = \dfrac{10}{5} = 2.00$

$1 - 0.9772 = 0.0228 = 2.28\%$

29. a)

Bear Sightings

b) Yes, positive

c)

x	y	x^2	y^2	xy
765	119	585,225	14,161	91,035
926	127	857,476	16,129	117,602
1145	150	1,311,025	22,500	171,750
842	119	708,964	14,161	100,198
1485	153	2,205,225	23,409	227,205
1702	156	2,896,804	24,336	265,512
6865	824	8,564,719	114,696	973,302

$$r = \frac{6(973,302) - 6865(824)}{\sqrt{6(8,564,719) - 47,128,225}\sqrt{6(114,696) - 678,976}} = \frac{183,052}{\sqrt{4,260,089}\sqrt{9200}} \approx 0.925$$

d) Yes, $|\,0.925\,| > 0.811$

e)
$$m = \frac{6(973,302) - 6865(824)}{6(8,564,719) - 47,128,225} = \frac{183,052}{4,260,089} \approx 0.04,$$

$$b = \frac{824 - \dfrac{183,052}{4,260,089}(6865)}{6} \approx 88.17, \quad y = 0.04x + 88.17$$

f) $y = 0.04(1500) + 88.17 = 148.17 \approx 148$ bears

30. a)

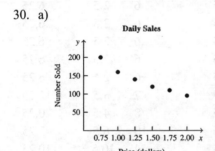

b) Yes; negative because generally as the price increases, the number sold decreases.

c)

x	y	x^2	y^2	xy
0.75	200	0.5625	40,000	150
1.00	160	1	25,600	160
1.25	140	1.5625	19,600	175
1.50	120	2.25	14,400	180
1.75	110	3.0625	12,100	192.5
2.00	95	4	9025	190
8.25	825	12.4375	120,725	1047.5

$$r = \frac{6(1047.5) - 8.25(825)}{\sqrt{6(12.4375) - 68.0625}\sqrt{6(120,725) - 680,625}} - \frac{-521.25}{\sqrt{6.5625}\sqrt{43,725}} \approx -0.973$$

d) Yes, $\mid -0.973 \mid > 0.811$

e) $m = \dfrac{6(1047.5)-8.25(825)}{6(12.4375)-68.0625} = \dfrac{-521.25}{6.5625} \approx -79.4$

$b = \dfrac{825 - \dfrac{-521.25}{6.5625}(8.25)}{6} \approx 246.7, \quad y = -79.4x + 246.7$

f) $y = -79.4(1.60)+246.7 = 119.66 \approx 120$ sold

31. 180 lb
33. 25%
35. 14%
37. $192 + (2)(23) = 238$ lb

32. 185 lb
34. 25%
36. $(100)(192) = 19,200$ lb
38. $192 - (1.8)(23) = 150.6$ lb

39. $\bar{x} = \dfrac{152}{43} \approx 3.53$

40. 2

41. 3

42. $\dfrac{0+14}{2} = 7$

43. $14 - 0 = 14$

44.

x	$x-\bar{x}$	$(x-\bar{x})^2$	x	$x-\bar{x}$	$(x-\bar{x})^2$	x	$x-\bar{x}$	$(x-\bar{x})^2$
0	−3.5	12.25	2	−1.5	2.25	4	0.5	0.25
0	−3.5	12.25	2	−1.5	2.25	5	1.5	2.25
0	−3.5	12.25	2	−1.5	2.25	5	1.5	2.25
0	−3.5	12.25	3	−0.5	0.25	5	1.5	2.25
0	−3.5	12.25	3	−0.5	0.25	6	2.5	6.25
0	−3.5	12.25	3	−0.5	0.25	6	2.5	6.25
1	−2.5	6.25	3	−0.5	0.25	6	2.5	6.25
1	−2.5	6.25	3	−0.5	0.25	6	2.5	6.25
2	−1.5	2.25	3	−0.5	0.25	6	2.5	6.25
2	−1.5	2.25	4	0.5	0.25	7	3.5	12.25
2	−1.5	2.25	4	0.5	0.25	8	4.5	20.25
2	−1.5	2.25	4	0.5	0.25	10	6.5	42.25
2	−1.5	2.25	4	0.5	0.25	14	10.5	110.25
2	−1.5	2.25	4	0.5	0.25			334.75
2	−1.5	2.25	4	0.5	0.25			

$\dfrac{334.75}{42} \approx 7.97, \quad s = \sqrt{7.97} \approx 2.82$

45.

# of Children	# of Presidents
0 - 1	8
2 - 3	16
4 - 5	10
6 - 7	6
8 - 9	1
10 - 11	1
12 - 13	0
14 - 15	1

46. and 47.

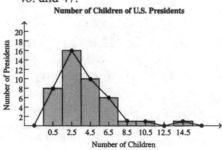

48. No, it is skewed to the right.
49. Answers will vary.
50. Answers will vary.

Chapter Test

1. $\bar{x} = \dfrac{210}{5} = 42$

2. 43

3. 43

4. $\dfrac{27 + 52}{2} = 39.5$

5. $52 - 27 = 25$

6.

x	$x - \bar{x}$	$(x - \bar{x})^2$
27	−15	225
43	1	1
43	1	1
45	3	9
52	<u>10</u>	<u>100</u>
	0	336

$\dfrac{336}{4} = 84, \; s = \sqrt{84} \approx 9.17$

7.

Class	Frequency
25 - 30	7
31 - 36	5
37 - 42	1
43 - 48	7
49 - 54	5
55 - 60	3
61 - 66	2

8.

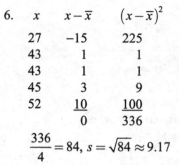

9.

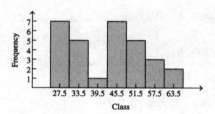

10. Mode = $815

11. Median = $790

12. $100\% - 25\% = 75\%$

13. 79%

14. $100(820) = \$82{,}000$

15. $\$820 + 1(\$40) = \$860$

16. $z_{36} = \dfrac{36 - 42}{5} = \dfrac{-6}{5} = -1.20$

$z_{53} = \dfrac{53 - 42}{5} = \dfrac{11}{5} = 2.20$

$0.9861 - 0.1151 = 0.8710 = 87.10\%$

17. $z_{35.75} = \dfrac{35.75 - 42}{5} = -1.25$

$1 - 0.1056 = 0.8944 = 89.44\%$

18. $z_{48.25} = \dfrac{48.25 - 42}{5} = 1.25$

    $1 - 0.8944 = 0.1056 = 10.56\%$

19. $z_{50} = \dfrac{50 - 42}{5} = 1.6$

    $0.9452 = 94.52\%$

20. a)

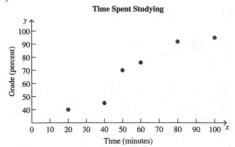

Time Spent Studying

b) Yes

c) The values in the last row of the calculation table are:

   $\Sigma x = 350, \ \Sigma y = 418,$

   $\Sigma x^2 = 24500, \ \Sigma y^2 = 31790,$

   $\Sigma xy = 27520$

   $r = \dfrac{6(27520) - (350)(418)}{\sqrt{6(24500) - (350)^2}\ \sqrt{6(31790) - (418)^2}} \approx 0.950$

d) Yes, $|\ 0.950\ | > 0.917$

e) $m = \dfrac{6(27520) - (350)(418)}{6(24500) - (350)^2} = 0.77$

   $b = \dfrac{418 - (0.77)(350)}{6} = 24.86, \quad y = 0.77x + 24.86$

f) $y = 0.77(75) + 24.86 \approx 83$

CHAPTER THIRTEEN

GRAPH THEORY

<u>Exercise Set 13.1</u>

1. Graph
3. Edge
5. Path
7. Degree

9.

11.

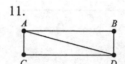

B and C are even. A and D are odd.

13.

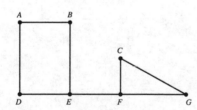

15. No. There is no edge connecting vertices C and D. Therefore, A, B, C, D, E is not a path.
17. Yes. One example is A, B, E, D, A, C, B.
19. No.

Copyright © 2017 Pearson Education, Inc.

21.

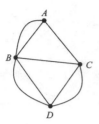

23.

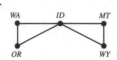

25.

27.

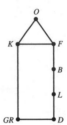

29.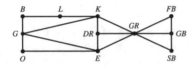

31.

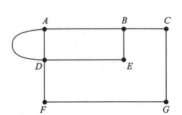

33. Disconnected. There is no path that connects
    A to *D*.
35. Connected
37. Edge *BC*
39. Edge *EF*
41. Answers will vary.

43. a) - c) Answers will vary.
    d) The sum of the degrees is equal to twice the number of edges. This is true since each edge
       must connect two vertices. Each edge then contributes two to the sum of the degrees.
45. a) and b)   Answers will vary.

Exercise Set 13.2
1. Euler
3. No
5. Odd
7. *A, B, D, E, C, A, D, C*; other answers are possible.
9. No. This graph has exactly two odd vertices. Each Euler path must begin with an odd vertex. *B* is an even
    vertex.
11. *E; B, D, E, F G, C, A, B, C, F,* other answers are possible.
13. No. A graph with exactly two odd vertices has no Euler circuits.
15. *A, B, C, D, E, F, B, D, F, A*; other answers are possible.
17. *C, D, E, F, A, B, D, F, B, C*; other answers are possible.
19. *E, F, A, B, C, D, F, B, D, E*; other answers are possible.
21. a) Yes. Each island would correspond to an odd vertex. According to item 2 of Euler's Theorem, a graph
       with exactly two odd vertices has at least one Euler path, but no Euler circuit.
    b) They could start on either island and finish at the other.

In Exercises 23-28, one graph is shown. Other graphs are possible.
23. a)

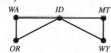

        b) Yes; *WA, ID, MT, WY, ID, OR, WA*
25. a)

    b) Vertices S and T are both odd. According to item 2 of Euler's Theorem, since there are exactly two odd
       vertices, at least one Euler path, but no Euler circuits exist. Yes; *S,T, N, A, P, N, Q, J, S, Q, T*
    c) No. (See part b.)

27. a)

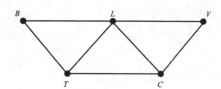

b) Vertices *T* and *C* are both odd. According to item 2 of Euler's Theorem, since there are exactly two odd vertices, at least one Euler path, but no Euler circuits exist.

Yes; *T, B, L, V, C, L, T, C*

c) No. (See part b.)

29. a) The graph representing the floor plan:

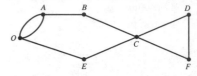

b) Yes

c) *O, A, B, C, D, F, C, E, O, A*

31. a) The graph representing the floor plan:

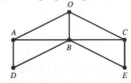

b) No; the graph has four odd vertices, and by Euler's theorem a graph with more than two odd vertices has neither an Euler path nor an Euler circuit.

33. a) Yes. The graph representing the map:

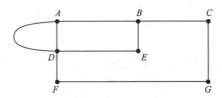

They are seeking an Euler path or an Euler circuit. Note that vertices *A* and *B* are both odd. According to item 2 of Euler's Theorem, since there are exactly two odd vertices, at least one Euler path, but no Euler circuits exist.

b) The residents would need to start at the intersection of Maple Cir., Walnut St., and Willow St. or at the intersection of Walnut St. and Oak St.

35. *A, B, E, D, C, A, D, B*, other answers are possible.
37. *A, B, H, E, B, C, E, F, H, G, F, D, C, A, G*; other answers are possible.
39. *A, B, C, D, I, H, G, B, E, H, C, E, G, F, A*; other answers are possible.

41. *A, C, D, G, H, F, C, F, E, B, A*; other answers are possible.
43. *A, B, C, E, B, D, E, F, I, E, H, D, G, H, I, J, F, C, A*; other answers are possible.
45. *UT, CO, NM, AZ, CA, NV, UT, AZ, NV*; other answers are possible.

47. *B, A, E, H, I, J, K, D, C, G, G, J, F, C, B, F, I, E, B*; other answers are possible.

49. *J, G, G, C, F, J, K, D, C, B, F, I, E, B, A, E, H, I, J*; other answers are possible.

51. a) Yes. There are no odd vertices.

    b) Yes. There are no odd vertices.

53. a) No. There are more than two odd vertices.

    b) No. There is at least one odd vertex.

55. a) No.

    b) California, Nevada, and Louisiana (and others) have an odd number of states bordering them. Since a graph of the United States would have more than two odd vertices, no Euler path and no Euler circuit exist.

57. a)   b)   c)

Exercise Set 13.3

1. Salesman

3. Hamilton

5. Euler

7. Force

9. *B,A,E,F,G,H,D,C* and *H,D,C,G,F,E,A,B*; other answers are possible.

11. *A, B, C, D, G, F, E, H* and *E, H, F, G, D, C, A, B*; other answers are possible.

13. *A,G,J,D,E,B,C,F,I,H* and *J,D,A,G,F,I,H,E,B,C*; other answers are possible.

15. *A, B, D, G, E, H, F, C, A* and *D, B, A, C, F, H, E, G, D*; other answers are possible.

17. *A, B, C, F, I, E, H, G, D, A* and *A, E, B, C, F, I, H, G, D, A*; other answers are possible.

19.

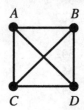

21. The number of unique Hamilton circuits within the complete graph with 6 vertices representing this situation is $(6-1)! = 5! = 5 \cdot 4 \cdot 3 \cdot 2 \cdot 1 = 120$ ways

23. The number of unique Hamilton circuits within the complete graph with thirteen vertices representing this situation is $(13-1)! = 12! = 12 \cdot 11 \cdot 10 \cdot 9 \cdot 8 \cdot 7 \cdot 6 \cdot 5 \cdot 4 \cdot 3 \cdot 2 \cdot 1 = 479,001,600$ ways

In Exercises 25-31, other graphs are possible.

25.a)

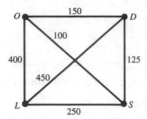

b)

Hamilton Circuit	First Leg/Distance	Second Leg/Distance	Third Leg/Distance	Fourth Leg/Distance	Total Distance
O, D, S, L, O	150	125	250	400	925 feet
O, D, L, S, O	150	450	250	100	950 feet
O, L, S, D, O	400	250	125	150	925 feet
O, L, D, S, O	400	450	125	100	1075 feet
O, S, D, L, O	100	125	450	400	1075 feet
O, S, L, D, O	100	250	450	150	950 feet

The shortest route is *O, D, S, L, O* or *O, L, S, D, O*

c) 925 feet (But note that if Mary goes back to her office from the library by revisiting the student center, her total trip will be 875 feet.)

27. a)

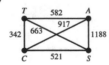

b)

Hamilton Circuit	First Leg/Distance	Second Leg/Distance	Third Leg/Distance	Fourth Leg/Distance	Total Distance
T, C, S, A, T	342	521	1188	582	2633 mi
T, C, A, S, T	342	917	1188	663	3110 mi
T, S, C, A, T	663	521	917	582	2683 mi
T, S, A, C, T	663	1188	917	342	3110 mi
T, A, C, S, T	582	917	521	663	2683 mi
T, A, S, C, T	582	1188	521	342	2633 mi

The shortest route is *T, C, S, A, T* or T, A, S, C, T

c) 2633 miles

29. a)

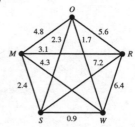
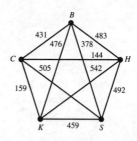

b) *O, W, S, M, R, O;* for $1.7 + 0.9 + 2.4 + 3.1 + 5.6 = 13.7$ miles
c) Answers will vary.

31. a)

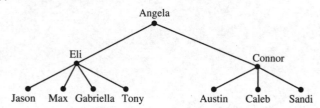

b) *S, B, C, H, K, S* for $378 + 431 + 144 + 542 + 459 = \$1954$
c) Answers will vary.

33. a) – d) Answers will vary.
35. *A, E, D, N, O, F, G, Q, P, T, M, L, C, B, J, K, S, R, I, H, A,* other answers are possible.

Exercise Set 13.4
1. Tree
3. Circuits
5. Minimum-cost
7.

9.

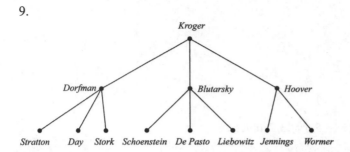

11.

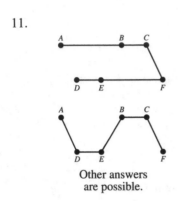

Other answers
are possible.

13.

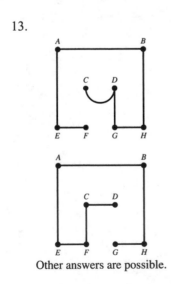

Other answers are possible.

15.

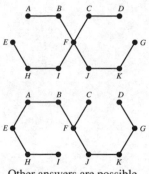

Other answers are possible.

17.

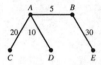

Choose edges in the following order:
AB, AD, AC, BE. Other answers are possible.

19.

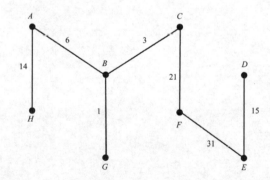

Choose edges in the following order:
GB, BC, BA, AH, DE, CF, FE

21.

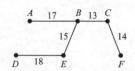

Choose edges in the following order:
BC, CF, EB, AB, DE

23.

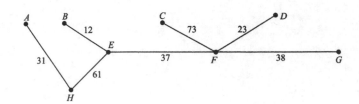

Choose edges in the following order: *BE, FD, AH, EF, FG, HE, CF*

25. a)

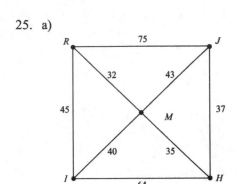

Other answers are possible.

b)

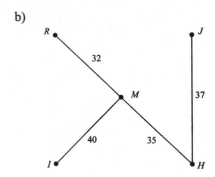

Choose edges in the following order:
RM, MH, JH, IM

c) $45(32 + 35 + 37 + 40) = 45(144) = \$6480$

27. a)

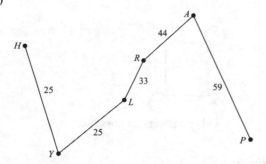

Choose edges in the following order: *HY, YL, LR, RA, AP*

b) 7100(25 + 25 + 33 + 44 + 59) = 7100(186) = $1,320,600

29. a)

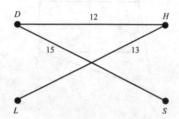

Choose edges in the following order:
DH, HL, DS

b) 3500 (12 + 13 + 15) = 3500 (40) = $140,000

31. a)

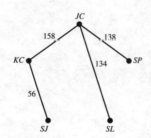

Choose edges in the following order: *KC-SJ, JC-SL, JC-SP, JC-KC*

b) 3700(56 + 134 + 138 + 158) = 3700(486) = $1,798,200

33. Answers will vary.
35. Answers will vary.

Review Exercises

1.

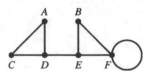

Other answers are possible.

2.

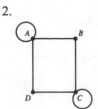

Other answers ae possible.

3. *E, C, D, F, E, G, A, B, H, G*; other answers are possible.

4. No. A path that includes each edge exactly one time would start at vertex *E* and end at vertex *G*, or vice-versa.

5.

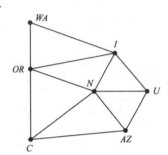

6.

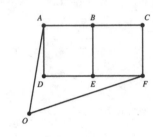

7. Connected

8. Disconnected. There is no path that connects *A* to *C*.

9. Edge *BD*

10. *C, A, B, F, H, G, C, D, E, D, E, F*; other answers are possible.

11. *F, E, D, E, D, C, G, H, F, B, A, C*; other answers are possible.

12. *A, B, E, G, F, D, C, A, D, E, A*; other answers are possible.

13. *E, D, C, A, E, B, A, D, F, G, E*; other answers are possible.

14. a)

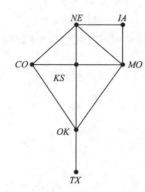

   b) Vertices *CO* and *TX* are both odd. According to item 2 of Euler's Theorem, since there are exactly two odd vertices, at least one Euler path, but no Euler circuits exist.

   Yes; *CO, NE, IA, MO, NE, KS, MO, OK, CO, KS, OK, TX*; other answers are possible.

   c) No. (See part b) above.)

15. a)

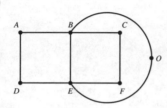

   b) Yes; the graph has no odd vertices, so there is at least one Euler path, which is also an Euler circuit.

   c) The person may start in any room or outside and will finish where he or she started.

16. a) Yes. The graph representing the map:

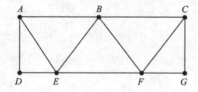

   The officer is seeking an Euler path or an Euler circuit. Note that vertices *A* and *C* are both odd. According to item 2 of Euler's Theorem, since there are exactly two odd vertices, at least one Euler path but no Euler circuits exist.

   b) The officer would have to start at either the intersection of Dayne St., Gibson Pl., and Alvarez Ave. or at the intersection of Chambers St., Fletcher Ct., and Alvarez Ave.

17. *C,A,B,G,F,A,D,C,F,D,B,E,D,G,E*; other answers are possible.
18. *A, B, D, E, I, J, O, N, L, K, G, H, L, M, I, H, D, C, G, F, A*; other answers are possible.
19. *C,A,E,K,I,F,G,J,L,H,B,D* and *D,B,H,L,J,G,F,C,A,E,K,I*; other answers are possible.
20. *A, B, E, F, J, I, L, K, H, G, C, D, A* and *I, J, F, E, B, A, D, C, G, H, K, L, I*; other answers are possible.

21.

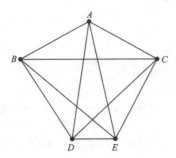

22. The number of unique Hamilton circuits within the complete graph with 5 vertices representing this situation is $(5-1)! = 4! = 4 \cdot 3 \cdot 2 \cdot 1 = 24$ ways

23. a)

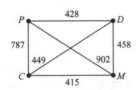

b)

Hamilton Circuit	First Leg/Cost	Second Leg/Cost	Third Leg/Cost	Fourth Leg/Cost	Total Cost
P, D, C, M, P	428	449	415	902	$2194
P, D, M, C, P	428	458	415	787	$2088
P, C, M, D, P	787	415	458	428	$2088
P, C, D, M, P	787	449	458	902	$2596
P, M, D, C, P	902	458	449	787	$2596
P, M, C, D, P	902	415	449	428	$2194

The least expensive route is *P, D, M, C, P* or *P, C, M, D, P*

c) $2088

24. a)

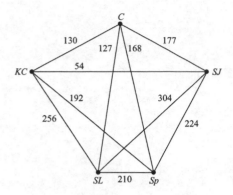

b) *SJ, KC, C, SL, Sp, SJ* traveling a total of 54 + 130 + 127 + 210 + 224 = 745 miles

c) *Sp, C, SL, KC, SJ, Sp* traveling a total of 168 + 127 + 256 + 54 + 224 = 829 miles

25.

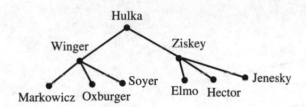

26.

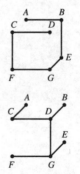

Other answers are possible.

27.

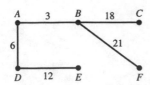

Choose edges in the following order:

AB, AD, DE, BC, BF

28.   a)

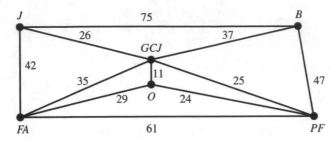

b)

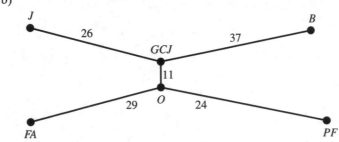

Choose edges in the following order:

O GCJ, O PF, J GCJ, FA O, GCJ B

c)   3.75 (11 + 24 + 26 + 29 + 37) = 3.75 (127) = $476.25

Chapter Test

1.

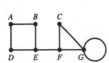

Edge *AB* is a bridge. There is a loop at vertex *G*.

Other answers are possible.

2.

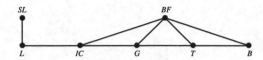

Edge *AB* is a bridge. There is a loop at vertex *G*.
Other answers are possible.

3.

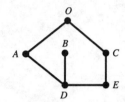

4. One example:

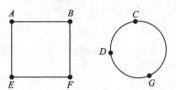

5.   *A, B, E, D, C, A, D, B*; other answers are possible.
6.   *A, B, D, G, H, F, C, B, E, D, F, E, C, A*; other answers are possible.
7.   Yes. The person may start in room *A* and end in room *B* or vice versa.
8.   *A, D, E, A, F, E, H, F, I, G, F, B, G, C, B, A*; other answers are possible.
9.   *B, A, D, E, F, C, G*; other answers are possible.
10. *A, B, C, G, E, D, H, I, K, J, F, A*; other answers are possible.

11.

12. The number of unique Hamilton circuits within the complete graph with G vertices representing
     this situation is $(6-1)! = 5! = 120$ ways

13.

14. *C, R, N, U, C* or *C, U, N, R, C* for $618
15. *C, R, U, N, C* for $662

16.

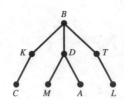

17.

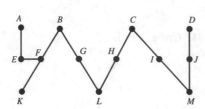

Other answers are possible.

18.

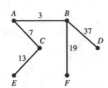

Choose edges in the following order: *AB,AC, CE, BF, BD*

19. Choose edges in the following order:
V2-V4, V3-V4, V4-V5, V1-V2

20. 1.25 (29 + 32 + 41 + 45) = 1.25 (147) = $183.75

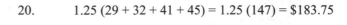

CHAPTER FOURTEEN

VOTING AND APPORTIONMENT

Exercise Set 14.1

1. Majority

3. $\dfrac{n(n-1)}{2}$

5. Plurality

7. Pairwise comparison

9. a) Li is the winner; he received the most votes using the plurality method.

   b) No. $\dfrac{102,503}{76,431+91,863+102,503} = \dfrac{102,503}{270,797} \approx 0.39$ is not a majority.   Majority is > 135,399 votes.

11.

Number of votes	3	1	2	2	1
First	P	K	L	L	K
Second	K	P	P	K	L
Third	L	L	K	P	P

13. $9 + 5 + 3 + 2 = 19$ employees

15. Sun wins with the most votes (9).

17. A majority out of 19 votes is 10 or more votes.
    First choice votes: (S) 9, (L) 8, (C) 2
    None receives a majority, thus C with the least
    votes is eliminated.
    Second round: (S) 9, (L) 5+3+2 = 10
    Lake wins with a majority of 10 votes.

19. W: 4 1st place votes = (4)(3) = 12
       2 2nd place votes = (2)(2) = 4
       3 3rd place votes = (3)(1) = 3
    O: 2 1st place votes = (2)(3) = 6
       4 2nd place votes = (4)(2) = 8
       3 3rd place votes = (3)(1) = 3
    V: 3 1st place votes = (3)(3) = 9
       3 2nd place votes = (3)(2) = 6
       3 3rd place votes = (3)(1) = 3
    W = 19 points; O = 17 points; V = 18 points
    Washington, D.C. wins with 19 points.

21. W vs. O: W = 3+2+1 = 6 O = 2+1 = 3
    W gets 1 pt.
    W vs. V: W = 3+1 = 4 V = 2+2+1 = 5
    V gets 1 pt.
    O vs. V: O = 3+2 = 5 V = 2+1+1 = 4
    O gets 1 pt.

    Three-way tie, no winner.

23. Votes: (L) 8+3+2 = 13 (S) 6+3 = 9
    (H) 4+3+2 = 9 (T) 1
    Los Angeles wins with the most votes.

25. A majority out of 32 votes is 17 or more votes.
    First choice votes: (L) 13, (S) 9, (H) 9, (T) 1
    None receives a majority, thus T with the least
    votes is eliminated.
    Second round: (L) 13, (S) 9, (H) 10
    No majority, thus eliminate S.
    Third round: (L) 16, (H) 16
    Since L and H tied, there is no winner.

27. B: 5 1ˢᵗ place votes = (5)(3) = 15
    4 2ⁿᵈ place votes = (4)(2) = 8
    3 3ʳᵈ place votes = (3)(1) = 3
    S: 1 1ˢᵗ place vote = (1)(3) = 3
    7 2ⁿᵈ place votes = (7)(2) = 14
    4 3ʳᵈ place votes = (4)(1) = 4
    M: 6 1ˢᵗ place votes = (6)(3) = 18
    1 2ⁿᵈ place vote = (1)(2) = 2
    5 3ʳᵈ place votes = (5)(1) = 5
    B = 26 points; S = 21 points; M = 25 points
    Brownstein wins with 26 points.

29. B vs. S: B = 5+4 = 9 S = 1+2 = 3
    B gets 1 pt.
    B vs. M: B = 5 M = 1+4+2 = 7
    M gets 1 pt.
    S vs. M: S = 5+1 = 6 M = 4+2 = 6
    S and M get 0.5 pt.
    B = 1 pt. S = 0.5 pt. M = 1.5 pts.
    Marquez wins with 1.5 points.

31. A majority out of 12 votes is 7 or more votes.
    Most last place votes: (B) 3, (S) 4, (M) 5
    Thus M with the most last place votes is
    eliminated.
    Second round using the most last place votes:
    (B) 1+2 = 3, (S) 4+5 = 9
    Brownstein wins with the least last place votes.

33. L: 5 1ˢᵗ place votes = (5)(3) = 15
    6 3ʳᵈ place votes = (6)(1) = 6
    E: 2 1ˢᵗ place votes = (2)(3) = 6
    9 2ⁿᵈ place votes = (9)(2) = 18

    O: 4 1ˢᵗ place votes = (4)(3) = 12
    2 2ⁿᵈ place votes = (2)(2) = 4
    5 3ʳᵈ place votes = (5)(1) = 5
    L = 21 points; E = 24 points; O = 21 points
    Erie Road wins with 24 points.

35. L vs. E: L = 5 E = 2+4 = 6 E gets 1 pt.
    L vs. O: L = 5 O = 2+4 = 6 O gets 1 pt.
    E vs. O: E = 5+2 = 7 O = 4 E gets 1 pt.
    Erie Road wins with 2 points.

37. a) Votes: (B): 12, (L): 2, (D): 1
Becker wins with the most votes.

   b) B: 12 1st place votes = (12)(4) = 48
   3 2nd place votes = (3)(3) = 9
   L: 2 1st place votes = (2)(4) = 8
   8 2nd place votes = (8)(3) = 24
   5 3rd place votes = (5)(2) = 10
   M: 10 3rd place votes = (10)(2) = 20
   5 4th place votes = (5)(1) = 5
   D: 1 1st place vote = (1)(4) = 4
   4 2nd place votes = (4)(3) = 12
   10 4th place votes = (10)(1) = 10
   B = 57 points; L = 42 points;
   M = 25 points, D = 26 points
   B wins with 57 points.

37. c) A majority out of 15 votes is 8 or more votes.
First choice votes: (B) 12, (L) 2
   (M) 0, (D) 1
Because B already has a majority, B wins.

   d) B vs. L: B = 8+4+1 = 13 L = 2
   B gets 1 pt.
   B vs. M: B = 8+4+2+1 = 15 B gets 1 pt.
   B vs. D: B = 8+4+2 = 14 D = 1
   B gets 1 pt.
   L vs. M: L = 8+4+2+1 = 15 L gets 1 pt.
   L vs. D: L = 8+2 = 10 D = 4+1 = 5
   L = gets 1 pt.
   M vs. D: M = 8+2 = 10 D = 4+1 = 5
   M gets 1 pt.
   B wins with 3 points.

39. a) G: 8 1st place votes = (8)(4) = 32
   3 2nd place vote = (3)(3) = 9
   4 3rd place votes = (4)(2) = 8
   14 4th place votes = (14)(1) = 14
   I: 3 1st place votes = (3)(4) = 12
   7 2nd place vote = (7)(3) = 21
   19 3rd place votes = (19)(2) = 38
   P: 14 1st place votes = (14)(4) = 56
   8 2nd place votes = (8)(3) = 24
   3 3rd place votes = (3)(2) = 6
   4 4th place votes = (4)(1) = 4
   Z: 4 1st place votes = (4)(4) = 16
   11 2nd place vote = (11)(3) = 33
   3 3rd place votes = (3)(2) = 6
   11 4th place votes = (11)(1) = 11
   G = 63 points; I = 71 points; P = 90 points;
   Z = 66 points
   P wins with 91 points.

39. b) Votes: (G): 8, (I): 3, (P): 14, (Z): 4
P wins with the most votes.

   c) A majority out of 29 votes is 15 or more votes.
   First choice votes: (G) 8, (I) 3
   (P) 14, (Z) = 4
   None receives a majority, thus I with the least
   votes is eliminated.
   Second round: (G) 11, (P) 14, (Z) 4
   No majority, thus eliminate Z.
   Third round: (G) 15, (P) 14
   G wins with 15 votes.

   d) G vs. P: G = 15 P = 14 G gets 1 pt.
   G vs. Z: G= 11 Z = 18 Z gets 1 pt.
   G vs. I: G = 8 I = 21 I gets 1 pt.
   P vs. Z: P = 25 Z = 4 P gets 1 pt.
   P vs. I: P = 22 I = 7 P gets 1 pt.
   Z vs. I: Z = 15 I = 14 Z gets 1 pt.
   P and Z tie with 2 points.

41. By ranking their choices, voters are able to provide more information with the Borda count method.

43. a) If there were only two columns then only two of the candidates were the first choice of the voters. If each of the 15 voters cast a ballot, then one of the voters must have received a majority of votes because 15 cannot be split evenly.

   b) An odd number cannot be divided evenly so one of the two first choice candidates must receive more than half of the votes.

45. a) C: 4 + 1 +1 = 6 R: 4 + 4 + 3 = 11
   W: 3 + 3 + 2 + 2 + 1 + 1 = 12
   T: 4 + 3 + 2 + 2 = 11
   The Warriors finished 1st, the Rams and the
   Tigers tied for 2nd, and the Comets were 4th.

   b) C: 5 + 0 = 5 R: 5 + 5 + 3 = 13
   W: 3 + 3 + 1 + 1 + 0 + 0 = 8
   T: 5 + 3 + 1 + 1 = 10
   Rams - 1st, Tigers - 2nd, Warriors - 3rd, and
   Comets - 4th.

47. a) Each voter casts $4+3+2+1 = 10$ votes.
$(15)(10) = 150$ votes
b) $150 - (35+40+25) = 150 - 100 = 50$ votes
c) Yes. Candidate D has more votes than each of the other 3 candidates.

49. Answers will vary.

Exercise Set 14.2

1. Majority
3. Head – to – head
5. Borda count
7. Plurality with elimination

9. The plurality method yields Orlando is the winner with a majority of 8 1st place votes. However, if the Borda count method is used:
Orlando $(8)(3) + (3)(2) + (4)(1) = 24 + 6 + 4 = 34$
Nashville $(4+3)(3) + (8)(2) = 21 + 16 = 37$
Portland $(4)(2) + (8+3)(1) = 8 + 11 = 19$
The winner is Nashville using the Borda count method, thus violating the majority criterion.

11. a) Total votes $= 2+4+2+3 = 11$
A vs. B: $A = 4+2 = 6$ $B = 2+3 = 5$
A gets 1 pt.
A vs. C: $A = 2+4 = 6$ $C = 2+3 = 5$
A gets 1 pt.
B vs. C: $B = 2+4 = 6$ $C = 2+3 = 5$
B gets 1 pt.
Plan A wins all head-to-head comparisons.
b) C wins by a plurality of 5 votes. No, the head-to-head criterion is not satisfied.

13. P: 4 1st place votes $= (4)(3) = 12$
2 2nd place votes $= (2)(2) = 4$
3 3rd place votes $= (3)(1) = 3$
L: 3 1st place vote $= (3)(3) = 9$
5 2nd place vote $= (5)(2) = 10$
1 3rd place vote $= (1)(1) = 1$
S: 2 1st place votes $= (2)(3) = 6$
2 2nd place vote $= (2)(2) = 4$
5 3rd place vote $= (5)(1) = 5$
P = 19 points; L = 20 points; S = 15 points
P vs. L: $P = 4+1 = 5$ $L = 4$ P gets 1 pt.
P vs. S: $P = 4+1 = 5$ $S = 4$ P gets 1 pt.
L vs. S: $L = 4+1+2 = 7$ $S = 2$ L gets 1 pt.
Because Parking wins its head-to-head comparisons and the Lounge Areas win by Borda count method, the head-to-head criterion is not satisfied.

15. A majority out of 21 votes is 11 or more votes.
Plurality with elimination:
First choice votes: $A = 6, B = 8, C = 7$
Eliminate A. $B = 8, C = 13$; C wins

Pairwise comparison:
A vs. B: $A = 7+6 = 13$ $B = 8$ A gets 1 pt.
A vs. C: $A = 8+6 = 14$ $C = 7$ A gets 1 pt.
B vs. C: $B = 8$ $C = 7+6=13$ C gets 1 pt.

No, because C wins by plurality with elimination but A wins using head-to-head comparison.

17. Votes: A: 12,   B: 4,   C: 9; thus, A wins.
    If B drops out, we get the following:
    Votes: A: 12,   C: 9 + 4 = 13, thus C would win.
    The irrelevant alternatives criterion is not satisfied.

19. A receives 53 points, B receives 56 points, and
    C receives 53 points. Thus, B wins using the
    Borda count method. If C drops out, we get the
    following: B receives 46 points, and C receives
    35 points. Thus, B still wins. The
    irrelevant alternatives criterion is satisfied.

21. A majority out of 32 voters is 17 or more votes.
    Votes: A: 8 + 3 = 11, B: 9, C: 12; none has a
    majority, thus eliminate B.
    Votes: A:8 + 3 = 11,   C: 9 +12 = 21, thus C
    wins. If the three voters who voted for A,C,B
    change to C,A,B, the new set of votes becomes:
    Votes: A: 8,   B: 9,   C: 15; none has a
    majority, thus eliminate A.
    Votes: B: 17, C:15, B wins.
    Thus, the monotonicity criterion is not satisfied.

23. A vs. B: A = 13   B = 13   A gets 0.5, B gets 0.5.
    A vs. C: A = 13   C = 13   A gets 0.5, C gets 0.5.
    A vs. D: A = 13   D = 13   A gets 0.5, D gets 0.5.
    B vs. C: B = 13   C = 13   B gets 0.5, C gets 0.5.
    B vs. D: B = 13   D = 13   B gets 0.5, D gets 0.5.
    C vs. D: C = 12   D = 14   D gets 1 pt.
    D wins with 2 pts.
    After the change in votes:
    A vs. B: A = 8   B = 18   B gets 1 pt.
    A vs. C: A = 13   C = 13   A gets 0.5, C gets 0.5.
    A vs. D: A = 13   D = 13   A gets 0.5, D gets 0.5.
    B vs. C: B = 18   C = 8   B gets 1 pt.
    B vs. D: B = 13   D = 13   B gets 0.5, D gets 0.5.
    C vs. D: C = 7   D = 19   D gets 1 pt.
    B wins with 2.5 pts.
    The monotonicity criterion is not satisfied.

25. A receives 2 points, B receives 3 point, C receives
    2 points, D receives 1 point, and E receives 2 pts.
    B wins by pairwise comparison.
    After A, C and E drop out, the new set of votes is
    B: 2   D: 3, thus D wins. The irrelevant
    alternatives criterion is not satisfied.

27. Total votes = 7   A wins with a majority of
    4 votes.
    A: 4 1st place votes = (4)(3) = 12
        3 3rd place votes = (3)(1) = 3
    B: 2 1st place vote = (2)(3) = 6
        5 2nd place vote = (5)(2) = 10
    C: 1 1st place vote = (1)(3) = 3
        2 2nd place votes = (2)(2) = 6
        4 3rd place votes = (4)(1) = 4
    A = 15 points;   B = 16 points;   C = 11 points
    B wins with 21 points. No. The majority
    criterion is not satisfied.

29. a) Washington, D.C., with 12 out of 23 votes
    b) Again, Washington, D.C., since it has a majority.
    c) N = (20)(3) + (3)(1) = 63
       W = (12)(4) + (3)(3) + (8)(1) = 65
       P = (8)(4) + (15)(2) = 62
       B = (3)(4) + (8)(2) + (12)(1) = 40
       Washington wins.
    d) Again, Washington, D.C., since it has a majority.

e) N vs. W: N = 8   W = 15, W gets 1 pt.
   N vs. P: N= 12   P = 11, N gets 1 pt.
   N vs. B: N= 20   B = 3, N gets 1 pt.
   W vs. P: W = 15   P = 8, W gets 1 pt.
   W vs. B: W= 12   B = 11, W gets 1 pt.
   P vs. B: P = 20   B = 3, P gets 1 pt.
   Washington, D.C., wins with 3 pts.
f) None of them.

31. a) A majority out of 82 votes is 42 or more votes.
    First choice votes: (J) 28, (L) 30, (C) 24
    None receives a majority, thus C with the least
    votes is eliminated.
    Second round: (J) 52, (L) 30
    Thus, LeBron James is selected..

    b) No majority on the 1st vote; L is eliminated
    with the fewest votes.
    Second round: (J) 38, (C) 44
    Bradley Cooper is chosen.

    c) Yes.

35. Answers will vary.

33. A candidate who holds a plurality will only gain
    strength and holds an even larger lead if more
    favorable votes are added.

37. Answers will vary.

Exercise Set 14.3

1. Divisor
3. Upper
5. Quota
7. Hamilton
9. a) Webster's   b) Adams'   c) Jefferson's

11. a) $\dfrac{8,000,000}{160} = 50,000$ = standard divisor

    b)

State	A	B	C	D	Total
Population	1,345,000	2,855,000	982,000	2,818,000	8,000,000
Standard Quota	26.9	57.1	19.64	56.36	

13. a) and b) Modified divisor: 49,300

State	A	B	C	D	Total
Population	1,345,000	2,855,000	982,000	2,818,000	8,000,000
Modified Quota	27.28	57.91	19.92	57.16	
Jefferson's Apportionment (round down)	27	57	19	57	160

15. a) and b) Modified divisor: 50,700

State	A	B	C	D	Total
Population	1,345,000	2,855,000	982,000	2,818,000	8,000,000
Modified Quota	26.53	56.31	19.37	55.58	
Adams' Apportionment (round up)	27	57	20	56	160

17. Standard divisor: 50,000

State	A	B	C	D	Total
Population	1,345,000	2,855,000	982,000	2,818,000	8,000,000
Standard Quota	26.9	57.1	19.64	56.36	
Webster's Apportionment (standard rounding)	27	57	20	56	160

19. a) Standard divisor $= \dfrac{\text{total}}{25} = \dfrac{675}{25} = 27$

b)

Hotel	A	B	C	Total
Amount	306	214	155	675
Standard Quota	11.33	7.93	5.74	

21. a) and b)

Hotel	A	B	C	Total
Amount	306	214	155	675
Modified Quota	11.86	8.29	6.01	
Jefferson's Apportionment (rounded down)	11	8	6	25

23. a) and b)

Hotel	A	B	C	Total
Amount	306	214	155	675
Modified Quota	10.55	7.38	5.34	
Adams' Apportionment (rounded up)	11	8	6	25

25. a) and b)

Store	A	B	C	Total
Amount	306	214	155	675
Standard Quota	11.33	7.93	5.74	
Webster's Apportionment (standard rounding)	11	8	6	25

27. a) A standard divisor = $\dfrac{\text{total}}{50} = \dfrac{600}{50} = 12$

   b) and c)

Resort	A	B	C	D	Total
Rooms	98	114	143	245	600
Standard Quota	8.17	9.50	11.92	20.42	
Lower Quota	8	9	11	20	48
Hamilton's Method	8	10	12	20	50

29. Using a divisor of 12.4:

Resort	A	B	C	D	Total
Rooms	98	114	143	245	600
Modified Quota	7.90	9.19	11.53	19.76	
Adams' Method	8	10	12	20	50

31. a) Standard divisor = $\dfrac{\text{total}}{250} = \dfrac{13000}{250} = 52$

   b) and c)

School	LA	Sci.	Eng.	Bus.	Hum	Total
Enrollment	1746	7095	2131	937	1091	13,000
Standard Quota	33.58	136.44	40.98	18.02	20.98	
Lower Quota	33	136	40	18	20	247
Hamilton's Apportionment	34	136	41	18	21	250

33. A divisor of 51.5 was used.

School	LA	Sci.	Eng.	Bus.	Hum	Total
Enrollment	1746	7095	2131	937	1091	13,000
Modified Quota	33.90	137.77	41.38	18.19	21.18	
Jefferson's Apportionment (round down)	33	137	41	18	21	250

35. a) A standard divisor = $\dfrac{\text{total}}{120} = \dfrac{10800}{120} = 90$

   b) and c)

Dealership	A	B	C	D	Total
Annual Sales	3840	2886	2392	1682	10,800
Standard Quota	42.67	32.07	26.58	18.69	120.00
Hamilton's Apportionment	43	32	26	19	120

37. A divisor of 90.3 was used.

Dealership	A	B	C	D	Total
Annual Sales	3840	2886	2392	1682	10,800
Modified Quota	42.52	31.96	26.49	18.63	
Webster's Apportionment	43	32	26	19	120

39. a) Standard divisor = $\dfrac{75,000}{100} = 750$

   b) and c)

Route	A	B	C	D	E	F	Total
Passengers	9070	15,275	12,810	5720	25,250	6875	75,000
Standard Quota	12.09	20.37	17.08	7.63	33.67	9.17	
Lower Quota	12	20	17	7	33	9	98
Hamilton's Method	12	20	17	8	34	9	100

41. The divisor 765 was used.

Route	A	B	C	D	E	F	Total
Passengers	9070	15,275	12,810	5720	25,250	6875	75,000
Modified Quota	11.86	19.97	16.75	7.48	33.01	8.99	
Adams' Method	12	20	17	8	34	9	100

43 a) Standard divisor = $\dfrac{\text{total}}{200} = \dfrac{2400}{200} = 12$

   b) and c)

Shift	A	B	C	D	Total
Room calls	751	980	503	166	2400
Standard Quota	62.58	81.67	41.92	13.83	
Lower Quota	62	81	41	13	197
Hamilton's Method	62	82	42	14	200

45. The divisor 11.9 was used.

Shift	A	B	C	D	Total
Room calls	751	980	503	166	2400
Modified Quota	63.11	82.35	42.27	13.95	
Jefferson's Apportionment (round down)	63	82	42	13	200

47. Standard divisor = $\dfrac{3615920}{105} = 34437.33$

   a) Hamilton's Apportionment: 7, 2, 2, 2, 8, 14, 4, 5, 10, 10, 13, 2, 6, 2, 18
   b) Jefferson's Apportionment: 7, 1, 2, 2, 8, 14, 4, 5, 10, 10, 13, 2, 6, 2, 19
   c) States that benefited: Virginia   States Disadvantaged: Delaware

49. One possible answer is A; 743, B: 367, C: 432, D: 491, E: 519, F: 388

Exercise Set 14.4

1. Population

3. Alabama

5. Small

7. New divisor = $\dfrac{1080}{61} = 17.71$

Office	A	B	C	D	E	Total
Employees	246	201	196	211	226	1080
Standard Quota	13.89	11.35	11.07	11.91	12.76	
Lower Quota	13	11	11	11	12	58
Hamilton's Apportionment	14	11	11	12	13	61

No. No office suffers a loss so the Alabama paradox does not occur.

9. a) Standard divisor = $\dfrac{900}{30} = 30$

State	A	B	C	Total
Population	161	250	489	900
Standard Quota	5.37	8.33	16.30	
Hamilton's Apportionment	6	8	16	30

b) New divisor = $\dfrac{900}{31} = 29.03$

State	A	B	C	Total
Population	161	250	489	900
Standard Quota	5.55	8.61	16.84	
Hamilton's Apportionment	5	9	17	31

Yes, state A loses 1 seat and states B and C each gain 1 seat.

11. a) Standard divisor = $\dfrac{25,000}{200} = 125$

City	A	B	C	Total
Population	8130	4030	12,840	25,000
Standard Quota	65.04	32.24	102.72	
Hamilton's Apportionment	65	32	103	200

b) New divisor = $\dfrac{25,125}{200} = 125.625$

City	A	B	C	Total
New Population	8150	4030	12,945	25,125
Standard Quota	64.88	32.08	103.04	
Hamilton's Apportionment	65	32	103	200

No. None of the Cities loses a bonus.

13. a) Standard divisor $= \dfrac{5400}{54} = 100$

Division	A	B	C	D	E	Total
Population	733	1538	933	1133	1063	5400
Standard Quota	7.33	15.38	9.33	11.33	10.63	
Lower Quota	7	15	9	11	10	52
Hamilton's Apportionment	7	16	9	11	11	54

b) New divisor $= \dfrac{5454}{54} = 101$

Division	A	B	C	D	E	Total
Population	733	1539	933	1133	1116	
Standard Quota	7.26	15.24	9.24	11.22	11.05	
Lower Quota	7	15	9	11	11	53
Hamilton's Apportionment	8	15	9	11	11	54

Yes. Division B loses an internship to Division A even though the population of division B grew faster than the population of division A.

15. a) Standard divisor $= \dfrac{4800}{48} = 100$

Tech. Data	A	B	Total
Employees	844	3956	4800
Standard Quota	8.44	39.56	
Lower Quota	8	39	47
Hamilton's Apportionment	8	40	48

b) New divisor $= \dfrac{5524}{55} = 100.44$

Tech. Data	A	B	C	Total
Employees	844	3956	724	5524
Standard Quota	8.40	39.39	7.21	
Lower Quota	8	39	7	54
Hamilton's Apportionment	9	39	7	55

Yes. Group B loses a manager.

17. a) Standard divisor $= \dfrac{3300}{33} = 100$

State	A	B	Total
Population	744	2556	3300
Standard Quota	7.44	25.56	
Lower Quota	7	25	32
Hamilton's Apportionment	7	26	33

b) New divisor $= \dfrac{4010}{40} = 100.25$

State	A	B	C	Total
Population	744	2556	710	4010
Standard Quota	7.42	25.50	7.08	
Lower Quota	7	25	7	39
Hamilton's Apportionment	7	26	7	40

No. The apportionment is the same.

Review Exercises

1. a) Comstock wins with the most votes (20).

   b) A majority out of 42 voters is 22 or more votes. Comstock does not have a majority.

2. a) Michelle MacDougal wins with the most votes (231).

   b) Yes. A majority out of 413 voters is 207 or more votes.

3.

# of votes	3	2	1	1	3
First	B	A	D	D	C
Second	A	C	C	A	B
Third	C	D	A	B	A
Fourth	D	B	B	C	D

4.

# of votes	2	2	2	1
First	C	A	B	C
Second	A	C	A	B
Third	B	B	C	A

5. Number of votes $= 5 + 3 + 1 + 2 = 11$

6. A tie between DVD player and digital camera with
    A plurality vote of 9 each.

7. C: 31 points,   D: 33 points,   B: 30 points,
    J: 16 points. Domino's Pizza wins with 33 points.

9. There is a three-way tie. There is no winner.

8. A majority out 11 voters is 6 or more votes.
    Votes: C: 5, D: 4, B: 2.
    None has a majority, thus eliminate B.
    Votes: C: 7, D: 4
    Chipotle Mexican Grill wins.

10. First round last place votes: Votes: J: 6, C: 3, D: 2,
     B: 0, so J is eliminated.
     Second round last place votes: Votes: B: 5, C: 4,
     D: 2 so B is eliminated.
     Third round first place votes: Votes: C: 7, D: 4, so
     C has a plurality.
     Chipotle Mexican Grill wins.

11. $38+30+25+7+10 = 110$ students voted

12. Volleyball wins with a plurality of 40 votes.

13. S: 223 pts., V: 215 pts., B: 222 pts.
    Soccer wins.

15. S: 1 pt., V: 1 pt., B: 1 pt. A 3-way tie

17. a) Yes. A majority out of 372 voters is 187 or more
       votes. American Music receives a majority.
    b) Votes: A: 161+134 = 295, T: 65, W: 12,
       S: 0 American Music wins.
    c) A: 1387 pts., T: 740 pts., W: 741 pts.,
       S: 852 pts. American Music wins.
    d) 187 or more votes is needed for a majority.
       Votes: A: 295, T: 45, W: 12, S: 0
       American Music wins.
    e) A: 3 pts., T: 1 pt., W: 1 pt., S: 1 pt.
       American Music wins.

19. a) A majority out of 16 voters is 9 or more votes.
       Votes: (C): 4+3+ = 7, (F): 1+1 = 2,
       (A): 0, (W): 6+1 = 7 None has a majority,
       thus eliminate A. Votes: (C): 4+3 = 7,
       (F): 1+1 = 2, (W): 6 + 1 = 7 None has a
       majority, thus eliminate F
       Votes: (C): 4+3+1 = 8, (W): 6+1+1 = 8.
       Thus, C and W tie.
    b) Use the Borda count method to break the tie.
       (C) = 46 points, (W) = 50 points;
       Where On Earth is Carmen San Diego wins.

22. a) A majority out of 50 voters is 26 or more votes.
       Votes: A: 14, B: 12+8 = 20, C: 16
       None has the majority, thus eliminate A.
       Votes: B :12+8 = 16, C: 16+14 = 30 C wins.

    b) The new preference table is

Number of votes	12	16	8	14
First	B	C	C	A
Second	A	B	B	C
Third	C	A	A	B

       Votes: A: 14, B: 12, C: 24; None has a
       majority, thus eliminate B.
       Votes: A: 26, C: 24 A wins. When the
       order is changed A wins. Therefore, the
       monotonicity criterion is not satisfied.

14. A majority out of 110 voters is 56 or more votes.
    Votes: S: 38, V: 40, B: 32; None has a majority,
    thus eliminate B. Votes: S: 45, V: 65
    Volleyball wins.

16. Votes: S: 38, V: 40, B: 32 None has a
    majority, thus eliminate V with the most last place
    votes. Votes: S: 68, B: 42. Soccer wins.

18. Votes: C: 25, S: 80, D: 45, L: 50
    a) A majority out of 200 voters is 101 or more
       votes. None of the cities has a majority.
    b) Seattle, with a plurality of 70 votes.
    c) C: 495, S: 410, D: 495, L: 600; Las Vegas wins.
    d) C: 30, S: 70, D: 45, L: 55; no city has a majority
       so eliminate C.
       S: 70, D: 45, L: 85; no city has a majority so
       eliminate D.
       S: 70, L: 130; Las Vegas wins.
       S: 80, L: 120; Las Vegas wins.
    e) Las Vegas wins with 3 points; Dallas has 2,
       Chicago 1 and Seattle has 0.

19. c) (C) vs. (W): C: 4+3+1 = 8 points,
       (W): 6+1+1 = 8 points.
       C and W tie again.

20. A: 49 pts., B: 59 pts, C: 52 pts., D: 30 pts.
    Using the Borda count, candidate B wins.
    However, A has a majority of first place votes,
    thus the majority criterion is not satisfied.

21. A wins all its head-to-head comparisons but B
    wins using the Borda count method.
    The head-to-head criterion is not satisfied.

22. c) If B drops out the new table is

Number of votes	12	16	8	14
First	A	C	C	A
Second	C	A	A	C

    Votes: A: 12+14 = 26, C: 16+8 = 25 A wins.
    Since C won the first election and then after B
    dropped out A won, the irrelevant criterion is
    not satisfied.

23. a) R to P: 59 to 55, R to N: 58 to 56,
    R to B: 58 to 56, so Ragu wins the
    head-to-head comparison.

    b) Prego with a plurality of 34.

    c) R; 289, P: 237, N: 307, B: 307; a tie between
    Newman's Own and Barilla.

    d) In the first round N is eliminated, then R is
    eliminated, and Barilla wins with
    80 votes.

    e) Ragu wins with 3 points.

    f) Plurality, Borda count, and plurality with
    elimination all violate the head-to-head criterion.

24. a) Yes. Fleetwood Mac is favored when
    compared to each of the other bands.

    b) Votes: R: 15, B: 34, J: 9+4 = 13,
    F: 25 Boston wins.

    c) R: 217 points, B: 198 points, J: 206 points,
    F: 249 points Fleetwood Mac wins.

    d) A majority out of 87 voters is 44 or more votes.
    Votes: R: 15, B: 34, J: 13, F:25
    None has a majority, thus eliminate J.
    Votes: R: 15+9+4 = 28, B: 34, F: 25
    None has a majority, thus eliminate F.
    Votes: R: 28+25 = 53, B: 34 REO wins.

    e) R = 2 pts., B = 0 pts., J = 1 pt., F = 3 pts.
    Thus, Fleetwood Mac wins.

    f) Plurality and plurality w/elimination methods

25. A majority out of 70 cotes is 36 or more votes, which A has. Using the Borda count method:
    $$A = 4(36) + 2(20) + 1(8) + 2(6) = 204$$
    $$B = 3(36) + 4(20) + 3(8) + 3(6) = 230$$
    $$C = 1(36) + 3(30) + 4(8) + 1(6) = 134$$
    $$D = 2(36) + 1(20) + 2(8) + 4(6) = 132$$
    B wins with 230 points.
    The majority criterion is not satisfied.

26. A majority out of 82 votes is 42 or more votes.
    C wins the first election with 20 + 10 = 30 votes.
    Using the plurality with elimination method:
    First election: A = 28, B = 24, C = 30.
    None has a majority, thus eliminate B.
    Second election: A = 36 + 24 + 2 = 62, which is a majority.
    The monotonicity criterion is not satisfied.

27. The Borda count method for the first election is:
    $$A = 4(24) + 4(16) + 4(16) + 1(16) + 5(8) + 3(4) + 3(5) = 307$$
    $$B = 5(24) + 5(16) + 1(16) + 2(16) + 2(8) + 4(4) + 2(5) = 290$$
    $$C = 1(24) + 2(16) + 3(16) + 5(16) + 1(8) + 1(4) + 1(5) = 201$$
    $$D = 3(24) + 1(16) + 2(16) + 4(16) + 3(8) + 5(4) + 5(5) = 253$$
    $$E = 2(24) + 3(16) + 5(16) + 3(16) + 4(8) + 2(4) + 4(5) = 284$$
    A wins with 307 points.
    If D drops out,
    $$A = 3(24) + 3(16) + 3(16) + 1(16) + 4(8) + 3(4) + 3(5) = 243$$
    $$B = 4(24) + 4(16) + 1(16) + 2(16) + 2(8) + 4(4) + 2(5) = 250$$
    $$C = 1(24) + 1(16) + 2(16) + 4(16) + 1(8) + 1(4) + 1(5) = 153$$
    $$E = 2(24) + 2(16) + 4(16) + 3(16) + 3(8) + 2(4) + 4(5) = 244$$
    B wins with 250 points.
    This violates the irrelevant alternatives criterion.
    Using the pairwise election methods for the first election:
    A = 3 pts., B = 2 pts., C = 1 pt., D = 2 pts., E = 2 pts.
    If D drops out,
    A = 2 pts, B = 2 pts., C = 0 pts., E = 2 pts. No winner.
    This violates the irrelevant alternative criterion.

28. Standard divisor $= \dfrac{6000}{10} = 600$

Region	A	B	C	Total
Number of Houses	2592	1428	1980	6000
Standard Quota	4.32	2.38	3.30	
Lower Quota	4	2	3	9
Hamilton's Apportionment	4	3	3	10

29. Using the modified divisor 500.

Region	A	B	C	Total
Number of Houses	2592	1428	1980	6000
Modified Quota	5.18	2.86	3.96	
Jefferson's Apportionment (rounded down)	5	2	3	10

30. Using the modified divisor 700.

Region	A	B	C	Total
Number of Houses	2592	1428	1980	6000
Modified Quota	3.70	2.04	2.83	
Adams' Apportionment (rounded up)	4	3	3	10

31. Using the modified divisor 575.

Region	A	B	C	Total
Number of Houses	2592	1428	1980	6000
Modified Quota	4.51	2.48	3.4	
Webster's Apportionment (normal rounding)	5	2	3	10

32. Yes. Hamilton's Apportionment becomes 5, 2, 4.   Region B loses one truck.

33. Standard divisor $= \dfrac{690}{23} = 30$

Course	A	B	C	Total
Number of Students	311	219	160	690
Standard Quota	10.37	7.30	5.33	
Lower Quota	10	7	5	22
Hamilton's Apportionment	11	7	5	23

34. Use the modified divisor 28.

Course	A	B	C	Total
Number of Students	311	219	160	690
Modified Quota	11.211	7.82	5.71	
Jefferson's Apportionment (round down)	11	7	5	23

35. Use the modified divisor 31.4

Course	A	B	C	Total
Number of Students	311	219	160	690
Modified Quota	9.90	6.97	5.10	
Adams' Apportionment (round up)	10	7	6	23

36. Use the modified divisor 29.5

Course	A	B	C	Total
Number of Students	311	219	160	690
Modified Quota	10.54	7.42	5.42	
Webster's Apportionment (standard rounding)	11	7	5	23

37. The new divisor is $\dfrac{698}{23} = 30.35$

Course	A	B	C	Total
Number of Students	317	219	162	698
Standard Quota	10.44	7.22	5.34	
Lower Quota	10	7	5	22
Hamilton's Apportionment	11	7	5	23

No. The apportionment remains the same.

38. The standard divisor $= \dfrac{50000}{50} = 1000$

State	A	B	Total
Population	4420	45,580	50,000
Standard Quota	4.42	45.58	
Hamilton's Apportionment	4	46	50

39. Using a divisor of 990:

State	A	B	Total
Population	4420	45,580	50,000
Modified Quota	4.46	46.04	
Jefferson's Apportionment	4	46	50

40. Using a divisor of 1020:

State	A	B	Total
Population	4420	45,580	50,000
Modified Quota	4.33	44.69	
Adams' Apportionment	5	45	50

41. Using a divisor of 1000:

State	A	B	Total
Population	4420	45,580	50,000
Modified Quota	4.42	45.58	
Webster's Apportionment	4	46	50

42. The new divisor is $\dfrac{55,400}{55} = 1007.27$

State	A	B	C	Total
Population	4420	45,580	5400	55,400
Standard Quota	4.39	45.25	5.36	
Hamilton's Apportionment	5	45	5	55

Yes. State B loses a seat.

Chapter Test

1. $4 + 3 + 3 + 2 = 12$ members voted.

2. No lunch has a majority of ≥ 7 votes.

3. Pizza wins with a plurality of 5 votes.

4. D: 25, P: 26, B: 21; pizza wins.

5. B is eliminated and D then has a plurality of 7; deli sandwiches wins.

6. D: 1.5 pts., P 1 pt., B 0.5 pt.; deli sandwiches wins.

7. Votes: $26 + 14 + 29 + 30 + 43 = 142$

8. Salamander wins with 43 votes.

9. (H) 1st $(40)(4) = 160$
   2nd $(59)(3) = 177$
   3rd $(0)(2) = 0$
   4th $(43)(1) = 43$ H receives 380 points.

   (I) 1st $(29)(4) = 116$
   2nd $(40)(3) = 120$
   3rd $(73)(2) = 146$
   4th $(0)(1) = 0$ I receives 382 points

   (L) 1st $(30)(4) = 120$
   2nd $(43)(3) = 129$
   3rd $(43)(2) = 86$
   4th $(26)(1) = 26$ L receives 361 points

   (S) 1st $(43)(4) = 172$
   2nd $(0)(3) = 0$
   3rd $(26)(2) = 52$
   4th $(73)(1) = 73$ S receives 297 points.

   The iguana (I) wins with the most points.

10. A majority out of 142 voters is 72 or more votes.
    Votes: H: 40, I: 29, L: 30, S: 43; None has a majority, thus eliminate I. Votes: H: 69, L: 30, S: 43   None has a majority, thus eliminate L. Votes: H: 99, S: 43
    The hamster wins.

11. H vs. I: I gets 1 pt.   H vs. L: L gets 1 pt.
    H vs. S: H gets 1 pt.   I vs. L: L gets 1 pt.
    I vs. S: I gets 1 pt.   L vs. S: L gets 1 pt.
    The lemming wins with 3 points.

12. Plurality: Votes: W: 86, X: 52+28 = 80, Y: 60,
    Z: 58   W wins.
    Borda count: W gets 594 points, X gets 760 points,
    Y gets 722 points, Z gets 764 points   Z wins
    Plurality with elimination: A majority out of 284
    voters is 143 or more votes.
    Votes: W: 86, X: 80, Y: 60, Z: 58
    None has a majority, thus eliminate Z.
    Votes: W: 86, X: 80+58 = 138, Y: 60
    None has a majority, thus eliminate Y.
    Votes: W: 86, X: 138+60 = 198   X wins.

12. Head-to-Head: When Y is compared to each of the
    others, Y is favored. Thus Y wins the
    head-to-head comparison.
    Plurality, Borda count and Plurality with elimination
    each violate the head-to-head criterion. The pairwise
    method never violates the head-to-head criterion.

13. A majority out of 35 voters is 18 or more votes.
    El Capitan (E) has a majority.
    However, the mule deer (M) wins using the Borda
    count method with 115 points. Thus the majority
    criterion is violated.

14. The standard divisor = $\dfrac{33,000}{30} = 1100$

State	A	B	C	Total
Population	6933	9533	16534	33,000
Standard Quota	6.30	8.67	15.03	
Hamilton's Apportionment	6	9	15	30

15. The divisor 1040 was used.

State	A	B	C	Total
Population	6933	9533	16534	33,000
Modified Quota	6.67	9.17	15.90	
Jefferson's Apportionment (round down)	6	9	15	30

16. The divisor used was 1160

State	A	B	C	Total
Population	6933	9533	16534	33,000
Standard Quota	5.98	8.22	14.25	
Adams' Apportionment	6	9	15	30

17. The divisor 1100 was used.

State	A	B	C	Total
Population	6933	9533	16534	33,000
Modified Quota	6.30	8.67	15.03	
Webster's Apportionment	6	9	15	30

18. The new divisor 1064.52

State	A	B	C	Total
Population	6933	9533	16534	33,000
Standard Quota	6.51	8.96	15.53	
Hamilton's Apportionment	6	9	16	31

The Alabama paradox does not occur, since none of the states loses a seat.

19. The divisor $= \dfrac{33,826}{30} = 1127.53$

State	A	B	C	Total
Population	7072	9724	17030	33,826
Standard Quota	6.27	8.62	15.10	
Hamilton's Apportionment	6	9	15	30

The Alabama paradox does not occur, since none of the states loses a seat.

20. The new divisor is $\dfrac{38,100}{35} = 1088.57$

State	A	B	C	D	Total
Population	6933	9533	16534	5100	38100
Standard Quota	6.37	8.76	15.19	4.69	
Hamilton's Apportionment	6	9	15	5	36

The new states paradox does not occur, since none of the existing states loses a seat.